XXII International School on Nuclear Physics, Neutron Physics and Applications 2017

Journal of Physics: Conference Series Volume 1023

Varna, Bulgaria
10 – 16 September 2017

ISBN: 978-1-5108-6323-1
ISSN: 1742-6588

Printed from e-media with permission by:

Curran Associates, Inc.
57 Morehouse Lane
Red Hook, NY 12571

CURRAN ASSOCIATES INC. proceedings.com

Printed by Curran Associates, Inc. (2018)

For permission requests, please contact the Institute of Physics
at the address below.

Institute of Physics
Dirac House, Temple Back
Bristol BS1 6BE UK

Phone: 44 1 17 929 7481
Fax: 44 1 17 920 0979

techtracking@iop.org

Additional copies of this publication are available from:

Curran Associates, Inc.
57 Morehouse Lane
Red Hook, NY 12571 USA
Phone: 845-758-0400
Fax: 845-758-2633
Email: curran@proceedings.com
Web: www.proceedings.com

Table of contents

Volume 1023

XXII International School on Nuclear Physics, Neutron Physics and Applications

10–16 September 2017, Varna, Bulgaria

Accepted papers received: 1 May 2018
Published online: 15 May 2018

Preface

Papers

IOP Conf. Series: Journal of Physics: Conf. Series **1023** (2018) 011001 doi:10.1088/1742-6596/1023/1/011001

Preface

The present volume contains the lectures and short talks given at the XXII International School on Nuclear Physics, Neutron Physics and Applications.

The meeting was held from 10 to 16 of September 2017 in "Grifid Encanto Beach Hotel", in the Golden Sands Resort near the city of Varna.

The School was organized by the Institute for Nuclear Research and Nuclear Energy of Bulgarian Academy of Sciences.

In keeping with a long-standing tradition, the School has been held every second year since 1973. The School's program has been restructured to encompass our new international links and today it is more similar to an international conference than to a classical nuclear physics school. Fifty participants from 12 countries enjoyed the scientific presentations and discussions during the pleasant evenings.

The program of the School spanned the latest results in fundamental areas such as nuclear structure and reactions as well as examples of the latest applications of nuclear methods.

The main topics of the presentations were:

- Exotic nuclei and their properties
- Nuclear structure
- Superheavy elements, synthesis and properties
- Rare processes and decays
- Radioactive beams, production and research programs
- Evolution of shell structure
- Collective modes of the nucleus
- Nuclear astrophysics
- Applications of exotic beams in materials research
- Current and future nuclear facilities

We would like to thank conference participants, members of the advisory and local organizing committees for their contribution to the success of the XXII International School on Nuclear Physics, Neutron Physics and Applications. We are also grateful to the reviewers for their positive contribution to the high standard of the articles presented in this volume.

Finally, we would like to thank the Bulgarian National Science Fund for their financial support.

Dimitar Tonev, Sevdalina Dimitrova

Chairs of the Organizing committee of
the XXII International School on Nuclear Physics,
Neutron Physics and Applications.

IOP Conf. Series: Journal of Physics: Conf. Series **1023** (2018) 011002 doi:10.1088/1742-6596/1023/1/011002

Peer review statement

All papers published in this volume of *Journal of Physics: Conference Series* have been peer reviewed through processes administered by the proceedings Editors. Reviews were conducted by expert referees to the professional and scientific standards expected of a proceedings journal published by IOP Publishing.

IOP Conf. Series: Journal of Physics: Conf. Series **1023** (2018) 012001 doi:10.1088/1742-6596/1023/1/012001

Influence of complex configurations on properties of pygmy dipole resonances

N N Arsenyev[1], A P Severyukhin[1], V V Voronov[1], Nguyen Van Giai[2]

[1] Bogoliubov Laboratory of Theoretical Physics, Joint Institute for Nuclear Research, 141980 Dubna, Moscow region, Russia

[2] Institut de Physique Nucléaire, CNRS-IN2P3 and Univ. Paris-Sud, 91405 Orsay, France

E-mail: voronov@theor.jinr.ru

Abstract. Starting from the quasiparticle random phase approximation based on the Skyrme interaction SLy5, we study the effects of phonon-phonon coupling (PPC) on the low-energy electric dipole responses in some spherical nuclei. The inclusion of the PPC results in the formation of low-energy 1^- states. There is an impact of the PPC effect on low-energy $E1$ strength. The PPC effect on the electric dipole polarizability is discussed. We predict a strong increase of the summed $E1$ strength below 10 MeV, with increasing neutron number from ^{48}Ca till ^{58}Ca.

1. Introduction

It is well known that collective dipole excitations in atomic nuclei can arise from out-of-phase oscillations of the proton and neutron "fluids" giving the giant dipole resonance (GDR) [1]. Systematic investigations established it to be a global feature of nuclei from the very light to the heaviest nuclei [2, 3]. In recent years, the interest is more focused on the low-lying dipole strength, that is located below the GDR energies. The concentration of the $E1$ strength around the particle separation energy is commonly called the pygmy dipole resonance (PDR) because of its weak strength in comparison with the GDR, which dominates the $E1$ strength distribution in nuclei [4]. In analogy to the GDR, the PDR has been interpreted as a collective oscillation of the neutron skin with respect to a $N \approx Z$ inert core (see Ref. [5] and references therein). The total sum of the measured energy-weighted sum rule of such $E1$ distributions is less than 1-2% of the Thomas-Reiche-Kuhn (TRK) sum rule value for stable nuclei and less than 5-6% for unstable neutron-rich nuclei [4]. Recent theoretical calculations indicate that such a low-energy peak is a common property of neutron-rich nuclei lying in different mass regions [6, 7]. The occurrence of non-negligible low-lying $E1$-strength can influence the radiative neutron capture cross section by orders of magnitude and, consequently, also the rate of the astrophysical r-process nucleosynthesis [8].

The strong proton shell closure at $Z = 20$ and the already good experimental knowledge of the chain of calcium isotopes makes [9, 10, 11, 12] calcium an attractive element for a PDR study. Indeed, indications of a trend for increasing low-energy dipole strength with increasing mass can be observed in the dipole excitation functions (above neutron separation energy) in the stable Ca isotopes [9, 10]. The results were generally consistent with the theoretical prediction regarding the shifting of dipole strength to lower energies, see, e.g., Refs. [6, 13, 14, 15]. Moreover, recent

IOP Conf. Series: Journal of Physics: Conf. Series **1023** (2018) 012001 doi:10.1088/1742-6596/1023/1/012001

experimental studies indicate $N = 32$ as a new magic number in Ca isotopes due to the high energy of the first 2^+ state in ^{52}Ca [16] and the trend obtained for the two-neutron separation energies [17]. The first experimental spectroscopic study of low-lying states in ^{54}Ca have been performed at RIKEN [18]: the 2_1^+ energy in ^{54}Ca was found to be only $\sim$ 500 keV below that in ^{52}Ca, suggesting a $N = 34$ new shell closure. Finally, we note that new progress in the production of neutron-rich Ca isotopes can be expected at the NSCL at Michigan State University [19]. Future measurements of excited states and masses for the neighboring Ca isotopes will further enhance our understanding of nuclear states in very neutron-rich systems. Thus, the spectroscopy of neutron-rich calcium isotopes provides a valuable information, with important tests of theoretical calculations.

A powerful microscopic approach is the quasiparticle-phonon model (QPM) [20, 21]. A special emphasis on a reliable description of the mean-field part, reproducing as close as possible the ground-state properties of nuclei along an isotopic chain has been done in [22, 23]. This is achieved by solving the ground-state problem in a semimicroscopic approach based on a Skyrme energy density functional (EDF) [22, 23]. The EDF+QPM calculations have been applied for the low-energy dipole strength [22, 23]. The QRPA with a self-consistent mean-field derived from Skyrme EDF is one of the most successful methods for studying the low-energy dipole strength, see e.g., [24, 14, 25, 26, 27]. Such an approach describes the properties of the low-lying states less accurately than more phenomenological ones, but the results are in a reasonable agreement with experimental data. On the other hand, due to the anharmonicity of vibrations there is a coupling between one-phonon and more complex states [20, 21]. The main difficulty is that the complexity of calculations beyond standard QRPA increases rapidly with the size of the configuration space, so that one has to work within limited spaces. Using a finite-rank separable approximation (FRSA) [28, 29, 30, 31] for the residual interaction one can overcome this difficulty. The FRSA was thus used to study the electric low-energy excitations and giant resonances within and beyond the QRPA [30, 31, 32, 33]. In this paper our approach applied for PDR features of neutron-rich nuclei. We will give an illustration of our approach for ^{48}Ca with closed neutron shell in comparison to the $N = 30$ isotope ^{50}Ca. Preliminary results of our studies for neutron-rich Sn isotopes are reported in Refs. [26, 34, 35].

2. The FRSA model

The starting point of the method is the Hartree-Fock (HF)-BCS calculation [36] of the ground state based on Skyrme interactions. Spherical symmetry is imposed on the quasiparticle wave functions. The continuous part of the single-particle spectrum is discretized by diagonalizing the Skyrme HF Hamiltonian on a harmonic oscillator basis. In the particle-hole (p-h) channel we use the Skyrme interaction with the tensor components and their inclusion leads to the modification of the spin-orbit potential [37, 38]. The pairing correlations are generated by a density-dependent zero-range force

$$V_{pair}(\mathbf{r}_1, \mathbf{r}_2) = V_0 \left[1 - \eta \left(\frac{\rho(r_1)}{\rho_0} \right)^{\gamma} \right] \delta(\mathbf{r}_1 - \mathbf{r}_2), \tag{1}$$

where $\rho(r_1)$ is the particle density in coordinate space, ρ_0 being the nuclear matter saturation density; γ, η, and V_0 are adjusted parameters. The parameters are determined by adjusting the empirical odd-even mass differences of the nuclei in the region under study.

To build the QRPA equations on the basis of HF-BCS quasiparticle states, the residual interaction is consistently derived from the Skyrme force in the p-h channel and from the zero-range pairing force in the particle-particle (p-p) channel [39]. The FRSA for the residual interaction enables us to find QRPA eigenvalues as the roots of a relatively simple secular equation [28, 31]. The cut-off of the discretized continuous part of the single-particle spectra is at the energy of 100 MeV. This is sufficient to exhaust practically all the sum rules.

IOP Conf. Series: Journal of Physics: Conf. Series **1023** (2018) 012001 doi:10.1088/1742-6596/1023/1/012001

To take into account the effects of the phonon-phonon coupling (PPC) we follow the basic QPM ideas [20, 21]. We construct the wave functions from a linear combination of one-phonon and two-phonon configurations

$$\Psi_\nu(\lambda\mu) = \left(\sum_i R_i(\lambda\nu)Q^+_{\lambda\mu i} + \sum_{\lambda_1 i_1 \lambda_2 i_2} P^{\lambda_1 i_1}_{\lambda_2 i_2}(\lambda\nu)\left[Q^+_{\lambda_1\mu_1 i_1}Q^+_{\lambda_2\mu_2 i_2}\right]_{\lambda\mu}\right)|0\rangle, \quad (2)$$

where λ denotes the total angular momentum and μ its z-projection in the laboratory system. The ground state is the QRPA phonon vacuum $|0\rangle$. The unknown amplitudes $R_i(\lambda\nu)$ and $P^{\lambda_1 i_1}_{\lambda_2 i_2}(\lambda\nu)$ are determined from the variational principle, which leads to a set of linear equations [30, 33]

$$(\omega_{\lambda i} - E_\nu)R_i(\lambda\nu) + \sum_{\lambda_1 i_1 \lambda_2 i_2} U^{\lambda_1 i_1}_{\lambda_2 i_2}(\lambda i)P^{\lambda_1 i_1}_{\lambda_2 i_2}(\lambda\nu) = 0, \quad (3)$$

$$2(\omega_{\lambda_1 i_1} + \omega_{\lambda_2 i_2} - E_\nu)P^{\lambda_1 i_1}_{\lambda_2 i_2}(\lambda\nu) + \sum_i U^{\lambda_1 i_1}_{\lambda_2 i_2}(\lambda i)R_i(\lambda\nu) = 0. \quad (4)$$

The rank of the set of linear equations (3) and (4) is equal to the number of one- and two-phonon configurations included in the wave function (2). Its solution requires to compute the Hamiltonian matrix elements coupling one- and two-phonon configurations:

$$U^{\lambda_1 i_1}_{\lambda_2 i_2}(\lambda i) = \langle 0|Q_{\lambda i}H\left[Q^+_{\lambda_1 i_1}Q^+_{\lambda_2 i_2}\right]_\lambda|0\rangle. \quad (5)$$

Equations (3) and (4) have the same form as the QPM equations [20, 21], where the single-particle spectrum and the residual interaction are derived from the same Skyrme EDF.

3. Details of calculations

We apply the above approach to study the influence of the PPC on the strength $E1$ distributions of the neutron-rich Ca isotopes. We use the Skyrme interactions SLy5 [40] and SLy5+T [41] in the p-h channel. The parameters of the force SLy5 have been adjusted to reproduce nuclear matter properties, as well as nuclear charge radii, binding energies of doubly magic nuclei. The force SLy5+T involves the tensor terms added without refitting of the parameters of the central interaction (the tensor interaction parameters are $\alpha_T = -170$ MeVfm^5 and $\beta_T = 100$ MeVfm^5). These parametrizations describe correctly the binding energies of even-even Ca isotopes [42]. It is shown [42] that the proton-neutron rms differences become larger when the neutron number increases. The same evolution is obtained with other Skyrme EDF's [6]. In the case of ^{48}Ca, the experimental neutron skin thickness ($0.14-0.20$ fm) has been determined from the $E1$ strength distribution which is extracted from proton inelastic scattering [43]. HF-BCS analysis gives the neutron skin ΔR_{np} of ^{48}Ca to be 0.16 fm and 0.14 fm with the SLy5 and SLy5+T EDF's, respectively. The theoretical "model-averaged" estimate for ΔR_{np} is 0.176$\pm$0.018 fm [44]. In addition, the *ab initio* calculations for the neutron skin in ^{48}Ca is $0.12 \leq \Delta R_{np} \leq 0.15$ fm [45].

For the interaction in the p-p channel, we use a zero-range volume force, i.e., $\eta = 0$ in Eq. (1). The strength V_0 is taken equal to -270 MeVfm^3. This value of the pairing strength is

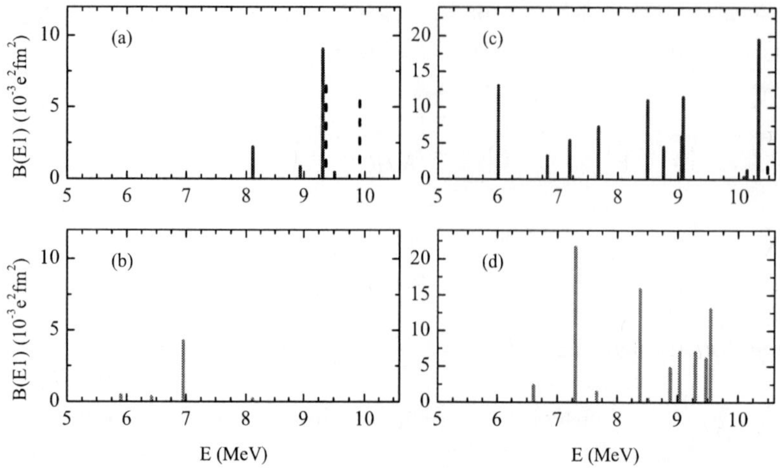

Figure 1. Low-energy $E1$ strength distribution of ^{40}Ca (resp. ^{48}Ca) is shown in the left (resp. right) panels. The dashed and solid lines correspond to the SLy5 calculations within the RPA and taking into account the PPC effects, respectively. Panels (b) and (d): experimental data are from Ref. [12].

fitted to reproduce the experimental neutron pairing energies of ^{50,52,54}Ca obtained from binding energies of neighboring isotopes. This choice of the pairing interaction has also been used for a satisfactory description of the experimental data of ^{70,72,74,76}Ni [46], ^{90,92}Zr and ^{92,94}Mo [33]. Thus, hereafter we use the Skyrme interaction SLy5 with and without tensor components in the particle-hole channel together with the volume zero-range force acting in the particle-particle channel.

In order to construct the wave functions (2) of the 1^- states, in the present study we take into account all two-phonon terms that are constructed from the phonons with multipolarities $\lambda \leq 5$ [26, 34, 35]. As it was shown in [42]the overall agreement of the calculated energies and $B(E\lambda)$ values with the experimental data looks reasonable.

All dipole excitations with energies below 35 MeV and 15 most collective phonons of the other multipolarities are included in the wave function (2). We have checked that extending the configuration space plays a minor role in our calculations.

4. Results and discussion

As a first step in the present analysis, we examine the PPC effects on the $E1$ strength distributions for doubly-magic ^{40,48}Ca isotopes. A comparison of such calculations with recent experimental data [12] demonstrates that the RPA approach cannot reproduce correctly the low-energy $E1$ strength distributions, see Fig. 1. For ^{40}Ca the PPC calculation predicts the first 1^- state significantly higher than the experimental two-phonon candidate [47] (see Fig. 1). It is worth pointing out that our results for ^{40,48}Ca are in good agreement with the RQTBA calculations taking into account the effects of coupling between quasiparticles and phonons [15]. In addition, we discuss the GDR energy region. For ^{48}Ca, the photo-absorption process is well studied experimentally. The photo-absorption cross section up to 27 MeV is displayed in Fig. 2(a). The cross section is computed by using a Lorentzian smearing with an averaging parameter $\Delta = 1.0$ MeV. The PPC effects yield a noticeable redistribution of the GDR strength in comparison with the RPA results. It is worth mentioning that the coupling increases the GDR width from 6.9 MeV to 7.3 MeV in the energy region $10 - 26$ MeV. The experimental GDR width and energy are 6.98 MeV and 19.5 MeV [48], respectively. The calculated characteristics

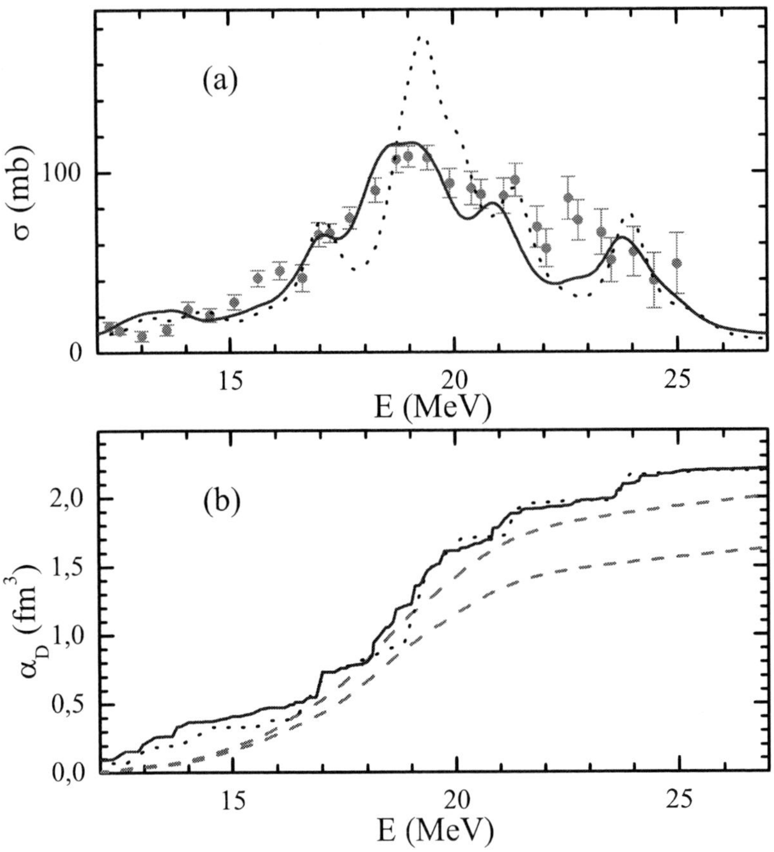

Figure 2. (a) The estimated photo-absorption cross section for ^{48}Ca (filled circles) are taken from Ref. [48]. The dotted and solid lines correspond to the calculations within the RPA with the SLy5 EDF and taking into account the PPC, respectively. (b) Running sum of the electric dipole polarizability for ^{48}Ca calculated within the RPA with the SLy5 EDF (dotted line) and the RPA plus PPC (solid line) in comparison to experimental determination of α_D (the two dashed lines indicate upper and lower limits) [43].

of the GDR are in agreement with the observed values. We conclude that the main mechanisms of the GDR formation in ^{48}Ca can be taken into account correctly and consistently in the PPC approach.

In order to perform further investigations on the ^{48}Ca nucleus we have extracted the electric dipole polarizability [1, 49, 50], which represents a handle to constrain the equation of state of neutron matter and the physics of neutron stars.

Running sums of α_D values for ^{48}Ca in the energy region below 27 MeV are given in Fig. 2(b). It is shown that the PPC does not affect the description of the electric dipole polarizability. The results differ insignificantly. Moreover, we have checked that inclusion of the tensor components does not change the value of α_D obtained by integrating the $E1$ strength up to 60 MeV: $\alpha_D = 2.28$ fm^3 in the case of the SLy5 EDF and $\alpha_D = 2.20$ fm^3 in the SLy5+T. Both effective interactions reproduce the experimental data $\alpha_D = 2.07{\pm}0.22$ fm^3 [43] and they are in good agreement with the "model-averaged" value of $2.306{\pm}0.089$ fm^3, which is predicted in Ref. [44]. Although the GDR strength dominates, contributions to α_D value at lower and higher excitation energies must be taken into account.

IOP Conf. Series: Journal of Physics: Conf. Series **1023** (2018) 012001 doi:10.1088/1742-6596/1023/1/012001

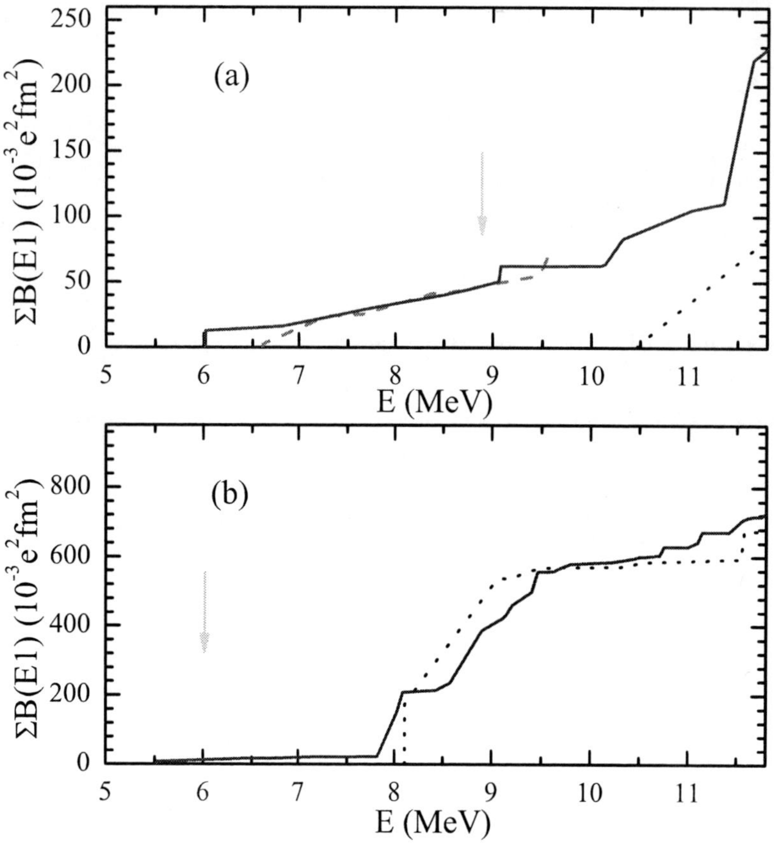

Figure 3. Running sums of the low-lying dipole strengths in ^{48}Ca (top panel) and ^{50}Ca (lower panel). The dotted and solid lines correspond to the calculations within the QRPA with the SLy5 EDF and taking into account the PPC, respectively. Experimental data (dashed line) are taken from Ref. [10]. The calculated S_{1n} energy is indicated by the solid arrow.

To complete the discussion we consider α_D as a function of the neutron number for Ca isotopes. The result of the SLy5 calculation with the PPC predicts a monotonic increase of the α_D value with neutron number, and only a small kink in the calculated excitation energies is found at the $N = 28$ shell closure. The calculated polarizabilities α_D of ^{40}Ca and ^{48}Ca are in excellent agreement with the experimental data [43]. We find that the correlation between the value of α_D and neutron skin ΔR_{np} is discerned, see [42]. With the increase of the neutron skin one observes a smooth increase of α_D. Thus, the α_D value and the neutron skin ΔR_{np} are correlated as predicted in Ref. [50]. Besides this, the α_D value can be measured in finite nuclei and, as a result, the ΔR_{np} value can be extracted.

Let us now discuss the strong increase of the summed $E1$ strength below 10 MeV ($\sum B(E1)$), with increasing the neutron number. In Fig. 3(a) the calculated running sum for ^{48}Ca is plotted as a function of the excitation energy. In the same plot the calculated S_{1n} values are shown. In the case of the RPA, there is no 1^- state below 10 MeV (see Fig. 1). The RPA calculations predict the first dipole state around 10.5 MeV. In contrast to the RPA case, the PPC results in the formation of low-lying 1^- states in this energy region. The dominant contribution in the wave function of the 1^- states comes from the two-phonon configurations ($> 60\%$). These states originate from the fragmentation of the RPA states above 10 MeV. As one can see in Fig. 3(a), the calculated running sum of the $\sum B(E1)$ value is close to the experimental $\sum B(E1)$ value. The PPC calculations give a total dipole strength of 0.063 e^{2}fm^2. The experimental value of $\sum B(E1)$ is 0.0687±0.0075 e^{2}fm^2 in the same interval [10]. The PPC effects produce a sizable impact on the low-energy $E1$ strength of ^{48}Ca. It is remarkable that the contributions of the low-lying 1^- states to the value of α_D is small (0.033 fm^3), three times as much as the value deduced experimentally [43]. It is shown that the α_D value is more sensitive to the fine structure of the $E1$ strength distribution. The PPC calculations reproduce the observed trend in light Ca isotopes, although the theoretical value of $\sum B(E1)$ for ^{44}Ca underestimates the experimental value by a factor of 2 [42].

In contrast to the case of ^{48}Ca, the PPC effects on the low-energy dipole spectrum of ^{50}Ca is weak (see Fig. 3(b)). Thus, the $\sum B(E1)$ values for ^{50}Ca results predominantly from the QRPA distribution of $E1$ strength. A similar result is observed in the case of ^{52,54,56,58}Ca isotopes. The separation energies decrease much faster than the value of $\sum B(E1)$. This means, of course, that the observation of the PDR in (γ, γ') experiments will be strongly hindered.

Our investigation [42] demonstrates that there is a correlation between the PDR properties and the neutron skin $\Delta R_{\rm np}$.

5. Conclusions

The electric dipole polarizability is a particularly important observable, as it can be measured in finite nuclei and it provides important information on the neutron skin thickness that can be extracted. In this study, Skyrme QRPA calculations including the phonon-phonon coupling have been performed for the $E1$ response in neutron-rich Ca isotopes, some of which should become experimentally accessible in the near future. Our calculations reproduce the data for the neutron skin thickness and neutron separation energies. It is shown that the phonon-phonon coupling has small influence on the dipole polarizability.

For ^{48}Ca, the PPC effect have a damping and smoothing action which yields a GDR cross section close to the experimental one in shape and magnitude. We find the impact of the shell closure $N = 28$ on the summed $E1$ strength below 10 MeV. The dipole response for $^{52-58}$Ca is characterized by the fragmentation of the strength distribution and its spreading into the low-energy region.

5.1. Acknowledgments

N.N.A., A.P.S., and V.V.V. thank the hospitality of IPN-Orsay where the part of this work was done. This work is partly supported by the CNRS-RFBR agreement No. 16-52-150003, the IN2P3-JINR agreement, and the RFBR grant No. 16-02-00228.

6. References

[1] Migdal A B 1944 *J. Phys. Acad. Sci. USSR* **8** 331; 1945 *J. Exp. Theor. Phys. USSR* **15** 81
[2] Berman and B L Fultz S C 1975 *Rev. Mod. Phys.* **47** 713
[3] Dietrich S S and Berman B L 1988 *At. Data Nucl. Data Tables* **38** 199
[4] Savran D *et al.* 2013 *Prog. Part. Nucl. Phys.* **70** 210
[5] Paar N*et al.* 2007 *Rep. Prog. Phys.* **70** 691
[6] Inakura T *et al.* 2011 *Phys. Rev.* C **84** 021302(R)
[7] Ebata S *et al.* 2014 *Phys. Rev.* C **90**, 024303 2015 **92**, 049902(E)
[8] Arnould M *et al.* 2007 *Phys. Rep.* **450** 97
[9] Hartmann T *et al.* 2000 *Phys. Rev. Lett.* **85** 274
[10] Hartmann T *et al.* 2004 *Phys. Rev. Lett.* **93** 192501
[11] Isaak J *et al.* 2011 *Phys. Rev.* C **83** 034304
[12] Derya V*et al.* 2014 *Phys. Lett.* B **730**288
[13] Soloviev V G *et al.* 1978 *Nucl. Phys.* A **304** 503
[14] Terasaki J and Engel J 2006 *Phys. Rev.* C **74** 044301
[15] Egorova I A and Litvinova E 2016 *Phys. Rev.* C **94** 034322

[16] Gade A *et al.* 2006 *Phys. Rev.* C **74** 021302(R)
[17] Wienholtz F *et al.* 2013 *Nature* **498** 346
[18] Steppenbeck D *et al.* 2013 *Nature* **502** 207
[19] Tarasov O B *et al.* 2009 *Phys. Rev. Lett.* **102** 142501
[20] Soloviev V G 1992 *Theory of Atomic Nuclei: Quasiparticles and Phonons*(Institute of Physics, Bristol and Philadelphia)
[21] Lo Iudice N *et al.* 2012 *J. Phys. G* **39** 043101
[22] Tsoneva N *et al.* 2014 *Phys. Lett.* B **586** 213
[23] Tsoneva N and Lenske H 2008 *Phys. Rev.* C **77** 024321
[24] Sarchi *et al.* 2004 *Phys. Lett.* B **601** 27
[25] Avdeenkov A *et al.* 2011 *Phys. Rev.* C **83** 064316
[26] Arsenyev N N *et al.* 2012 *EPJ Web of Conf.* **38** 17002
[27] Repko A *et al.* 2013 *Phys. Rev.* C **87** 024305
[28] Nguyen Van Giai *et al.* 1998 *Phys. Rev.* C **57** 1204
[29] Severyukhin A P *et al.* 2002 *Phys. Rev.* C **66** 034304
[30] Severyukhin A P *et al.* 2004 *Eur. Phys. J.* A **22** 397
[31] Severyukhin A P *et al.* 2008 *Phys. Rev.* C **77** 024322
[32] Severyukhin A P *et al.* 2009 *Phys. At. Nucl.* **72** 1149
[33] Severyukhin A P *et al.* 2012 *Phys. Rev.* C **86** 024311
[34] Arsenyev N N*et al.*2015 *Acta Phys. Pol.* B **46** 517
[35] Arsenyev N N *et al.* 2016 *EPJ Web of Conf.* **107** 05006
[36] Ring P and Schuck P 1980 *The Nuclear Many Body Problem* (Berlin: Springer)
[37] Stancu *et al.* 1977 *Phys. Lett.*B **68** 108
[38] Lesinski T *et al.* 2007 *Phys. Rev.* C **76** 014312
[39] Terasaki J *et al.* 2005 *Phys. Rev.* C **71** 034310
[40] Chabanat E *et al.* 1998 *Nucl. Phys.* A **635** 231
[41] Colò G *et al.* 2007 *Phys. Lett.* B **646** 227 ; 2008 **668** 457(E)
[42] Arsenyev N N *et al.* 2017 *Phys. Rev.* C **95** 054312
[43] Birkhan J *et al.* *arXiv:1611.07072v1 [nucl-ex].*
[44] Piekarewicz J *et al.* 2012 *Phys. Rev.* C **85** 041302(R)
[45] Hagen G *et al.*2016 *Nature Physics* **12** 186
[46] Severyukhin A P 2014 *Phys. Rev.* C **90** 044320
[47] Derya V *et al.*] 2016 *Phys. Rev.* C **93** 034311
[48] O'Keefe G J *et al.* 1987 *Nucl. Phys.* A **469** 239
[49] Bohigas O *et al.* 1981 *Phys. Lett.* B **102**
[50] Satuła W*et al.* 2006 *Phys. Rev.* C **74** 011301(R)

Weak-interaction rates in stellar conditions

Pedro Sarriguren

Instituto de Estructura de la Materia, IEM-CSIC, Serrano 123, E-28006 Madrid, Spain

E-mail: p.sarriguren@csic.es

Abstract. Weak-interaction rates, including β-decay and electron captures, are studied in several mass regions at various densities and temperatures of astrophysical interest. In particular, we study odd-A nuclei in the pf-shell region, which are involved in presupernova formations. Weak rates are relevant to understand the late stages of the stellar evolution, as well as the nucleosynthesis of heavy nuclei. The nuclear structure involved in the weak processes is studied within a quasiparticle proton-neutron random-phase approximation with residual interactions in both particle-hole and particle-particle channels on top of a deformed Skyrme Hartree-Fock mean field with pairing correlations. First, the energy distributions of the Gamow-Teller strength are discussed and compared with the available experimental information, measured under terrestrial conditions from charge-exchange reactions. Then, the sensitivity of the weak-interaction rates to both astrophysical densities and temperatures is studied. Special attention is paid to the relative contribution to these rates of thermally populated excited states in the decaying nucleus and to the electron captures from the degenerate electron plasma.

1. Introduction

Weak β-decay and electron-capture (EC) have long been recognized [1] as key processes to understand the late stages of the stellar evolution. In particular, the presupernova stellar structure, as well as the nucleosynthesis of heavy nuclei, are determined to a large extent by those mechanisms. pf-shell nuclei are specially important in these scenarios because they are the main constituents of the stellar core in presupernova formations leading to core-collapse (type II) or thermonuclear (type Ia) supernovae.

From the theoretical side, the first extensive calculations of stellar weak rates in relevant ranges of densities (ρ) and temperatures (T) were done in Ref. [2] under severe assumptions on the energy distribution of the Gamow-Teller (GT) strength. Concerning the nuclear structure involved, different approaches are found in the literature that can be roughly divided into shell model (SM) [3] and proton-neutron quasiparticle random-phase approximation (QRPA) [4] categories. Although QRPA calculations cannot reach the detailed spectroscopic accuracy achieved by present state-of-the-art SM calculations, the global performance of QRPA is quite satisfactory. Moreover, one clear advantage of the QRPA method is that it can be extended to heavier nuclei, which are beyond the present capability of full SM calculations, without increasing much the complexity of the calculation.

It is also important to realize that there are clear differences between terrestrial and stellar decay rates caused by the effect of high ρ and T conditions. One effect of T is directly

IOP Conf. Series: Journal of Physics: Conf. Series **1023** (2018) 012002 doi:10.1088/1742-6596/1023/1/012002

related to the thermal population of excited states in the decaying nucleus, accompanied by the corresponding depopulation of the ground states. The weak rates of excited states can be significantly different from those of the ground state and a case by case consideration is needed. Another distinctive effect comes from the fact that atoms in stellar scenarios are completely ionized, and consequently electrons are no longer bound to the nuclei, but forming a degenerate plasma that obeys a Fermi-Dirac distribution. This opens the possibility for continuum EC, in contrast to the orbital EC caused by bound electrons in the atom under terrestrial conditions. These two effects make weak-interaction rates in the stellar interior sensitive functions of T and ρ.

In the case of even-even nuclei the eventual contributions from excited states could be safely neglected because of the high energy excitation of the first excited states, typically 2^+ states beyond 1 MeV that can hardly be excited within the range of temperatures considered in this work. However, low-lying excited states would contribute to the rates in the case of even-even well deformed nuclei [5], where the rotational states drop easily below 1 MeV, as well as in odd-A nuclei where quasiparticle states are found at very low excitation energies. In this work we study contributions to the weak rates coming from excited states in odd-A nuclei. To perform this study, a set of nuclei has been chosen according to their interest in presupernova models. Here we focus on ^{55}Mn and ^{55}Co.

2. Theoretical Formalism

The weak-interaction rate of a nucleus can be expressed as follows,

$$\lambda = \sum_i \lambda_i P_i(T); \quad P_i(T) = \frac{2J_i+1}{G} e^{-E_i/(k_B T)}, \tag{1}$$

where $P_i(T)$ is the probability of occupation of the excited state i in the parent nucleus. Assuming thermal equilibrium, it is given by a Boltzmann distribution. $G = \sum_i (2J_i+1)\, e^{-E_i/(k_B T)}$ is the partition function and $J_i(E_i)$ is the angular momentum (excitation energy) of the parent nucleus state i. In principle, the sum extends over all excited states in the parent nucleus up to the proton separation energy. However, because of the range of temperatures considered in this work (T=1-10 GK), only a few low-lying excited states are expected to contribute significantly. The scale of excitation energies to consider is determined by $k_B T$, which for maximum T around 10 GK is given by $k_B T = 0.862$ MeV. The weak-interaction rate corresponding to the parent state i is given by

$$\lambda_i = \sum_f \lambda_{if} = \frac{\ln 2}{D} \sum_f B_{if} \Phi_{if}(\rho, T), \tag{2}$$

where the sum extends over all the states in the final nucleus that can be reached in the process and $D = 6146$ s. This expression is decomposed into a phase space factor Φ_{if}, which is a function of ρ and T and a nuclear structure part B_{if} that contains the transition probabilities for allowed GT transitions.

In the astrophysical scenarios of our study, nuclei are fully ionized and continuum ECs from the degenerate electron plasma are possible. The phase space factor for continuum EC is given by

$$\Phi_{if}^{EC} = \int_{\omega_\ell}^{\infty} \omega p (Q_{if}+\omega)^2 F(Z,\omega) \times S_e(\omega) \left[1 - S_\nu(Q_{if}+\omega)\right] d\omega \,. \tag{3}$$

and the corresponding phase space factor for β^+ decay is given by

$$\Phi_{if}^{\beta^+} = \int_{1}^{Q_{if}} \omega p (Q_{if}-\omega)^2 F(-Z+1,\omega) \times \left[1 - S_{e^+}(\omega)\right] \left[1 - S_\nu(Q_{if}-\omega)\right] d\omega \,. \tag{4}$$

In these expressions ω is the total energy of the electron, $p = \sqrt{\omega^2 - 1}$ is the momentum, and Q_{if} is the total energy available

$$Q_{if} = \frac{1}{m_e c^2} \left(M_p - M_d + E_i - E_f \right) , \tag{5}$$

written in terms of the nuclear masses of parent (M_p) and daughter (M_d) nuclei and their excitation energies E_i and E_f, respectively. $F(Z, \omega)$ is the Fermi function that takes into account the distortion of the electron wave function due to the Coulomb interaction.

S_e, S_{e^+}, and S_ν, are electron, positron, and neutrino distribution functions, respectively. Its presence inhibits or enhances the phase space available. In the stellar scenarios considered here the commonly accepted assumption is that $S_{e^+} = 0$ because electron-positron pair creation becomes important only at higher energies and $S_\nu = 0$ because neutrinos and antineutrinos can escape freely from the interior of the star at these densities. The electron distribution is described by a Fermi-Dirac distribution

$$S_e = \frac{1}{\exp\left[(\omega - \mu_e) / (k_B T) \right] + 1} , \tag{6}$$

where μ_e is the chemical potential. The phase space factor for EC in Eq. (3) is therefore a sensitive function of both ρ and T, through the electron distribution S_e. On the other hand, the phase space factor for β^+ decay in Eq. (4) under the assumptions $S_{e^+} = S_\nu = 0$, does not depend on ρ and T. The only dependence of the positron decay rates to ρ and T appears indirectly through the population of excited states.

The nuclear structure part of the problem is described within the QRPA formalism. The method starts with a selfconsistent deformed Hartree-Fock mean field calculation with Skyrme interactions including pairing correlations. The single-particle energies, wave functions, and occupation probabilities are generated from this mean field. The Skyrme force SLy4 is used in this work. The solution of the HF equation is found by assuming time reversal and axial symmetry. The single-particle wave functions are expanded in terms of the eigenstates of an axially symmetric harmonic oscillator in cylindrical coordinates, using twelve major shells. The method also includes pairing between like nucleons in BCS approximation with fixed gap parameters for protons and neutrons, which are determined phenomenologically from the odd-even mass differences involving the experimental binding energies. Calculations for GT strengths are performed for the equilibrium shape of each nucleus, that is, for the configuration that minimizes the energy.

To describe GT transitions, a spin-isospin residual separable interaction is added to the Skyrme mean field and treated in a deformed proton-neutron QRPA. This interaction contains a *ph* and a *pp* part with coupling strengths χ^{ph}_{GT} and κ^{pp}_{GT}. In previous works [6] we studied the sensitivity of the GT strength distributions to the various ingredients contributing to the deformed QRPA calculations, namely to the nucleon-nucleon effective force, to pairing correlations, and to residual interactions. We found different sensitivities to them. In this work, all of these ingredients have been fixed to the most reasonable choices found previously. In particular we use the coupling strengths $\chi^{ph}_{GT} = 0.10$ MeV and $\kappa^{pp}_{GT} = 0.05$ MeV. The method has been successfully applied in the past to the study of the decay properties in different mass regions from proton-rich [5] to stable [7, 8], and to neutron-rich nuclei [9].

The GT strength for a transition from an initial state i to a final state f is given by

$$B_{if}(GT^\pm) = \frac{1}{2J_i + 1} \left(\frac{g_A}{g_V} \right)^2_{\text{eff}} \langle f || \sum_j^A \sigma_j t_j^\pm || i \rangle^2 , \tag{7}$$

IOP Conf. Series: Journal of Physics: Conf. Series **1023** (2018) 012002 doi:10.1088/1742-6596/1023/1/012002

where $(g_A/g_V)_{\rm eff} = 0.7(g_A/g_V)_{\rm bare}$ is an effective ratio of axial and vector coupling factors that takes into account in an effective manner the observed quenching of the GT strength.

When the parent nucleus has an odd nucleon, the ground state can be expressed as a one quasiparticle state in which the odd nucleon occupies the single-particle orbit of lowest energy. Quasiparticle excitations correspond to configurations with the odd nucleon in an excited state. Then, two types of transitions are possible. One type is due to phonon excitations in which the odd nucleon acts only as a spectator. These are three-quasiparticle states (3qp). In the intrinsic frame, the transition amplitudes are basically the same as in the even-even case, but with the blocked spectator excluded from the calculation. The other type of transitions are those involving the odd nucleon state. These are one-quasiparticle states (1qp), which are treated by taking into account phonon correlations in the quasiparticle transitions in first order perturbation. The transition amplitudes for the correlated states can be found in Refs. [4, 6].

3. Results

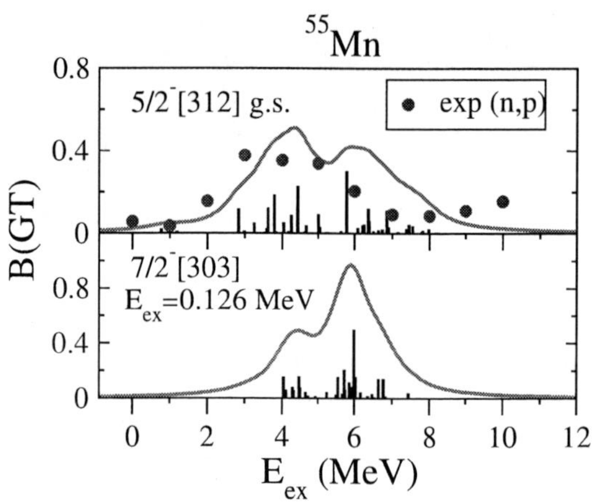

Figure 1. QRPA GT strength distributions $B(GT^+)$ for the transition of the ground state $5/2^-$ and excited state $7/2^-$ of ^{55}Mn going to ^{55}Cr, plotted versus the excitation energy of the daughter nucleus. Experimental data extracted from (n,p) are from Ref. [10].

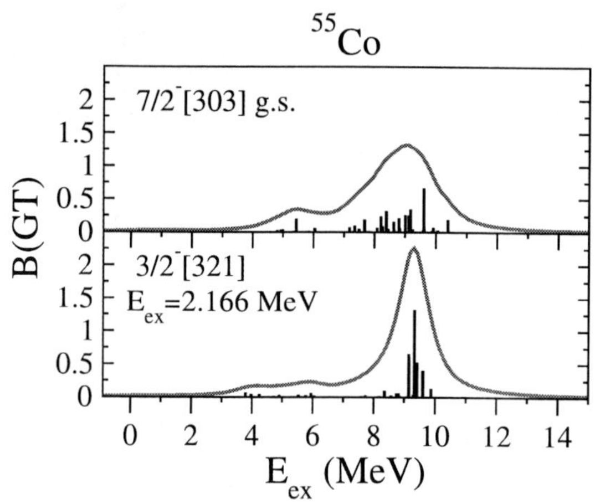

Figure 2. Same as in Figure 1, but for the ground state $7/2^-$ and excited state $3/2^-$ of ^{55}Co going to ^{55}Fe.

We start with the study the GT strength distributions corresponding to the ground state and the first excited states. In the case of ^{55}Mn (Figure 1) the profile of the ground state shows a broad peak centered a 5 MeV with a two-peaked substructure containing somewhat more strength in the lower one. The GT distribution of the excited state is close to that one with the strength shifted to the peak at higher energy. The distributions in the ground and excited states of ^{55}Co (Figure 2) are quite similar with a peak centered at about 9 MeV.

In the following figures we present the weak-interaction rates for these nuclei as a function of the temperature and for various densities. The range of T considered varies from $T_9 = 1$ up to $T_9 = 10$, in T_9 (GK) units, whereas the range in ρY_e varies from $\rho Y_e = 10^6$ mol/cm^3 up to $\rho Y_e = 10^{10}$ mol/cm^3. This grid of ρ and T includes those ranges relevant for astrophysical scenarios related to the silicon-burning stage in a presupernova star ($\rho Y_e = 10^7$ mol/cm^3 and $T_9 = 3$), as well as scenarios related to pre-collapse of the core and thermonuclear runaway type Ia supernova ($\rho Y_e = 10^9$ mol/cm^3 and $T_9 = 10$).

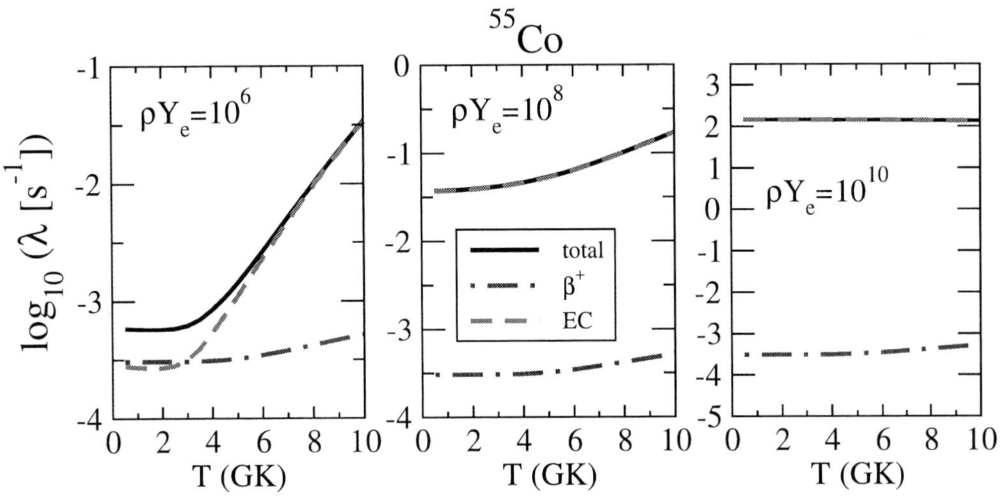

Figure 3. Weak-interaction rates as a function of T in ^{55}Co, decomposed into their EC and β^+ contributions for various densities.

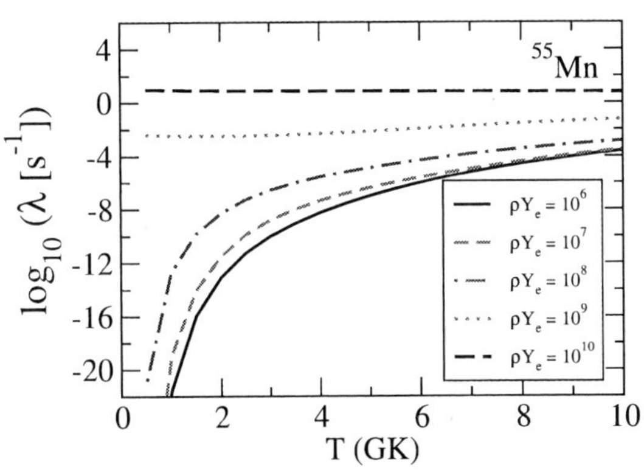

Figure 4. Total weak-interaction rates for ^{55}Mn as a function of T for various densities.

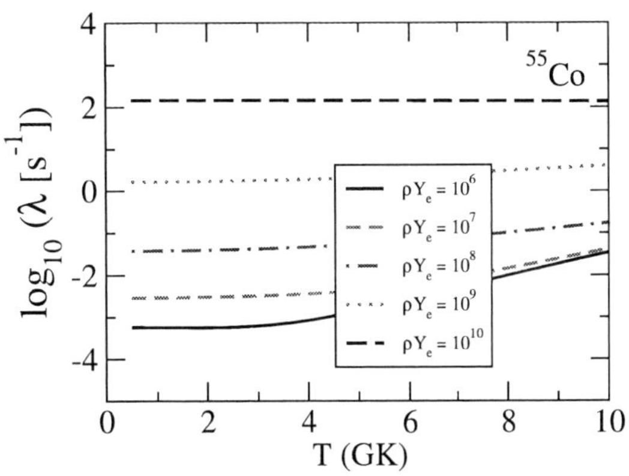

Figure 5. Same as in Figure 4, but for ^{55}Co.

At these ranges of densities and temperatures, weak-interaction rates are dominated by EC. β^+-decays will always contribute for sufficiently high ρ and T because of thermal population of excited states beyond Q_{EC} energies. However, for ρ and T values in this work, positron-decay contributions can be neglected. The most favored cases for β^+ contributions are those with positive Q_{EC} values, such as ^{55}Co ($Q_{EC} = 3.451$ MeV). We can see in Figure 3 for ^{55}Co the decomposition of the total weak-interaction rates into their EC and β^+ components. At $\rho Y_e = 10^6$ there is a competition between EC and β^+ at low T. At higher T, EC clearly dominates because it increases while β^+ remains almost constant. At higher densities, $\rho Y_e = 10^8 - 10^{10}$, β^+ is always very small in comparison with EC and can be safely neglected. The dominant EC determines the total rates in practically all the cases. Therefore, only the total rates will be shown in the following figures.

Figures 4 and 5 show the total rates as a function of the temperature for various densities from $\rho Y_e = 10^6$ up to $\rho Y_e = 10^{10}$ mol/cm^3. In these figures one can see that the rates always increase when either ρ or T increases. This is a simple consequence of the phase factors that increase accordingly. The sensitivity of the EC rates to the environmental (ρ,T) conditions is obvious. At

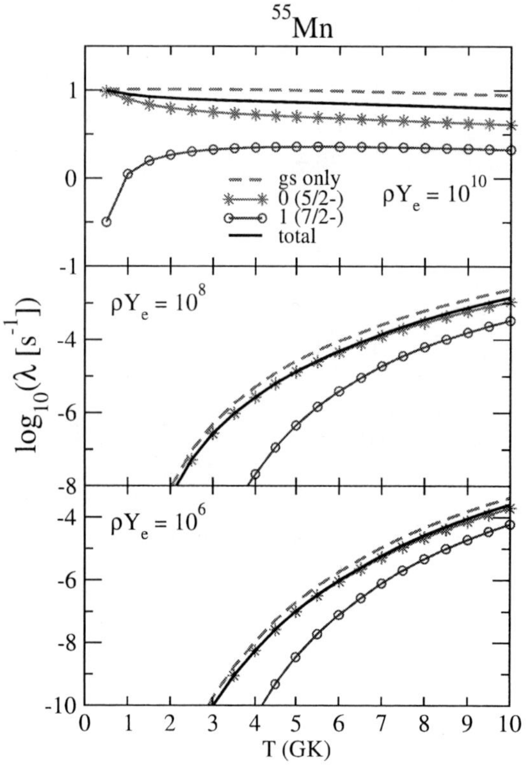

Figure 6. Weak-interaction rates for ^{55}Mn from SLy4-QRPA calculations as a function of the temperature T (GK) for various densities. Separate contributions from the ground and excited states are shown.

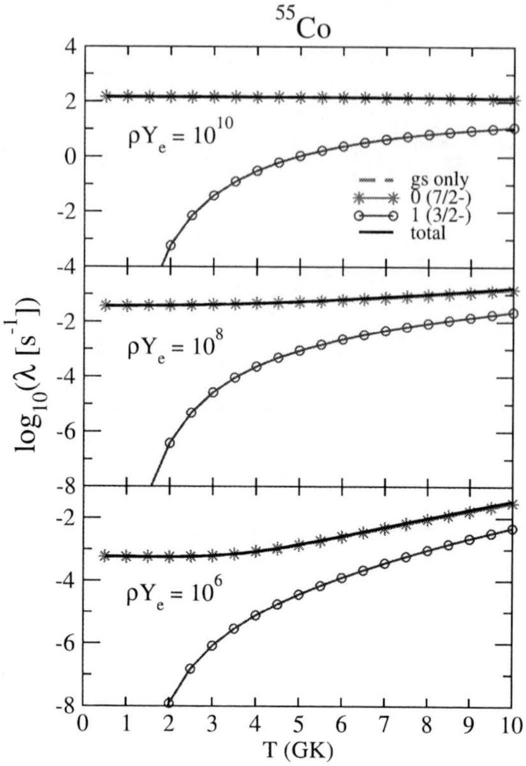

Figure 7. Same as in Figure 6, but for ^{55}Co.

low ρ (low Fermi energies) and low T (sharp shape of the energy distribution of the electrons S_e), the rates are low and very sensitive to details of the GT strength of the low-lying excitations and therefore to nuclear structure model calculations. On the other hand, when ρ increases (larger Fermi energies) and T increases (smearing of the electron distribution functions), EC rates also increase and become sensitive to all the spectrum. Then, the whole description of the GT strength distribution is more important than a detailed description of the low-lying spectrum. This is also the reason why all the rates become flat and roughly independent of T at high ρ. The dependence on the Q_{if} values is also apparent. These energies determine the lower integration limit of the EC phase factor in Eq. (3) and thus, the minimum energy of electrons in the stellar plasma to suffer a nuclear capture. The effect is more pronounced at low densities and one can roughly distinguish two types of patterns. Those that start with very low rates at low T and increase rapidly with T, such as the rates in ^{55}Mn and those exhibiting a rather flat behavior with much higher rates, as in the case of ^{55}Co. This global behavior can be understood in terms of the Q_{EC} energies. Large negative values, as in the cases of ^{55}Mn lead to very low rates. On the other hand, positive values of Q_{EC} as in ^{55}Co cause much higher rates.

In the next figures, Figures 6 - 7, the relative contributions of the ground and excited states to the total rates are studied. In addition, total rates and rates obtained from the ground state with full population labeled 'gs only' in the figures, are included. The relative importance of the various excited states is determined by their thermal population at a given T, by the phase factor, and by the structure of the GT strength distribution.

4. Conclusions

We have evaluated the weak-interaction rates of two odd-A pf-shell nuclei (^{55}Mn and ^{55}Co), which are representative of the constituents in presupernova formations. The nuclear structure involved in the calculation of the energy distribution of the Gamow-Teller strength is described within a selfconsistent deformed HF+BCS+QRPA formalism with density-dependent effective Skyrme interactions and spin-isospin residual interactions.

First, GT strength distributions for ground states and excited states have been studied in those nuclei and compared with the available data from charge-exchange reactions. Secondly, weak-interaction rates have been calculated at densities and temperatures holding in astrophysical scenarios of interest, such as silicon burning and Ia supernova pre-collapse stages. Although ECs are always dominant, positron decays have been also included because some nuclei are unstable $Q_{EC} > 0$ and because excited states above Q_{EC} values can be thermally populated.

The contributions of the excited states to the rates depend certainly on the density and temperature through the population of these states and their phase factors, but depend also on their GT structure. All in all, we find quite similar results for the total rates and for the rates considering only the contribution from the ground states fully populated. The reason is that population of excited states leads to a depopulation of ground states and both effects are compensated to some extent when the GT strength distributions are not very different from each other.

Acknowledgments

This work was supported in part by MINECO (Spain) under Research Grant No. FIS2014–51971–P.

References

[1] Burbidge EM *et al.* 1957 *Rev. Mod. Phys.* **29** 547
[2] Fuller GM, Fowler WA, and Newman MJ 1980 *Ap. J. Suppl.* **42** 447
[3] Langanke K and Martinez-Pinedo G 2000 *Nucl. Phys.* **A673** 481; Suzuki T *et al.* 2011 *Phys. Rev.* C **83** 044619
[4] Krumlinde J and Möller P 1984 *Nucl. Phys.* **A417** 419; Muto K *et al.* 1992 *Z. Phys.* A **341** 407
[5] Sarriguren P *et al.* 2005 *Eur. Phys. J.* A **24** 193; Sarriguren P 2009 *Phys. Rev.* C **79** 044315; Sarriguren P 2009 *Phys. Lett.* B **680** 438; Sarriguren P 2011 *Phys. Rev.* C **83** 025801
[6] Sarriguren P *et al.* 1998 *Nucl. Phys.* **A635** 55; Sarriguren P *et al.* 1999 *Nucl. Phys.* **A658** 13; Sarriguren P *et al.* 2001 *Nucl. Phys.* **A691** 631; Sarriguren P *et. al.* 2001 *Phys. Rev.* C **64** 064306
[7] Sarriguren P *et al.* 2003 *Nucl. Phys.* **A716** 230; Sarriguren P 2013 *Phys. Rev.* C **87** 045801; Sarriguren P 2016 *Phys. Rev.* C **93** 054309
[8] Sarriguren P *et al.* 2005 *Phys. Rev.* C **72** 054317; Moreno O *et al.* 2006 *Phys. Rev.* C **73** 054302; Boillos JM *et al.* 2015 *Phys. Rev.* C **91** 034311
[9] Sarriguren P and Pereira J 2010 *Phys. Rev.* C **81** 064314; Sarriguren P, Algora A, and Pereira J 2014 *Phys. Rev.* C **89** 034311; Sarriguren P 2015 *Phys. Rev.* C **91** 044304; Sarriguren P 2017 *Phys. Rev.* C **95** 014304
[10] El-Kateb S *et al.* 1994 Phys. Rev. C **49** 3128

IOP Conf. Series: Journal of Physics: Conf. Series **1023** (2018) 012003 doi:10.1088/1742-6596/1023/1/012003

Some aspects of self-consistent higher-order interactions

Hideo Sakamoto
Faculty of Engineering, Gifu University, Gifu 501-1193, Japan
E-mail: sakamoto@gifu-u.ac.jp

Abstract. After a brief review of the formalism of the self-consistent higher-order interactions, applications of a $([Q_3Q_3]^{(2)} \cdot Q_2)$ type of three-body interaction to the quadrupole moment of the 3^- state in ^{208}Pb and the energy splitting of the septuplet of states $(h_{9/2}3^-)I$ with $I = 3/2, 5/2, \cdots, 15/2$ in ^{209}Bi are discussed. It is shown that if the contribution of the three-body interaction is included, the theoretical value of the $Q_{el}(3^-)$ moment becomes rather small compared to the experiment, but the observed small energy splitting of the septuplet can essentially be understood within the particle-vibration coupling model. Roles of non-linear field couplings provided by the self-consistent higher-order interactions are also discussed.

1. Introduction

In a nuclear system, couplings between various modes of motion are not always weak but often become quite strong and can even affect fundamental modes themselves. Therefore, the construction of fundamental modes and the study of the couplings between them have been the important subjects for the theory of nuclear structure [1].

Through the study of giant resonances and low-lying collective states in deformed nuclei, it has been recognized that the concept of *nuclear self-consistency* (i.e. nuclear saturation and the self-consistency between an average potential and a density distribution) is essential to understand nuclear collective motions [1, 2]. By use of this remarkable concept as an important guiding principle, a general theory of self-consistent effective interactions in nuclei has been developed. In particular, higher-order effective interactions (e.g. a three-body interaction, etc.) are derived for general multipole modes under the condition that the system must observe the nuclear self-consistency in higher-order accuracy even when and especially when more than one mode is simultaneously excited in the system [3, 4, 5]. For the case of quadrupole mode, essentially the same type of many-body interactions were derived by Mrashalek [6].

The self-consistent higher-order interactions have been expected to modify coupling terms significantly compared with the cases when one uses only the two-body multipole-multipole interactions [3, 4, 5]. Among the self-consistent higher-order interactions, however, the higher-order quadrupole interactions have been investigated and applied most extensively [7, 8, 9, 10, 11, 12], while for other higher-order multipole interactions there have been relatively few discussions and realistic applications [5].

In this paper, after a brief review of the formalism of the self-consistent higher-order interactions, applications of a $([Q_3Q_3]^{(2)} \cdot Q_2)$ type of three-body interaction to the quadrupole moment of the 3^- state in ^{208}Pb and the energy splitting of the septuplet of states $(h_{9/2}3^-)I$ with

$I = 3/2, 5/2, \cdots, 15/2$ in ^{209}Bi are presented, and roles of non-linear field couplings originated from the self-consistent higher-order interactions are discussed.

2. Formalism of the self-consistent higher-order interactions

2.1. Improvement of the multipole interaction model

In the 1970s, difficulties of ordinary multipole interactions were pointed out by several authors [2, 13, 14]. For example, as shown by Kishimoto *et al.* [2], splittings of GR in deformed nuclei can not be explained by the conventional multipole-multipole (abbreviated as QQ) interaction model. Then an improved interaction model, the so-called doubly-stretched multipole-multipole (abbreviated as Q"Q") interaction model, was proposed. Since then, the doubly-stretched multipole interaction model has been successfully applied to describe deformed nuclei [2, 3, 5, 14, 15, 16, 17].

The Q"Q" interaction model has several advantages over the conventional QQ interaction model [3, 5]: (i) It guarantees exact elimination of spurious modes. (ii) It satisfies the self-consistency condition and the saturation condition more rigorously in deformed nuclei. (iii) It has the natural form of the effective interaction responsible for a fluctuation mode about a deformed equilibrium. (iv) The Q"Q" with the same strength for each K component can be approximately converted into the form of the QQ with K-dependent strengths. Such different strengths for each K agree very well with those used empirically in the QQ model for the analyses of low-lying collective quadrupole and octupole states in deformed nuclei.

Furthermore, for deformed nuclei, it has been shown that the double-stretched 2^λ-pole interaction in general is nothing but an improved 2^λ-pole interaction which effectively includes the two-body interaction part of the self-consistent three-body interactions to satisfy the nuclear self-consistency in higher-order accuracy [4, 5].

2.2. Higher-order quadrupole interactions

Concerning the origin of the conventional quadrupole-quadrupole interaction in nuclei, there is a famous explanation by Mottelson in terms of the core polarization and effective charge phenomena [18]. Along the line of it, higher-order quadrupole interactions have been derived to guarantee the nuclear self-consistency in higher-order accuracy for quadrupole deformed nuclei [3, 4, 5, 6]. In the harmonic oscillator model, since the analytic expression of the total energy functional and that of the nuclear self-consistency condition (Mottelson's relation) are available, by employing the Landau theory of Fermi systems, the effective n-body interaction can be derived by carrying out the nth order derivative of the total energy functional with respect to occupation probabilities. The resultant quadrupole interactions up to the fourth-order are expressed as

$$
\begin{aligned}
V^{(2)}_{\lambda=2} &= -\frac{\chi_{\lambda=2}}{2}(\hat{Q}_{\lambda=2}\cdot\hat{Q}_{\lambda=2}) \quad \text{with} \quad \chi_\lambda = \frac{4\pi}{2\lambda+1}\frac{m\omega_0^2}{A\langle r^{2\lambda-2}\rangle_0}, \\
V^{(3)}_{\lambda=2} &= \frac{1}{3!}\frac{\chi_2}{A\langle r^2\rangle_0}\left\{3(\hat{Q}_2\cdot\hat{Q}_2)\hat{R}_0 - \sqrt{\frac{56\pi}{5}}\left(\left[\hat{Q}_2\hat{Q}_2\right]^{(2)}\cdot\hat{Q}_2\right)\right\}, \\
V^{(4)}_{\lambda=2} &= -\frac{1}{4!}\frac{\chi_2}{(A\langle r^2\rangle_0)^2}\left\{12(\hat{Q}_2\cdot\hat{Q}_2)\hat{R}_0^2 - 8\sqrt{\frac{56\pi}{5}}\left(\left[\hat{Q}_2\hat{Q}_2\right]^{(2)}\cdot\hat{Q}_2\right)\hat{R}_0 + \frac{48\pi}{5}(\hat{Q}_2\cdot\hat{Q}_2)^2\right\}.
\end{aligned}
$$

Applications of the higher-order quadrupole interactions to realistic nuclei have been worked out by several authors [7, 8, 9, 10, 11, 12].

2.3. Higher-order interactions for general multipole modes

By use of the nuclear self-consistency as an important guiding principle, a theory of higher-order effective interactions for general multipole modes has been developed, and for shape oscillation modes in a harmonic oscillator potential model, an analytical expression of the self-consistent three-body interactions for general multipole modes has been presented [5]. For example, the three-body interaction for the dipole mode and that for the octupole mode are expressed as

$$V^{(3)}_{\lambda=1} = \frac{1}{3!}\frac{\chi_1}{A\langle r^2\rangle_0}\left\{3(\hat{Q}_1\cdot\hat{Q}_1)\hat{R}_0 + 3\sqrt{\frac{24\pi}{5}}\left(\left[\hat{Q}_1\hat{Q}_1\right]^{(2)}\cdot\hat{Q}_2\right)\right\}, \quad (1)$$

$$V^{(3)}_{\lambda=3} = \frac{1}{3!}\frac{\chi_3}{A\langle r^2\rangle_0}\left\{3(\hat{Q}_3\cdot\hat{Q}_3)\hat{R}_0 + \frac{18}{5}\sqrt{\frac{84\pi}{5}}\left(\left[\hat{Q}_3\hat{Q}_3\right]^{(2)}\cdot\hat{Q}_2\right)\right\}. \quad (2)$$

For the $V^{(3)}_{\lambda=1}$ interaction, it has been shown that the $([\hat{Q}_1\hat{Q}_1]^{(2)}\cdot\hat{Q}_2)$ term plays a role to recover the translational invariance of a quadrupole vibrating system, while the $(\hat{Q}_1\cdot\hat{Q}_1)\hat{R}_0$ term is necessary for the particle-vibration coupling model to resolve the problem of divergence in the self-energy of a single-particle due to the coupling to the translational mode [5]. Some aspects of the $([\hat{Q}_3\hat{Q}_3]^{(2)}\cdot\hat{Q}_2)$ term in the $V^{(3)}_{\lambda=3}$ interaction will be discussed in the next section.

3. Quadrupole-octupole coupling in ^{208}Pb region

In the nucleus ^{209}Bi, there exists a septuplet of states $(h_{9/2}3^-)I$ with $I = 3/2, 5/2, \cdots, 15/2$ at about 2.6 MeV which is almost equal to the excitation energy of the 3^- state of ^{208}Pb. The observed small energy splitting of the septuplet components implies that the particle-octupole coupling here is rather weak.

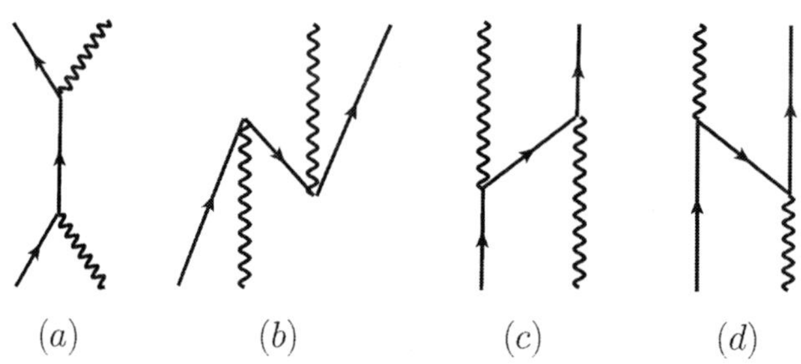

Figure 1. The second-order diagrams of the two-body interaction contributing to the energy of a particle-phonon multiplet [1].

From the late 1960's to the 1970's, detailed calculations were performed by Bohr and Mottelson [1, 18, 19] and Hamamoto [20, 21] from the perturbational point of view. They employed the particle-vibration coupling Hamiltonian $H_{coupl} = -R_0\frac{\partial V}{\partial r}(Y_3\cdot\alpha_3)$, where the amplitude of the octupole vibration was determined from the experimental $B(E3)$ value. Then it was shown that the structure of the above small energy splitting can essentially be reproduced by estimating the second-order particle-octupole coupling illustrated by the diagrams in Fig.1, with a correction for the $I = 3/2$ level which reflects the coupling to the $(d^{-1}_{3/2}0^+)$ configuration. They also evaluated the quadrupole moment of the 3^- state in ^{208}Pb by counting the diagrams illustrated in Fig.2. In the evaluation the polarization effect was included. As the result, Hamamoto got the calculated quadrupole moment of -0.13b (-0.14b) for the case of $e^p_{eff}(E2) = 1.5e$ $(2e)$ and $e^n_{eff}(E2) = e$ [20], and in the text book by Bohr and Mottelson the theoretical estimation was reported as $Q_{el}(3^-) \sim -0.1$b [1]. Then the effect of the quadrupole moment of the 3^- state on the energy splitting of the septuplet in ^{209}Bi was evaluated by

IOP Conf. Series: Journal of Physics: Conf. Series **1023** (2018) 012003 doi:10.1088/1742-6596/1023/1/012003

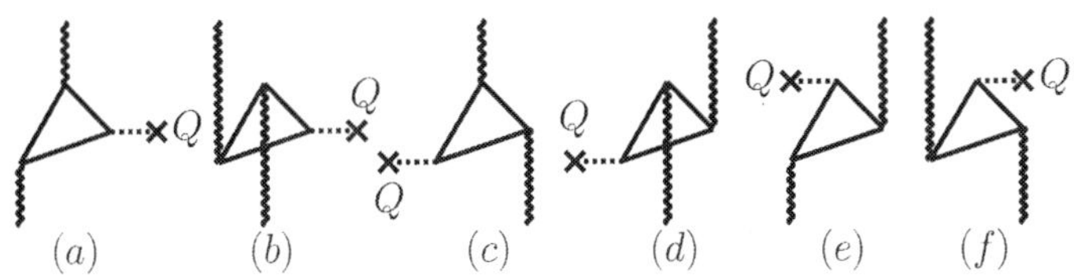

Figure 2. Diagrams contributing to the static quadrupole moment of a phonon [1].

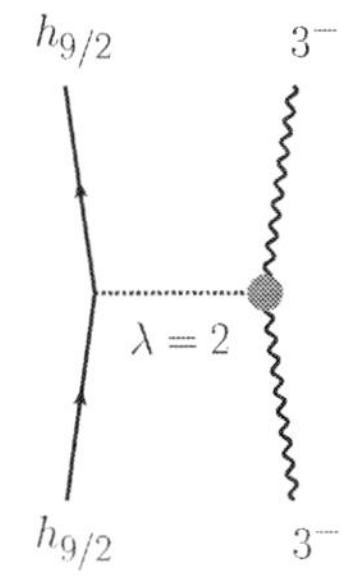

Figure 3. The diagonal matrix element of the interaction through the quadrupole moments of the phonon and the particle [20].

calculating the diagonal matrix element of the interaction through the quadrupole moments of the phonon and the particle shown in Fig.3. The resultant correction to the energy shift was estimated as

$$\delta E((h_{9/2}3^-)I) = 1.4Q_{el}(3^-)Q_{el}(h_{9/2})(-1)^{I-1/2}W(3\frac{9}{2}3\frac{9}{2};I2) \quad \text{keV} \cdot \text{fm}^{-4} \tag{3}$$

for the case of $e_{pol}(E2)$=0.5e, and it was indicated that if the value of $Q_{el}(3^-) \sim -1.0$b is used referring to the reported experimental values at that time [22, 23], the energy splitting of the septuplet becomes too big.

In 1984, Vermerr *et al.* published an updated value for the quadrupole moment of the 3^- state in ^{208}Pb as $Q_{el}^{exp}(3^-) = -0.34 \pm 0.15$ b [24, 25], and the value was used to reestimate the energy splitting of the septuplet in ^{209}Bi using coupling constants and energy shifts calculated by Bohr and Mottelson [1] and Hamamoto [21]. As the result, they concluded that the experimental value of -0.34 ± 0.15 b is somewhat larger than theoretical predictions and a value of $Q_{el}(3^-) = 0$ would give a better fit, but can in principle be accommodated within standard nuclear theories.

In the estimation of Fig.3 by Bohr and Mottelson, by Hamamoto, and probably also by Vermerr *et al.* accordingly, the value of $Q_{el}(h_{9/2}) \sim -0.4$b was used considering the experimental data known at that time [26]. Nowadays, however, there are various experimental data accumulated for the $Q_{el}(h_{9/2})$ ranging from $-0.37(4)$ to $-0.71(1)$b [27, 28], which should be confirmed [29]. Therefore, in case the experimental $Q_{el}(h_{9/2})$ is not so small, the above conclusion of Vermerr *et al.* needs to be reconsidered.

Furthermore, from our present point of view, also the particle-vibration coupling has to be reexamined by including non-linear field couplings coming from the self-consistent higher-order interactions. For the quadrupole-octupole coupling in ^{208}Pb region under consideration, the three-body $([Q_3Q_3]^{(2)} \cdot Q_2)$ interaction appearing in Eq.(2) is expected to play important roles together with the two-body $(Q_2 \cdot Q_2)$ and $(Q_3 \cdot Q_3)$ interactions [30]. The first-order diagrams of the three-body interaction contributing to the quadrupole moment of the 3^- state are illustrated in Fig.4.

I have calculated the contribution of the first-order diagrams by use of the modified oscillator model [31]. In the numerical calculation, all the particle-hole excitations within $3 \le N_p^{osc} \le 7$ and $4 \le N_n^{osc} \le 8$ have been included for the quadrupole mode, and for the amplitude of the collective octupole phonon the value of $\alpha_3 = 0.045$, which corresponds to $B(E3) = 39B_{sp}(E3)$ [1], has been adopted. As the result, the correction from the three-body interaction to the $Q_{el}(3^-)$

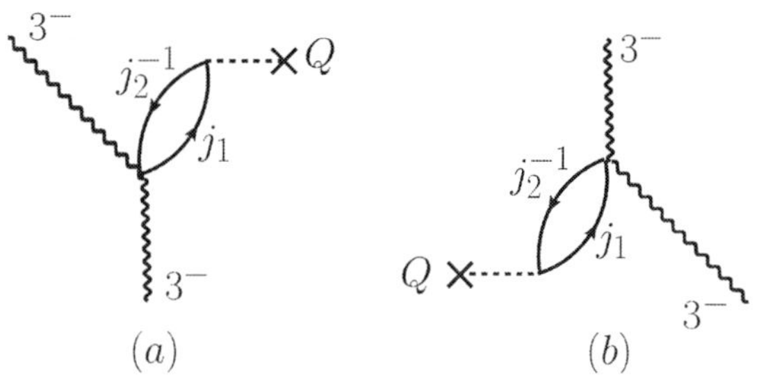

Figure 4. The first-order diagrams of the three-body interaction ($[Q_3Q_3]^{(2)}\cdot Q_2$) for the quadrupole moment of the 3^- state.

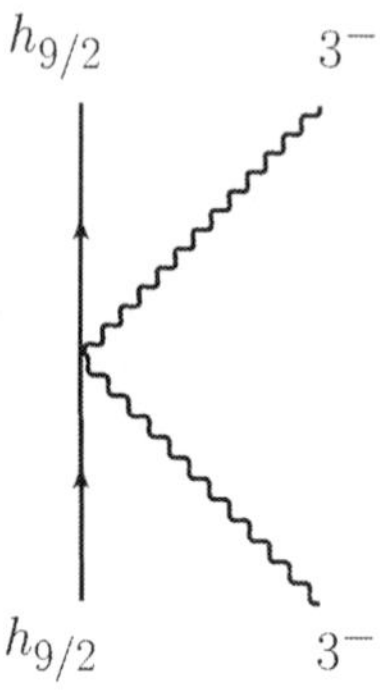

Figure 5. The diagonal matrix element of the three-body interaction ($[Q_3Q_3]^{(2)}\cdot Q_2$) for the energy splitting of the septuplet of states $(h_{9/2}3^-)I$.

moment has been estimated as $\delta Q_{el}(3^-) \sim 0.09$b. The estimation is somewhat preliminary because of the uncertainties involved in the calculation, but if the correctional value is added, for example, to the theoretical value estimated by Bohr-Mottelson [1] and Hamamoto [20, 21], the total $Q_{el}(3^-)$ moment becomes rather small. In this case, the theoretical moment might be too small compared to the experiment, but the observed small energy splitting of the septuplet can essentially be understood within the particle-vibration coupling model.

Thus, for the quadrupole moment of the octupole phonon under consideration, there seems to be a partial cancellation between the contribution of the second-order diagrams of the linear field-coupling, i.e. H_{coupl}, coming from the two-body interaction (Fig. 2) and that of the first-order diagrams of the quadratic field-coupling coming from the three-body interaction (Fig.4). Such a cancellation can also be seen in the energy shift of the septuplet. The leading-order contribution of the three-body interaction to the energy shift of the septuplet is illustrated in Fig.5, and is calculated as

$$\delta E'((h_{9/2}3^-)I) = -\chi_3' \langle 3^- \| \left[\hat{Q}_3\hat{Q}_3\right]^{(2)} \| 3^-\rangle \left\langle h_{9/2} \| \hat{Q}_2 \| h_{9/2}\right\rangle (-1)^{I-1/2} W(3\tfrac{9}{2}3\tfrac{9}{2};I2) \quad (4)$$

with

$$\chi_3' = \frac{3}{5}\sqrt{\frac{84\pi}{5}}\frac{\chi_3}{A\langle r^2\rangle_0}.$$

δE and $\delta E'$ have the same Racah coefficient, and are roughly the same order of magnitude with opposite signs with each other. Thus the contribution of Fig.3 to the energy splitting of the septuplet in ^{209}Bi is reduced by including the correction coming from the three-body interaction illustrated in Fig.5. As the result, the beautiful explanation for the structure of the energy splitting in terms essentially of the second-order diagrams in Fig.1 can be revived [32].

4. Summary and conclusions

Some properties of the self-consistent higher-order interactions, which have been derived to satisfy the *nuclear self-consistency* in higher-order accuracy, are sketched. The higher-order interactions have been expected to modify coupling terms significantly compared with the cases when one uses only the two-body multipole-multipole interactions.

IOP Conf. Series: Journal of Physics: Conf. Series **1023** (2018) 012003 doi:10.1088/1742-6596/1023/1/012003

To study roles especially of the $([Q_3Q_3]^{(2)} \cdot Q_2)$ type of the self-consistent three-body interaction, the quadrupole moment of the 3^- state in ^{208}Pb and the energy splitting of the septuplet of states $(h_{9/2}3^-)I$ with $I = 3/2, 5/2, \cdots, 15/2$ in ^{209}Bi are discussed from the perturbational point of view. It is shown that if the contribution of the three-body interaction is included, the theoretical value of the $Q_{el}(3^-)$ moment becomes rather small compared to the experiment, but the observed small energy splitting of the septuplet can essentially be understood within the particle-vibration coupling model.

The above cancellation mechanism seems to be one of the examples of the important role for the higher-order interactions. In general, it can be said that the self-consistent higher-order interactions provide mechanisms to restore the order in the system when more than one mode is simultaneously excited in a system.

References

[1] A. Bohr and B. R. Mottelson, *Nuclear Structure*,Vol. II (Benjamin, New York, 1975).
[2] T.Kishimoto, J.M.Moss, D.H.Youngblood, J.D.Bronson, C.M.Rozsa, D.R.Brown and A.D.Bacher, Phys. Rev. Lett. **35**, 552 (1975).
[3] T. Kishimoto, *Proc. Int. Conf. on Highly Excited States in Nuclear Reaction* (RCNP, Osaka 1980) p.145; p.377
[4] T. Kishimoto, T. Tamura and T. Kammuri, Prog. Theor. Phys. Suppl. **74&75**, 170 (1983).
[5] H. Sakamoto and T. Kishimoto, Nucl. Phys. **A501**, 205 (1989); **A501**, 242 (1989).
[6] E. R. Marshalek, Phys. Rev. Lett. **51**, 1534 (1983); Phys. Rev. C **29**, 640 (1984); Phys. Lett. **244B**, 1 (1990).
[7] M. Matsuo and K. Matsuyanagi, Prog. Theor. Phys. **78**, 591 (1987).
[8] H. Sakamoto and T. Kishimoto, Nucl. Phys. **A486**, 1 (1988); **A528**, 73 (1991).
[9] H. Aiba, Prog. Theor. Phys. **84**, 908 (1990).
[10] K. Yamada, Prog. Theor. Phys. **89**, 995 (1993).
[11] H. Sakamoto, Phys. Rev. C **52**, 177 (1995); **64**, 024303 (2001); **93**, 034315 (2016).
[12] H. Sakamoto, Acta Phys. Pol. B **48**, 573 (2017).
[13] Z. Bochnacki, Phys. Lett. **31B** 175 (1970).
[14] T. Suzuku and D.J. Rowe, Nucl. Phys. **A289**, 461 (1977).
[15] T. Suzuki and H. Sagawa, Prog. Theor. Phys. **65**, 565 (1981).
[16] Y. R. Shimizu and K. Matsuyanagi, Prog. Theor. Phys. **70**, 144 (1983); **71**, 960 (1984);**72**, 799 (1984).
[17] S. Åberg, Phys. Lett. **157B**, 9 (1985).
[18] B. R. Mottelson, *Nikko Summer School Lectures*, (NORDITA, Copenhagen, 1967) pub.288.
[19] B. R. Mottelson, *Nobel Lecture*, Dec.11, 1975; Science **193**, 287 (1976).
[20] I. Hamamoto, Nucl. Phys. **A126**, 545 (1969); **A135**, 576 (1969); **A141**, 1 (1970); **A155**, 362 (1970); **A232**, 445 (1974); Phys. Rep. **10**, 63 (1974).
[21] I. Hamamoto, in *Elementary modes of excitation in nuclei* Eds A. Bohr and R. A. Broglia (North-Holland, Amsterdam 1977) p.252.
[22] A. R. Barnett and W. R. Phillips, Phys. Rev. **186**, 1205 (1969).
[23] A. R. Barnett, S. F. Biagi, D. K. Olsen and W. R. Phillips, J. Phys. Soc. Jpn. Suppl. **34**, 389 (1973).
[24] W. Vermeer, M. Esat, J. Kuehner, R. Spear, A. Baxter, and S. Hinds, Aust. J. Phys., **37**, 123 (1984).
[25] M. J. Martin, Nucl. Data Sheets **108**, 1583 (2007).
[26] G. Eisele, I. Koniordos, G. Müller, and R. Winkler, Phys. Lett. **28B**, 256(1968).
[27] J. Bieroń and P. Pyykkö, Phys. Rev. Lett. **87**, 133003 (2001).
[28] J. Chen and F. G. Kondev, Nucl. Data Sheets **126**, 373 (2015).
[29] K. V. Koshelev, L. N. Labzowsky, and I. I. Tupitsyn, Phys. Lett. A **323**, 260 (2004).
[30] H. Sakamoto and T. Kishimoto, *Annual Report 1987* (Tandem Accelerator Center, Univ. of Tsukuba), UTTAC-**54** (1988) p.58.
[31] S. G. Nilsson, C. F. Tsang, A. Sobiczewski, Z. Szymański, S. Wycech, C. Gustafson, I.-L. Lamm, P.Möller and B. Nilsson, Nucl. Phys. **A131**, 1 (1969).
[32] I. Hamamoto, Phys. Scr. **91**, 23004 (2016).

IOP Conf. Series: Journal of Physics: Conf. Series **1023** (2018) 012004 doi:10.1088/1742-6596/1023/1/012004

Heavy neutron rich nuclei: production and investigation

S Zemlyanoy[1], K Avvakumov[1], N Kazarinov[1], V Fedosseev[2], R Bark[3], Z Blazczak[4] and Z Janas[5]

[1]Joint Institute for Nuclear Research, Dubna 141980, Russia
[2]CERN, Switzerland
[3]iThemba LABS, Nat. Research Foundation, South Africa
[4]Faculty of Physics, A. Mickiewicz University, Umultowska 85, 61-614 Poznan, Poland
[5]Faculty of Physics, University of Warsaw, Hoza 69, PL-00681 Warsaw, Poland

E-mail: zemlya@jinr.ru

Abstact. For production and investigation of heavy neutron rich nuclei devoted the new setup, which is under construction at Flerov Laboratory for Nuclear Reactions (FLNR) - JINR, Dubna now. This setup is planned to exploit available beams from the U-400M cyclotron in low energy multi-nucleon transfer reactions to study exotic neutron-rich nuclei located in the "north-east" region of nuclear map. Products from 4.5 to 9 MeV/nucleon heavy-ion collisions, such as ^{136}Xe on ^{208}Pb, are to be captured in a gas cell and selectively laser-ionized in a sextupole (quadrupole) ion guide extraction system.

PACS: 25.60.Je, 28.60.+s, 42.62.Fi

1. Introduction

In order to produce and study multinucleon transfer reaction products the GAs cell based Laser ionization an Separation setup (GALS) is designed at Flerov Laboratory for Nuclear Reactions (JINR). This setup is intended for production of heavy nuclei from the north-east part of nuclear map – near to or above the magic neutron number $N = 126$. At last years [1-5] heavy neutron-rich nuclides are of great interest for nuclear astrophysics. This becomes obvious when one looks at the predicted path for r-process nuclei and their decay (see Fig. 1 in [1] and [2]). In addition, study of the nuclear structure of extremely neutron rich nuclei is important for the basic nuclear spectroscopy. New information can be obtained about the changes of nuclear ground state properties, e.g. appearance of new or disappearance of the classical magic numbers (as in the case of light neutron-rich nuclei) as well as the occurrence of rapid nuclear structural changes, due to sudden onset of collectivity.

It should be noted that the region of heavy neutron rich nuclei along the neutron line $N = 126$ is still a blank field of the nuclide chart. Recently new isotopes were unambiguously identified in reactions with a ^{238}U beam impinging on a Be target at 1 GeV/nucleon [6]. The production cross-section for the new isotopes has been measured down to the pico-barn level and compared with predictions of different model calculations. Nevertheless the magic neutron number $N = 126$ for atoms with $Z = 75$ - 78 is as yet not achieved. This is because production and study of heavier neutron rich nuclei with $A > 160$ encounters both physical and technical problems (for details see [1, 4]). The production of neutron-rich heavy and superheavy nuclei using low-energy multinucleon transfer reactions has been discussed by Zagrebaev and Greiner [4]. The high predicted yield, shown in Fig. 1, opens a new opportunity in the studies at heavy ion facilities.

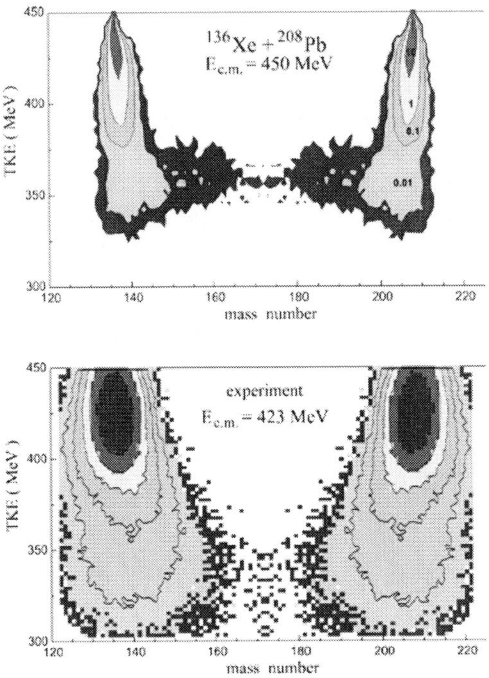

Figure. 1 Dependence of the products yield of transfer reactions on the total kinetic energy and mass number. The figure compares the theoretical predictions of [4] (above) and the experimental results (below) from the JINR cyclotron U-400M (courtesy of V. Zagrebaev and E. Kozulin). The numbers in the right-hand side of the theoretical predictions present the cross-section in mb.

2. Experimental techniques

The experimental technique is already described in detail in [1-3, 7, 8]. Here we present only a brief,

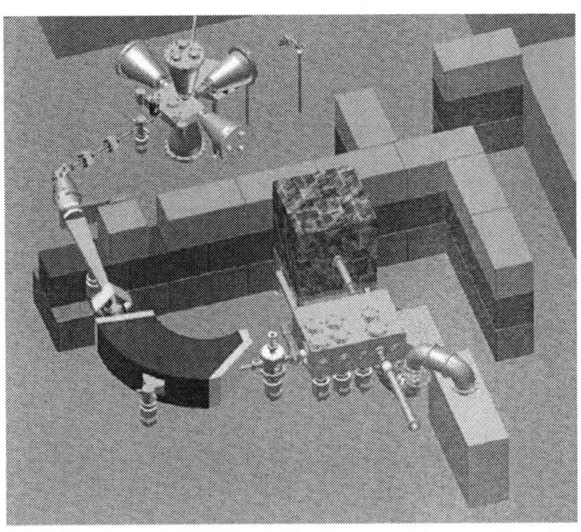

Figure 2. The allocation of the whole GALS facility in experimental cave near the hall of the cyclotron U-400M is shown in real scale.

overview of the experimental method that the GALS setup will use. The method, as proposed and applied in [7, 8], combines simultaneous *Z* and *A* separation based on stopping nuclei in gas cellsubsequent resonance laser ionization and mass-separation. The scheme of the GALS facility can be found in [3].

The operation principle is described in a number of papers (see e.g. [9] and the references therein). The first stage of laser system based on three Syra dye lasers pumped by two EgewaveYAG lasers and laser lab preparatory is almost finished. Now we concentrate on the GALS subsystems, located in the cave of Cyclotron U-400M experimental hall. Fig. 2 presents how these subsystems (Front-End, Enzel lenz, mass-separator, focal chamber and registration system) of GALS setup looks like. These subsystems are now in the advanced stage of realization. Mass separator could be a standard magnet separator similar to GPS at ISOLDE II [2, 10]) but nevertheless substantial efforts are needed for its realization. The same refers to the ion extraction and gas purification systems.

3D simulation of the beam dynamics in the GALS mass-separator was carried out in [11]. This simulation used 3D map of the AM magnetic field calculated using OPERA 3D code [12]. The number of particles used in this simulation was equal to 2 10^4. The computational model of the analyzing magnet is shown in Fig. 3.

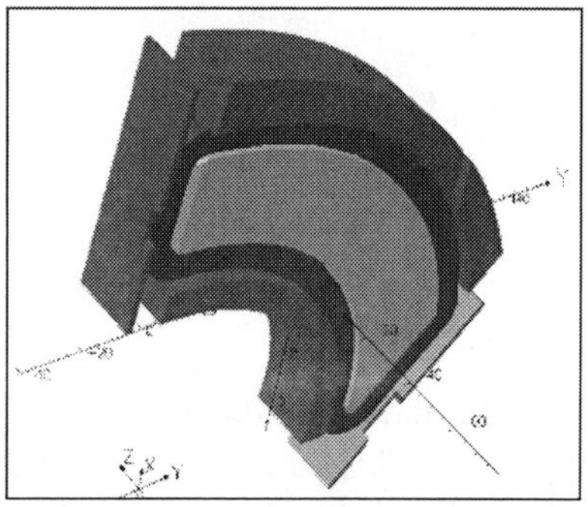

Figure 3. Computational model of the analyzing magnet

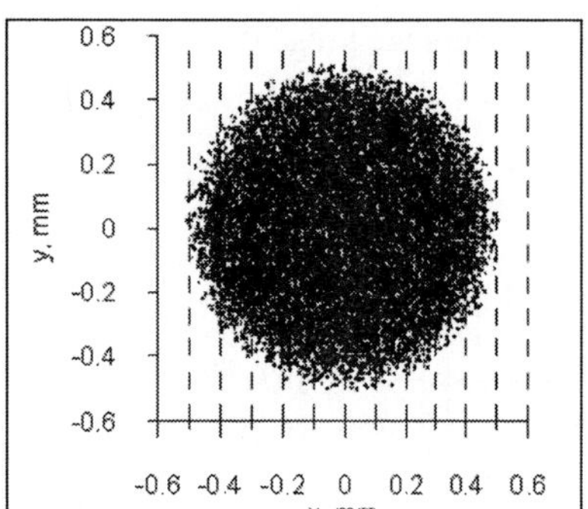

Figure 4. Initial particle distribution in plane {x,y}

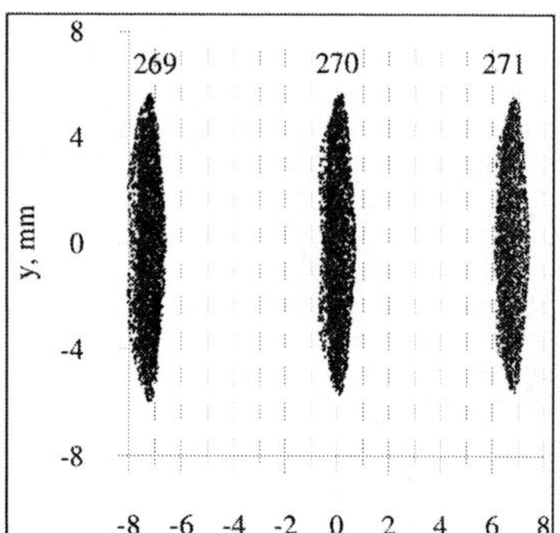

Figure 5. Ion distribution in the magnet focal plane

Fig. 6 shows the calculated distribution of the particles in the analysing magnet focal plane for the ions with A=269, 270, and 271. The estimated mass resolution of the spectrometer R_m = 1400.

3. The ion guide simulation

The GALS front end vacuum chamber is designed to be used with different types of gas cells, the experiments will be started with the dual chamber gas cell (Fig. 6), and a horn-type gas cell is planned

IOP Conf. Series: Journal of Physics: Conf. Series **1023** (2018) 012004 doi:10.1088/1742-6596/1023/1/012004

to be used later for better evacuation efficiency of atoms of interest and for the experiments with in-gas-jet ionization [13]. The second option means that an S- or L-shaped segmented ion guide could be used in the gas cell chamber, so there should be enough space for it. Thus, the distance between the cyclotron beam and the extraction electrode is quite long (approximately 740 mm). On the first stage of the GALS setup a single long (about 630 mm) sextupole ion guide (SPIG) is planned to be used for guiding the ion beam from the gas cell to the extraction and acceleration electrostatics. In order to estimate possible losses of the ions interacting with the gas jet and background gas the simulations of the system containing the gas cell exit and the SPIG were carried out.

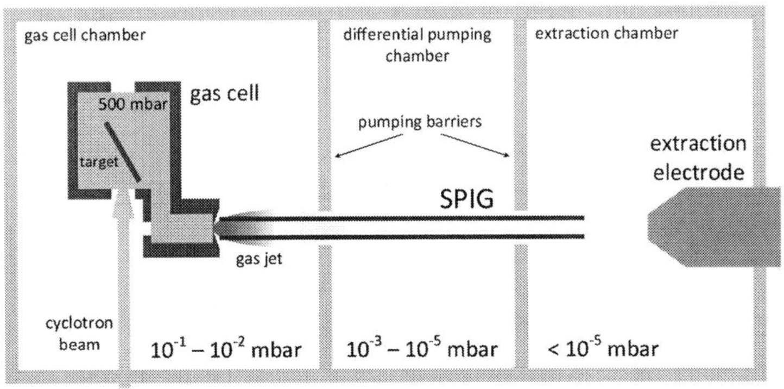

Figure 6. A schematic layout of GALS front-end.

Properties of a supersonic jet in a gas expansion from a gas cell to a vacuum chamber through a round orifice were discussed in detail in [14]. The simulations have been performed in the SIMION software package [15] using the hard sphere collision model and the additional code describing the gas jet [16]. In the model, the thermal ions propagate in a filled-cone direction distribution of half angle 30° out of the gas cell orifice, which has a diameter of 0.5 mm. The SPIG with an inner diameter of 3 mm consists of six rods (with a diameter of 1.5 mm and 630 mm long) cylindrically mounted on a sextupole structure, the distance from the orifice was set to 2 mm. The radiofrequency (4.7 MHz) voltage with peak-to-peak amplitude of 300 V is applied to the neighboring rods in the opposite phase, and the whole SPIG has a voltage of –210 V compared to the gas cell [17]. The pressure of gas (argon) was set to 500 mbar in the gas cell, $2 \cdot 10^{-2}$ mbar in the gas cell chamber, $2 \cdot 10^{-4}$ mbar in the differential pumping chamber and $2 \cdot 10^{-6}$ mbar in the extraction chamber.

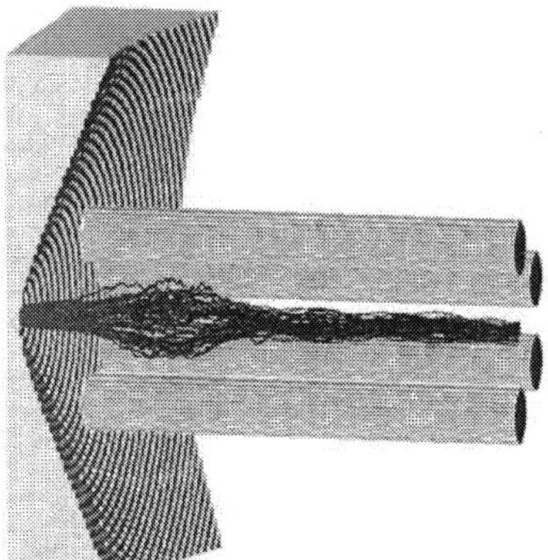

Figure 7. Simulated trajectories of the ions at SPIG entrance. The collisions with the buffer gas are marked with red

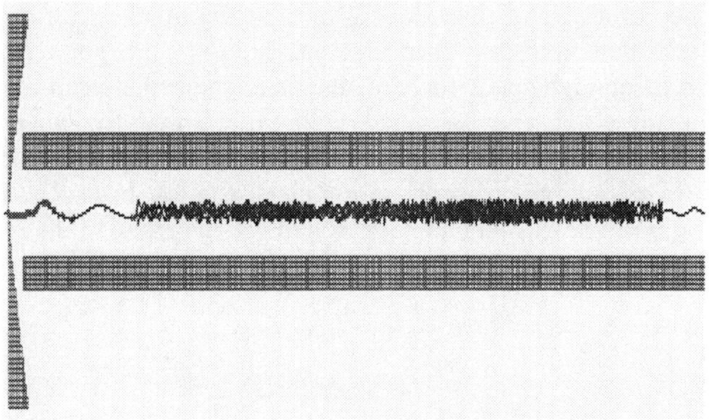

Figure 8. A trajectory of a single ion caught by a buffer gas inside the SPIG

Due to the simulations, almost no losses occur at SPIG entrance, but when the ions proceed from the gas jet to the background, they quickly lose most of the kinetic energy, which lead to high losses or at least very high time of flight. The trajectories of the ions interacting with the gas jet at the SPIG entrance are shown on Fig. 7. The ions' mean time of flight, mean longitudinal kinetic energy KE_x, their energy dispersion σ_{KE_x} and SPIG efficiency depending on the distance from the gas cell to the skimmer between the first and second cameras of the vacuum chamber are shown in Table 1, statistics of 200 ions used for each case, the ions' parameters were captured at the end of SPIG. As one can see when the distance is higher than 60 mm, the time of flight rises drastically, with decreasing of the kinetic energy along the beam axis. At such distances, the ions are being caught by buffer gas and it takes them about 5 ms or longer to diffuse to the SPIG exit, which is comparable with the time of evacuation from the gas cell. Fig. 8 represents an example a trajectory of such an ion inside the SPIG. The other option is losing such ions in collisions with the SPIG rods, but as the simulations show, the probability of this is rather low.

Based on the simulations results, one could assume the distances up to 60 mm to be acceptable for using experimentally, but the model used in the simulations does not describe in detail the slow pressure change in the skimmer aperture area, which means that experimental energy losses should be higher than indicated in the table. Thus, obtaining maximal efficiency and minimal time of flight is expected in case of skimmer location of 45-50 mm from the gas cell. These results are planned to be tested by further calculations and experiments.

Table 1. The parameters of the ions at the end of SPIG in dependence of the distance between the skimmer and the gas cell

Distance from the gas cell to the skimmer	Mean time of flight, μs	Mean longitudinal kinetic energy KE_x, eV	Longitudinal kinetic energy dispersion σ_{KEx}, eV	Efficiency, %
40 mm	1182.8	0.3031	0.1264	99
45 mm	1220.6	0.2901	0.1140	100
50 mm	1187.8	0.2980	0.1164	99.5
55 mm	1246.6	0.2935	0.1074	99.5
60 mm	1192.7	0.2986	0.1008	99.5
65 mm	6747.3	0.0497	0.0515	99
70 mm	9417.6	0.0394	0.0581	95

4. Planned experiments

We plan to perform the first online test on Os I. As already shown in [6], Os is very perspective candidate to achieve the neutron magic line $N = 126$. The last calculations of A.V. Karpov [15] predicted sufficiently high reaction cross section for Os isotope even above this magic line.

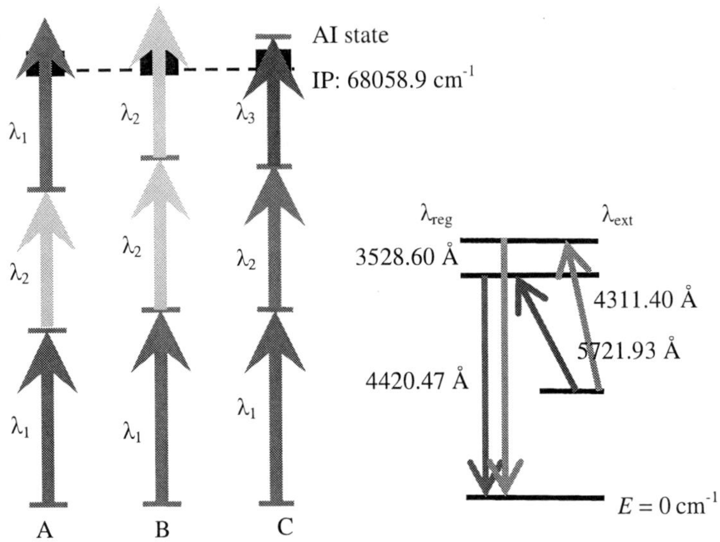

Fig. 9 *Left*: possible ionization schemes of Os I (the wavelengths (in air) A: $\lambda_1 = 3267.94$Å $\lambda_2 = 5509.33$Å, B: $\lambda_1 = 3018.04$Å $\lambda_2 = 5580.66$Å, C[20]: $\lambda_1 = 2909.06$Å $\lambda_2 = 4752.16$Å $\lambda_3 = 6043.15$Å). *Right*: transitions for the off-line test experiments

Many spectroscopic data of Os I have been studied to find appropriate two- or three-steps transitions for laser ionization. Although there are many observed spectral lines as yet the atomic spectra information is not complete: most of the transitions are not assigned [16 - 19] and the atomic strengths are not known. Nevertheless several of the possible RIS scheme for Os have been previously studied [20, 21]. Three of them that seem to be more effective (see e.g. [21]) are shown in Fig. 9. The most ionization schemes are two colored three steps, the 3rd step being non-resonant.

By the end of 2018 off-line experiments with Os I will be started. Their aim is to check the quality of the reference cell using excitation and registration with different wavelengths (see right hand side of Fig. 7).

The most favorable ionization scheme is expected to be the resonance three steps one via an auto-ionizing state. At present no literature information is available about such states except the one at 71032 cm^{-1} corresponding to the third step of case C [20]. For this reason at the second stage of the GALS activities a search for auto-ionizing states is planned. At the next phase of experiments an offline study of the optimal ionization scheme will be conducted. It supposes a detailed experimental study the efficiency of the different ionization schemes.

5. Conclusion

A new GALS facility at FLNR cyclotron U-400M is in a stage of build-up. GALS will apply the highly selective and efficient technique of stepwise resonant ionization in a gas cell (of the element of interest) with subsequent mass separation. The most efficient method to produce such nuclei, as motivated in [4], are the multi-nucleon transfer reactions. Our estimations show that at target thickness 0.3 mg/cm, ion beam of 0.1 pμA and setup efficiency of 10% we would be able to measure decay properties of 1 new isotope per day.

IOP Conf. Series: Journal of Physics: Conf. Series **1023** (2018) 012004 doi:10.1088/1742-6596/1023/1/012004

References

[1] Zagrebaev V I, Zemlyanoy S G, Kozulin E M, Kudryavtsev Yu, Fedosseev V, Bark R, Othman HA 2013 *Hyperfine Interact* **216,** 109

[2] Zagrebaev V I, Zemlyanoy S G, Kozulin E M *et al.* 2014 *Hyperfine Interact* **227,** 181

[3] Zagrebaev V I, Zemlyanoy S G, Kozulin E M *et al* 2016 *J. Phys.Conf. Series* Series **724,** 012057

[4] Zagrebaev V and Greiner W 2008 *Phys. Rev. Lett.* **101**, 122701

[5] Long Zh, Zhao-Qing Feng, Feng-Shou Zhang 2015 *J. Phys. G: Nucl. Part. Phys.* **42,** 085102

[6] Kurcewicz, J., Farinon, F., Geissel, H. et al. 2012 *Phys. Lett.* B **717** 371

[7] Kudryavstev Y *et al.* 2003 *Nucl. Instrum. Methods Phys. Res.* B **204**, 336

[8] Moore I D *et al* 2005 *J. Phys. G: Nucl. Part. Phys.* **31**, S1499

[9] Campbell P, Moore I D and Pearson M R 2016 *Prog. Particle Nucl. Phys.* **86,** 127

[10] Ravn H L and Allardyce B W 1989 In: *Treatise on Heavy-Ion Science*, **8**, 363

[11] Kazarinov N Yu 2016 *Physics of Particles and Nuclei Letters*, **13**, No. 7, 868

[12] Opera3D 2012 *Oxford OX5 1JE*, England

[13] Ferrer R *et al* 2013 *Nucl. Instrum. Methods Phys. Res.* B **317**, 570

[14] Kudryavtsev Y *et al* 2013 *Nucl. Instrum. Methods Phys.* Res. B **297**, 7

[15] http://simion.com/

[16] Traykov E (GANIL, CEA/DSM-CNRS/IN2P3, Caen, France) private communication

[17] Zemlyanoy S, Avvakumov A, Fedosseev V et al. 2017 *Hyperfine Interact* **238**, 31

[18] Karpov, A.V. EXON 2016 to be published

[19] Corliss, Charles H., Bozman, William R. Experimental transition probabilities for spectral lines of seventy elements, *NBS Monograph* **53** (1962)

[20] Meggers, William F. , Corliss, Charles H., Bourdon F. Scribner: Tables of spectral lines intensities (second edition), *NBS Monograph* **145** (1975)

[21] https://www.nist.gov/pml/atomic-spectra-database (version 5, updated October 2015)

[22] Sansonetti, J E, Martin W C 2005 *J. Phys. Chem. Reference Data* **34**, 1559

[23] Blum Joel D, Pellin M J, Calawa, W F et al. 1990 *Anal. Chem.* **62**, 209

[24] Calaway W F, Wien R C, Burnett D S et al. 1995 *J. Vac. Sci. Technol.* A **13**, 1310

IOP Conf. Series: Journal of Physics: Conf. Series **1023** (2018) 012005 doi:10.1088/1742-6596/1023/1/012005

Analysis of Nuclear Lifetimes Using the Gamma-ray Induced Doppler Shift Attenuation Method

F.C.L. Crespi

Università degli Studi di Milano and INFN, Milano, I-20133, Italy

fabio.crespi@mi.infn.it

Abstract. Lifetime measurements allow extraction of fundamental information on the nature of the excited states of a nuclear system. Since nuclear lifetimes cover many orders of magnitude, a number of experimental techniques and detection setups have been developed depending on the range of the lifetime of interest. The Gamma-ray Induced Doppler Shift Attenuation (GRIDSA) Method presented here is applied to the measurement of very short lifetimes, in the femtosecond range. It allows determining the nuclear lifetime by measuring the Doppler shift of a gamma ray emitted from the state of interest, in different directions with respect to a coincident preceding gamma ray, populating the same state and inducing a recoil of the nucleus in the target material with velocities of the order of 10^4-10^5 m/s. We realized an experiment in order to test the GRIDSA technique for the measurement of fs lifetimes after (n,γ) reactions. The measurement was performed at the Institut Laue-Langevin (ILL) with the 8 Ge-clover detectors of the FIPPS array. Preliminary results are discussed.

1. Introduction

It is well known that lifetime measurements allow us to extract fundamental information on the nature of the excited states of a nuclear system [1]. The lifetimes associated to gamma decay span many orders of magnitude, consequently a number of different experimental techniques has been developed. Each one of these is more appropriate for a particular range of lifetimes and/or a particular range of nuclei. The Gamma-ray Induced Doppler Shift Attenuation (GRIDSA) Method presented here is applied to the measurement of very short lifetimes, in the femtosecond range.

Since many years the attenuation of Doppler shift and broadening has been used in gamma spectroscopy to study very short lifetimes of nuclear states. In this context one of the most popular techniques is the Doppler Shift Attenuation Method (DSAM). In its application the interaction of accelerated ions with target nuclei generates recoiling ions with high velocity (several percent of c). The DSAM is used for measuring lifetimes in the 10^{-13} s < τ < 10^{-15} s interval.

Actually, the Gamma-Ray Induced Doppler Broadening (GRID) has been also used to measure lifetimes since a long time. This technique was developed at the Institut Laue Langevin (ILL) in Grenoble, making use of the ultra-high resolution double flat crystal spectrometers GAMS [2]. The GRID technique is based on observing Doppler broadened line shapes of gamma-rays after (n,γ) reactions. In this case the Doppler broadening is induced by a preceding gamma ray, the energy of the nuclear recoil is small (several hundreds of eV only). Nuclear state lifetime can be extracted when comparing these line shapes to computer simulations of the atomic slowing down. However, some features limit its applicability, for example: it requires massive samples (several grams), strong de-

IOP Conf. Series: Journal of Physics: Conf. Series **1023** (2018) 012005 doi:10.1088/1742-6596/1023/1/012005

excitation transitions and a good knowledge of the complete feeding scenario populating the level of interest. These limitations are not present in the Gamma-ray induced Doppler shift attenuation (GRIDSA) method, which is described in the next section.

2. The GRIDSA Method

The GRIDSA method [3] shares the same basic principles with the GRID technique, however in this case the Doppler- shift instead of -broadening is measured. In its application, following a (n,γ) reaction, a primary gamma ray with energy E_γ is emitted and consequently (for momentum conservation) the emitting nucleus must recoil in the opposite direction. Considering a simple two-step cascade to the ground state, the second emitted gamma ray can be Doppler shifted. This Doppler shift is the result of two effects on different time scales: the lifetime of the nuclear state the atomic time scale of the slowing down of the recoiling atoms in the lattice of the target material. One of these two features must be known in order to extract the other. In the present case, we aim to extract information on the nuclear level lifetime and the slowing down process of the nuclei in the target can be simulated starting from some physical assumptions on the type of collisions and interaction between atoms, and generating a velocity trend, as a function of time.

We applied the GRIDSA method using an array of 8 Ge-clover detectors: the FIPPS array operated at the ILL laboratory [see panel a) of figure1]. Let's consider gamma rays of a two-step cascade, where the level of interest (i.e. the one of which we are interested in measuring the lifetime) is between the two transitions. For the second among these two gamma rays in cascade, the events, which contribute to the peak can be separated in different spectra on the basis of the emission angle between the first and the second Ge-clover detector that fired in coincidence, then the shift between the peak of each spectrum and a reference peak that it is not Doppler shifted is measured. This difference is expected (provided that the lifetime of the level of interest is sufficiently short) to follow the trend illustrated in the panel b) of figure 1 and it is given by the formula: $\Delta E_{\gamma 2} \approx E_{\gamma 2} \cdot v_R/c \cdot \cos(\theta)$. Provided to have sufficient statistics, a way to improve the angular resolution consists in using the position information associated to the single crystals inside the clover. The centres of the crystals in the same clover can be at 24 degrees in case they are one next to each other, or at 35 degrees if they are on the diagonal. In this way the granularity in the angular coordinate improves and the curve indicated in the panel b) of figure 1 can be sampled with more experimental points.

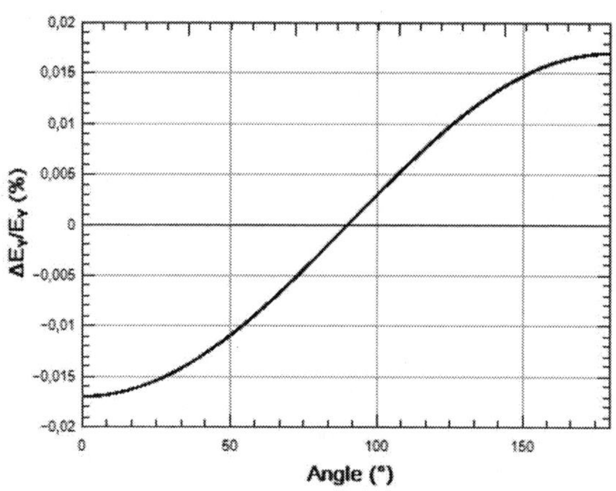

Figure 1. Panel a): picture of the 8 clover HPGe detectors of the FIPPS array operated at ILL. Panel b): Representation of the expected Doppler shift divided by the energy of the second gamma ray emitted, as a function of the angle between the first and the second gamma ray.

IOP Conf. Series: Journal of Physics: Conf. Series **1023** (2018) 012005 doi:10.1088/1742-6596/1023/1/012005

3. Experimental test of the GRIDSA and Preliminary Results

We proposed an experiment in order to test the GRIDSA technique to measure fs lifetimes after (n,γ) reactions. The experiment was performed in January 2017 at ILL with the 8 clover detectors of the FIPPS array [see panel a) of figure 1]. Different targets were used in order to scan over a certain range of isotope masses, lifetimes and recoil velocities. We present here preliminary results associated with the case of NaCl target. The studied case is the one of levels in the ^{36}Cl isotope, recoiling in the target after the reaction ^{35}Cl(n,γ)^{36}Cl realized with thermal neutrons. The maximum shift induced by the Doppler effect was expected to be ~0,5 keV, with v_R=5,1·104 m/s, (maximum velocity, when the nucleus begins to recoil).
In particular the first studied cascade was the one that, from the capture level at 8580 keV, de-excites the nucleus to the ground state with two gamma rays of 5715keV and 2864keV [figure 3 panel a)]. This case was chosen because the mean lifetime of the intermediate state is very short (about 23fs), the first gamma ray is rather energetic, so it can induce a sufficiently relevant recoil of the nucleus, and the second gamma ray is also rather energetic; in fact, the energy shift that we want to measure depends linearly on the energy of the second gamma ray. Moreover, the second gamma ray has to be emitted in the very short period of time in which the nucleus is still recoiling ($\sim 10^{-13}$s). A shift was therefore expected to be observed and, in fact, comparing the peaks relative to the events in which the two Ge-clover detectors that fired in coincidence are at 45 deg and 135 deg respectively [see figure 2 panel a)], a separation between the centroids is clearly visible as it is shown in the spectra of the panel b) in figure 2. As expected the 2864 keV peak is shifted towards lower energies in the case the two Ge-clover detectors that fired have a smaller relative angle (45 deg), since in this case the ^{36}Cl nucleus is recoiling moving away from the crystal that detected the 2864 keV gamma ray. The opposite is true for the case of the two Ge-clover detectors having a larger relative angle of 135 deg.
The centroids of the 2864 keV peak were calculated for each gamma spectrum obtained imposing a gating condition on the angle between the two HPGe crystals that fired in coincidence (detecting respectively the 2864 keV and the 5715 keV gamma rays from the excited ^{36}Cl nucleus). These centroids are plotted as a function of the angles between the two HPGe crystals in the plot displayed in the panel b) of figure 3. The trend obtained reflects that expected from the calculated curve shown in the panel b) of figure 1.

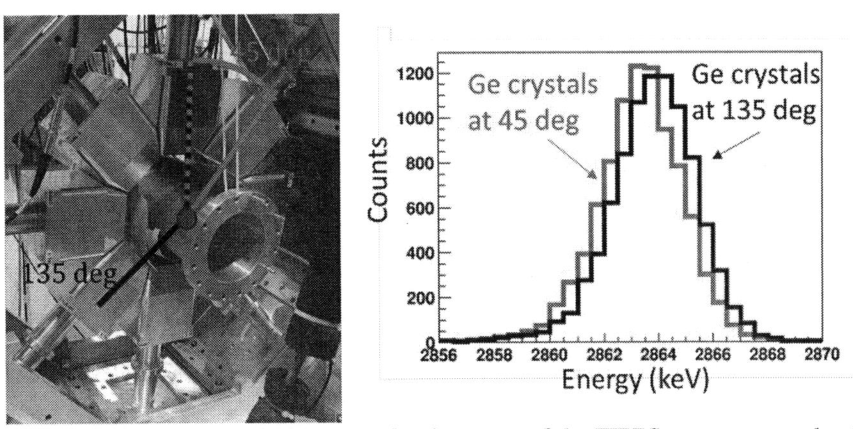

Figure 2. Panel a): picture of the 8 clover HPGe detectors of the FIPPS array operated at ILL. Panel b): peaks relative to the events in which the two Ge-clover detectors that fired in coincidence are at 45 degrees and 135 degrees respectively, a separation between the centroids is clearly visible.

IOP Conf. Series: Journal of Physics: Conf. Series **1023** (2018) 012005 doi:10.1088/1742-6596/1023/1/012005

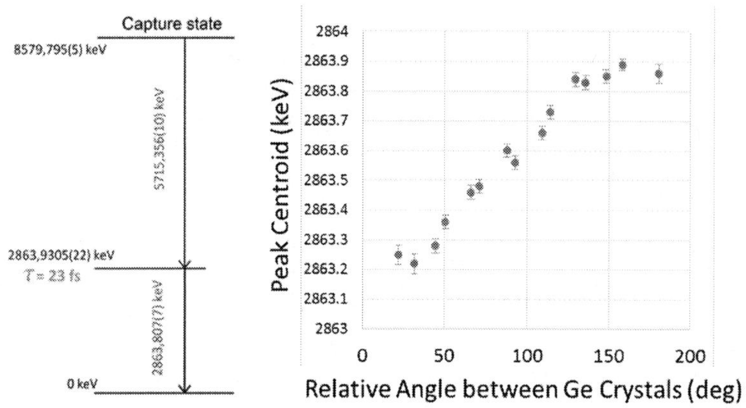

Figure 3. Panel a): Scheme of the 2-step transition trough the intermediate state of 2864 keV , with short mean lifetime: τ = 23fs. Panel b): the centroids of the 2864 keV peak plotted as a function of the angle between the two HPGe crystals that fired in coincidence.

However, in order to extract quantitative information on the nuclear level lifetimes the slowing down process of the recoiling nuclei in the target has to be under control. A simulation was performed using the SLOWDOWN code [4]. In the model the nuclei are considered to be placed in a lattice. After the neutron capture reaction the nucleus emits a high-energy gamma ray, which induces the nucleus itself to recoil. Then the nucleus gradually slows down, due to the interaction with the other atoms in the lattice. During this slowing down process other gamma rays can be emitted. This simulation allows us to obtain a velocity distribution at an instant of time corresponding to the decay time of the considered nucleus. This is shown in figure 4. The next step (which is not yet completed since the data analysis is still ongoing) consists of using the results of this simulation to produce calculated data to be compared with the experimental ones (e.g. like those displayed in the panel b) of figure 3) and then extract the measured value of the lifetime.

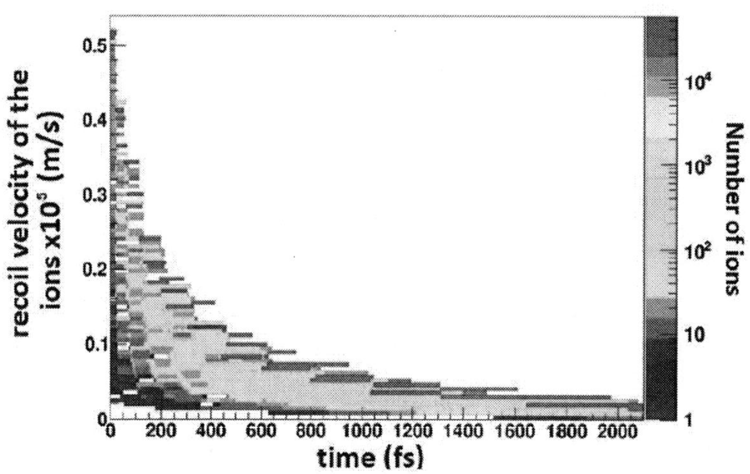

Figure 4. Two-dimensional histogram representing the simulated recoil velocity as a function of time for the case of ^{36}Cl nuclei: the different colors represent the probability, expressed in number of ions with a particular velocity at a specific instant of time.

IOP Conf. Series: Journal of Physics: Conf. Series **1023** (2018) 012005 doi:10.1088/1742-6596/1023/1/012005

References

[1] Nolan P J and Sharpey Schafer J 1979 Rep. Prog Phys 42 1
[2] H.G. Börner and J. Jolie, J. Phys. G 19 (1993) 217.
[3] T. Khan, T. von Egidy, F.J. Hartmann, J. Ott, M. Jentschel, Nucl. Instr. Meth. A 385 (1997) 100-107
[4] M. Jentschel, *Description of SLOWDOWN*, private communication.

Experimental challenges for the measurement of the ^{116}Cd(^{20}Ne,^{20}O)^{116}Sn double charge exchange reaction at 15 AMeV

D. Carbone[1], F. Cappuzzello[1,2], C. Agodi[1], M. Cavallaro[1], L. Acosta[3], D. Bonanno[4], D. Bongiovanni[1], T. Borello[5], I. Boztosun[6], S. Calabrese[1,2], D. Calvo[7], E.R. Chávez Lomelí[3], N. Deshmukh[1], P.N. de Faria[8], P. Finocchiaro[1], M. Fisichella[7], A. Foti[2,4], G. Gallo[2,4], A. Hacisalihoglu[1,9], F. Iazzi[7,10], R. Introzzi[7], G. Lanzalone[1,11], R. Linares[8], F. Longhitano[4], D. Lo Presti[2,4], N. Medina[5], A. Muoio[1], J.R.B. Oliveira[5], A. Pakou[12], L. Pandola[1], F. Pinna[7,10], S. Reito[4], G. Russo[2,4], G. Santagati[1], O. Sgouros[12], S.O. Solakcı[6], V. Soukeras[12], G. Souliotis[12], A. Spatafora[1,2], D. Torresi[1], S. Tudisco[1], A. Yildirim[6], V.A.B. Zagatto[8] for the NUMEN collaboration

[1] INFN-Laboratori Nazionali del Sud, Catania, Italy
[2] Università degli Studi di Catania, Catania, Italy
[3] Universidad Nacional Autònoma de México, Ciudad de México, Mexico
[4] INFN-Sezione di Catania, Catania, Italy
[5] Universidade de Sao Paulo, Sao Paulo, Brazil
[6] Akdeniz University, Antalya, Turkey
[7] INFN-Sezione di Torino, Turin, Italy
[8] Universidade Federal Fluminense, Niteroi, Brazil
[9] Istitute of Natural Science, Karadeniz Teknik Universitesi, Trabzon, Turkey
[10] Politecnico di Torino, Turin, Italy
[11] Università di Enna "Kore", Enna, Italy
[12] University of Ioannina, Ioannina, Greece

Abstract. The knowledge of the nuclear matrix elements (NME) entering in the expression of the half-life of the neutrinoless double beta decay is fundamental for neutrino physics. Information on the nuclear matrix elements can be obtained by measuring the absolute cross section of double charge exchange nuclear reactions. The two processes present some similarities, the initial and final-state wave functions are the same and the transition operators are similar. The experimental measurements of double charge exchange reactions induced by heavy ions present a number of challenging aspects, since such reactions are characterized by very low cross sections. Such difficulties are discussed for the measurement of the ^{116}Cd(^{20}Ne,^{20}O)^{116}Sn reaction at 15 AMeV.

1. Introduction

The physics of neutrinoless double beta (0νββ) decay has fundamental implications on particle physics, cosmology and fundamental physics. In particular it is considered one of the most promising ways to probe the Majorana or Dirac nature of neutrino and to have access to its effective mass. Furthermore, the observation of 0νββ would signal that the total lepton number is not conserved. For these and other fundamental implications, this physics case is presently one of the most important researches Beyond the Standard Model and might guide the way towards a Grand Unified Theory of

fundamental interactions. Since the 0νββ decay process involves nuclei, its analysis necessarily implies nuclear structure items. The 0νββ decay rate can be expressed as a product of three independent factors: the phase-space factor, the Nuclear Matrix Element (NME) and a term containing the effective neutrino masses. Thus, even if the decay rate will be measured, the knowledge of the NME is mandatory to extract information on the neutrino masses.
The evaluation of the NMEs is presently limited to state of the art model calculations based on different approaches (QRPA, shell-model, IBM etc.) [1-4]. However, significant differences in the obtained values are still found, due to the ambiguities in the models and the lack of strong constraints. Moreover, possible common approximations can correspond to systematic uncertainties.
The possibility of using heavy-ion induced Double Charge Exchange (DCE) reactions as tools to have experimentally driven information on the NMEs is at the basis of the NUMEN [5,6] and the NURE [7-8] projects. Despite the two processes are mediated by different interactions, there are some important similarities between them: i) the initial and final state wave functions in the two processes are the same, ii) the transition operators are similar, in both cases Fermi, Gamow-Teller and rank-two tensor components are present, iii) a large linear momentum (~100 MeV/c) is available in the virtual intermediate channel, iv) the two processes are non-local and are characterized by two vertices localized in a pair of valence nucleons, v) they take place in the same nuclear medium, vi) a relevant off-shell propagation through virtual intermediate channels is present.
The description of NMEs extracted from DCE and 0νββ presents the same degree of complexity, with the advantage for DCE to be "accessible" in laboratory. However a simple relation between DCE cross sections and ββ-decay half-lives is not trivial and needs to be explored.
In this context, an experimental campaign has started at the INFN-Laboratori Nazionali del Sud in Catania using the MAGNEX large acceptance magnetic spectrometer [9] focused on DCE reactions involving the nuclei of interest for 0νββ decay. These reactions represent an important experimental challenge since they are characterized by very low cross-sections and require a high energy resolution to distinguish the transitions in the region of the ground state. Both constraints are guaranteed by the use of the MAGNEX spectrometer, a tool with high performance and flexibility. In particular, the ^{116}Cd(^{20}Ne,^{20}O)^{116}Sn DCE reaction at 15 AMeV was recently measured for the first time. Some details about the experimental issues of this measurement are discussed in this paper.

2. Heavy-ion double charge exchange reactions

Due to the possibility to obtain information on the NMEs of ββ processes, the experimental study of nuclear transitions where the nuclear charge is changed by two units leaving the mass number unvaried was explored in the past. A large effort was done to study pion double charge exchange reactions (π^+,π^-) [10-11]. A factorization of the cross section in terms of a nuclear structure part, which is related to the nuclear matrix element, and a reaction part has been deduced in these cases [12-13]. However, suggestions that pion double charge exchange might be used to probe 0νββ decay NMEs were abandoned, due to the large differences in the momentum transfers and in the nature of the operators, as reported in [14]. Early studies of heavy-ion induced DCE reactions were also inconclusive. The reason was the lack of zero-degree data and the poor yields in the measured energy spectra and angular distributions, due to the very low cross sections involved, ranging from about 5–40 nb/sr [15] to 10 μb/sr [16]. Actually, this wide range of observed cross sections has never been deeply discussed. An additional complication in the interpretation of the data was due to possible contributions of multi-nucleon transfer reactions leading to the same final states [17].
Nowadays these experimental limitations are almost overcome, as we demonstrated in a pioneering experiment performed at the INFN-LNS laboratory. In particular, we studied the DCE reaction ^{40}Ca(^{18}O,^{18}Ne)^{40}Ar at 15 AMeV, with the aim to measure the cross section at zero degrees [18]. The key elements in the experiment were the high resolution Superconducting Cyclotron (CS) beams and the use of MAGNEX, a modern high resolution and large acceptance magnetic spectrometer characterized by high resolution in energy, mass and angle [19-20]. The high-order solution of the equation of motion is the key feature of MAGNEX, which guarantees the above mentioned performances and its relevance in the research of heavy-ion physics [21-25]. In this experiment we

have shown that high resolution and statistically significant experimental data can be measured for DCE processes and that precious information towards NME determination could be at our reach.

3. The NUMEN and NURE projects

To move towards nuclei candidates for $0\nu\beta\beta$ decay important experimental limits need to be overcome. The challenge is to measure a rare nuclear transition under a very high rate of heavy ions produced by the beam-target interaction. The exploration of nuclei of interests for $0\nu\beta\beta$ is particularly stimulating as well as challenging. We consider that:
a) The Q-value for DCE reactions on nuclei of interest for $0\nu\beta\beta$ is normally more negative than in the case of ^{40}Ca explored in ref. [18]. This could strongly reduce the cross section at very forward angles.
b) The (^{18}O,^{18}Ne) reaction is particularly advantageous, due to the large value of the B(GT) strengths. However, this reaction is of $\beta^+\beta^+$ kind, while most of the research on $0\nu\beta\beta$ is in the opposite side. None of the reactions of $\beta^-\beta^-$ kind looks like as favourable as the (^{18}O,^{18}Ne). For example, the (^{18}Ne,^{18}O) requires a radioactive beam, which cannot be available with comparable intensity. The proposed (^{20}Ne,^{20}O) has smaller B(GT), so a sensible reduction of the yield is expected;
d) In some cases gas or implanted targets are necessary, e.g. ^{136}Xe or ^{130}Xe, which are normally much thinner than solid state ones, with a consequent reduction of the collected yield;
e) In some cases the energy resolution (about half MeV) is not enough to separate the ground from the excited states in the final nucleus. Thus, the coincident detection of γ-rays from the de-excitation of the populated states is mandatory, but at the price of the collected yield.
As a consequence, the present limits of beam power (~100 W) for the CS accelerator and acceptable rate for the MAGNEX focal plane detector (few kHz) allow us to concentrate on only few cases, which are planned in the NURE project [7-8] (e.g. ^{116}Cd, ^{130}Te, ^{76}Ge). In order to start a systematic exploration of all the nuclei of interest for $0\nu\beta\beta$ decay, an upgraded set-up, able to work with at least two orders of magnitude more luminosity than the present, is necessary. This goal can be achieved by a substantial change in the technologies implemented in the beam extraction [26], in the control of the beam induced radioactivity, in the detection of the ejectiles [27-30] and in the power dissipation of the thin targets [31]. In addition, the project demands for an enhancement of the maximum accepted magnetic rigidity, preserving the geometry and field uniformity of the magnetic field [32-35] in order to keep the high-precision of the present trajectory reconstruction.
Finally, the development of a specific theory program to allow an accurate extraction of nuclear structure information from the measured cross sections is an important pillar of the NUMEN project. Relying on the use of the DWBA approximation for the cross section, the theory is focused on the development of microscopic models for DCE reactions, employing several approaches (QRPA, shell model, IBM) for inputs connected to nuclear structure quantities. We are also investigating the possible link between the theoretical description of the $0\nu\beta\beta$ decay and DCE reactions.

4. First experimental measurements on the ^{20}Ne + ^{116}Cd system

We performed the first experimental investigation of the (^{20}Ne,^{20}O) DCE reaction on a ^{116}Cd target, which is a candidate for the $0\nu\beta\beta$ decay. This is the first measurement of such a reaction, there are no data available in literature. A ^{20}Ne^{10+} cyclotron beam at 15 AMeV was delivered by the CS of INFN-LNS and impinged on a ^{116}Cd rolled target of 1370 μg/cm^2 thickness. The thickness was carefully chosen in order to obtain an energy resolution which allows to distinguish the transition to the residual ^{116}Sn ground state from its first excited state at 1.293 MeV. The MAGNEX spectrometer was placed at forward angles including zero degree in the full acceptance mode (~50 msr). The total covered angular range was $0° \leq \theta_{lab} \leq 8°$.
Usually when a measurement is performed at zero degree, the beam enters in the spectrometer and it is guided by the magnetic fields to a place in the focal plane region away from the detectors and finally collected by a faraday cup [18]. The fields are thus set in order to transport the ^{20}Ne^{10+} ions towards the faraday cup position. However, when the beam passes through the ^{116}Cd target a charge state distribution is originated. The maximum amount corresponds to the fully stripped ^{20}Ne^{10+} (~ 99%) but

a sizeable amount of beam in the 9^+ and 8^+ charge states is also produced. These lower charge state components have a magnetic rigidity similar to that of the ejectiles of interest: $^{20}F^{9+}$ for the Single Charge Exchange (SCE) and $^{20}O^{8+}$ for DCE. Consequently, they enter in the FPD acceptance, causing a limitation in the detector tolerable rate. For this reason, a second target was placed downstream of the ^{116}Cd to be used as a post-stripper material in order to minimize the amount of $^{20}Ne^{9+}$ and $^{20}Ne^{8+}$ beams. Different materials where tested and the final choice was a thick C foil of 987 μg/cm^2. With this configuration the charge state distribution is ~ 99.1 % of 10^+, ~ $9.0 \cdot 10^{-3}$ % of 9^+ and ~ $2.0 \cdot 10^{-5}$ % of 8^+ [36]. This solution allowed only partially to reduce the background and thus a system of shields before the FPD entrance was also equipped to stop such ejectiles.
Despite these experimental limitations, we were able to measure energy spectra and absolute cross sections for the DCE reaction channel. Moreover, we measured also other reaction channels (one- and two-proton transfer, one- and two-neutron transfer and single charge exchange), in order to estimate the role of the sequential multi-nucleon transfer routes on the diagonal DCE process. The data reduction and analysis are almost completed and the results will be published soon.

Acknowledgments
This project has received funding from the European Research Council (ERC) under the European Union's Horizon 2020 research and innovation programme (grant agreement No 714625).

References
[1] Vergados J D, Ejiri H, and Simkovic F 2012 *Rep. Prog. Phys.* **75** 106301
[2] Vogel P 2012 *Jour. of Phys. G: Nucl. and Part. Phys.* **39** 124002
[3] Engel J and Menéndez J (2017) *Rep. Prog. Phys.* **80** 046301
[4] Dell'Oro S, Marcocci S, Viel M and Vissani F 2016 *Adv. High Energy Phys.* **2016** 2162659
[5] Cappuzzello F et al. 2015 *J. Phys.: Conf. Ser.* **630** 12018
[6] Agodi C et al. 2015 *Nucl. Part. Phys. Proc.* **265** 28
[7] Cavallaro M et al. 2017 *PoS*(BORMIO2017)
[8] Cavallaro M 2017 *PoS*(NEUTEL2017) 031
[9] Cappuzzello F et al. 2016 *Eur. Phys.* J. A **52** 167
[10] Fazely A, Liu L C 1986 *Phys. Rev. Lett.* **57** 968
[11] Wu H, Wang R and Zhao E 1992 *Phys. Rev.* C **45** 1644
[12] Mordechai S et al. 1988 *Phys. Rev. Lett.* **61** 531
[13] Auerbach N, Zamick L, Zheng D C 1989 *Ann. Phys.* **192** 77
[14] Report to the Nuclear Science Advisory Committee, Neutrinoless Double Beta Decay, 2014.
[15] Blomgren J et al. 1995 *Phys. Lett.* B **362** 34
[16] Drake D M et al. 1980 *Phys. Rev. Lett.* **45** 1765
[17] Dasso C H and Vitturi A 1986 *Phys. Rev.* C **34** 743
[18] Cappuzzello F et al. (2015) *Eur. Phys. J.* A **51** 145
[19] Cunsolo A et al. 2007 *Eur. Phys. Jour.-Special Topics* **150** 343
[20] Cappuzzello F et al. (2014) *Nucl. Instr. Methods* A **763** 314
[21] Cappuzzello F et al. (2015) *Nature Communications* **6** 6743
[22] Pereira D et al. (2012) *Phys. Lett.* B **710** 426
[23] Carbone D et al. (2014) *Phys. Rev.* C **90** 064621
[24] Carbone D (2015) *Eur. Phys. J Plus* **130** 143
[25] Soukeras V (2015) *Phys. Rev.* C **91** 057601
[26] Calabretta L et al. (2017) *Modern Physics Letters* A **32** 17
[27] Cortesi M et al. 2017 *Review of Scientific Instruments* **88** 013303
[28] Muoio A et al. 2016 *EPJ Web of Conferences* **117** 10006
[29] Carbone D et al. 2016 *Results in Physics* **6** 863
[30] De Geronimo G et al. 2013 *IEEE Transactions on Nuclear Science* **60** 2314
[31] Iazzi F et al. 2017 *WIT Transactions on Engineering Sciences* DOI: 10.2495/MC170071
[32] Lazzaro A et al. 2008 *Nucl. Instr. and Methods* A **591** 394

[33] Lazzaro A et al. 2008 *Nucl. Instr. and Methods* A **585** 136
[34] Lazzaro A et al. 2007 *Nucl. Instr. and Methods* A **570** 192
[35] Lazzaro A et al. 2009 *Nucl. Instr. and Methods* A **602** 494
[36] Shima K et al. 1992 *Atom. Data and Nucl. Data Tables* **51** 2

IOP Conf. Series: Journal of Physics: Conf. Series **1023** (2018) 012007 doi:10.1088/1742-6596/1023/1/012007

In-beam test of the RPC architecture foreseen to be used for the CBM-TOF inner wall

M. Petriş*, D. Bartoş, M. Petrovici, L. Rădulescu, V. Simion
National Institute of Physics and Nuclear Engineering (IFIN-HH), Bucharest-Magurele, P.O.Box MG-6, RO-077125, Romania

I. Deppner, N. Herrmann, C. Simon
Physikalisches Institut, University of Heidelberg, Germany

J. Frühauf, M. Kiš, P-A. Loizeau
Gesellschaft für Schwerionenforschung, Darmstadt, Germany

E-mail: * mpetris@nipne.ro

Abstract. The Time Of Flight (TOF) subsystem is one of the main detectors of the CBM experiment. The TOF wall in conjunction with Silicon Tracking System (STS) is foreseen to identify charged hadrons, i.e. pions, kaons and protons, with a full azimuthal coverage at 2.5^0 - 25^0 polar angles. A system time resolution of at least 80 ps, including all contributions, such as electronics jitter and the resolution of the time reference system, is required. Such a performance should be maintained up to a counting rate larger than 30 kHz/cm^2 at the most inner region of TOF wall.

Our R&D activity has been focused on the development of two-dimensional position sensitive Multi-gap Resistive Plate Counter (MRPC) prototypes for the forward region of the CBM-TOF subdetector, the most demanding zone in terms of granularity and counting rate.

The in-beam tests using secondary particles produced in 30 GeV/u Pb ion collisions on a Pb target at SPS - CERN aimed to test the performance of these prototypes in conditions similar to the ones expected at SIS100 at FAIR. The performance of the prototypes is studied in conditions of exposure of the whole active area of the chamber to high multiplicity of reaction products.

The results show that this type of MRPC fulfill the challenging requirements of the CBM-TOF wall. Therefore, such an architecture is recommended as basic solution for CBM-TOF inner zone.

1. Introduction

Compressed Baryonic Matter (CBM) experiment at Facility for Antiproton and Ion Research (FAIR) in Darmstadt is a fixed target experiment with the mission to investigate the high net-baryon density matter produced in nucleus-nucleus collision in an energy range between 2 - 11 GeV/u delivered by SIS100 accelerator. The CBM experiment is designed for precision measurements of multidimensional correlations among different observables including particles with very low production cross sections, hyperons, heavy flavor hadrons, hypernuclei and strange objects, using the high-intensity heavy-ion beams provided by FAIR accelerators [1]. The beam extracted to the CBM cave will reach intensities up to 10^9 Au ions per second. The interaction rate will reach up to 10^7 reactions per second, with a charged particle multiplicity up to 1000 per event. In order to meet the challenging physics goal of measuring rare probes at such high interaction rate and multiplicity, the experimental set-up has to identify leptons and hadrons in a high counting rate environment. This requires fast and radiation hard detectors and front-end electronics, self triggered data acquisition and fast on-line event selection.

The Time Of Flight (TOF) subsystem is one of the core detectors of the CBM experiment. The TOF wall in conjunction with Silicon Tracking System (STS) is foreseen to identify charged hadrons, i.e. pions, kaons and protons, in the full acceptance of the system (the angular range covered by the STS detector of 2.5^0-25^0). It covers an active area of about 120 m^2 approximately rectangular in shape. A system time resolution of at least 80 ps including all possible contributions, such as electronics jitter and the resolution of the time reference system is needed [2]. Such a performance should be maintained up to a counting rate which, very close to the beam pipe, exceeds 30 kHz/cm^2 [2]. In order to fulfill these chalenging requirements at afordable costs, the CBM - TOF wall will be built based on state-of-the-art Multigap Resistive Plate Chambers (MRPC) [3].

2. A basic architecture for the inner zone of the CBM-TOF wall

Our R&D activity has been focused on the development of a Multi-Gap RPC with Multi-Strip readout (MSMGRPC) [4] for high counting rate and multiplicity environment, as it is anticipated for the inner zone of the CBM-TOF. Very good results in terms of efficiency, time and two-dimensional position resolutions were reported [5], for in-beam tests with minimum ionizing particles. Beased on these results we define a basic architecture of a module for CBM-TOF wall, assembling four MSMGRPC counters (strip pitch of 7.4 mm, 5.6 mm width and a strip length of 96 mm) in a staggered configuration on both x and y directions, in order to obtain a continuous coverage of the active area. Construction details of this basic architecture prototype (called further RPC2012) and its performance in the in-beam tests at CERN-PS accelerator using minimum ionizing particles and at CERN-SPS facility using reaction products produced by an Ar beam of 13 GeV/u on a Pb target, were reported in [6, 7]. The average cluster size was 1.6 strips per hit and the time resolution, including the electronics, was of ~60 ps [7]. For the signal processing in the tests mentioned above, a front-end electronics (FEE) based on 8 channels NINO chip [8] was used.

In the present contribution are reported the results obtained using the RPC2012 prototype in a heavy-ion in-beam test performed at CERN-SPS accelerator, in conditions closer to those foreseen for SIS100. In this test the signals delivered by the MSMGRPC were processed by a the FEE developed for the CBM-TOF wall.

3. In-beam tests of the basic architecture prototype

The in-beam test was performed at CERN-SPS with reaction products resulted from 30 GeV/u Pb ions collided on a Pb target. The signals delivered by the four counters of RPC2012 were processed by a FEE based on a 32 channel motherboard [2] containing 4 PADI chips[10], the front-end electronics anticipated to be used for the CBM-TOF wall. The signals were digitized by 32-channel FPGA-TDCs [11] and readout via TRB3 [12] data hubs. A comprehensive description of the whole setup is given in [9]. The experimental set-up was positioned at a polar angle of about 3^0 relative to the beam axis. The RPC2012 prototype was 'sandwiched' between two MSMGRPC prototypes called RPC2015SS [15] and RPC2015DS [14] (SS- Single Sided, DS-Double-Sided), housed in RPC2015 box and the RPCref prototype [5], as is shown in Fig. 1. The time difference between one of the two diamond detectors positioned in front of the target and the RPCref positioned downstream, at 4 m distance from the target, was used for particle

Figure 1. Photo of the experimental set-up based on MSMGRPC prototypes used in the in-beam test.

velocity correction. The relative spatial position of the seven MSMGRPCs and the operated

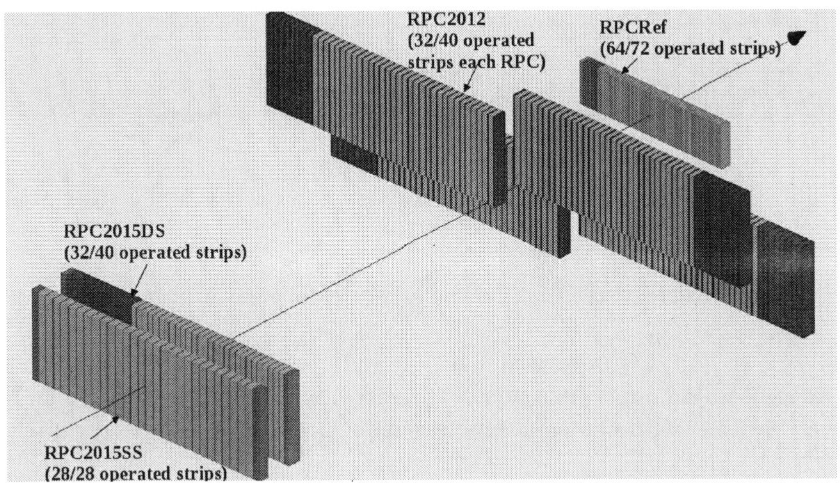

Figure 2. Sketch of the relative spatial positions of the seven MSMGRPCs in the experimental set-up at CERN-SPS.

area (green color) of each of them are presented in Fig. 2. One could observe in the figure that there is a partial overlap of the active area of each of the four RPC2012 counters, considered as detectors under study (DUT) with the the active area of DSRPC2015 positioned upstream, considered as reference (REF) counter for estimation of the time resolution and efficiency in the performed analysis.

4. Results

Details on the data analysis starting from unpacking to calibration and corrections for slewing effect, position and reaction product velocity spread, are described in [16]. In addition, due to the partial overlap of the active area between each DUT and REF, position cuts on both x and y directions were applied in the REF counter. In this way, hits which are focused on the middle of overlapped active area of both DUT and REF counters, were selected.

The efficiency is calculated as the ratio of the best matched hits between the diamond detector, RPC2015DS (REF), one of the RPC2012 (DUT) and RPCref, relative to the number of hits when only diamond detector, RPC2015DS, and RPCref are present. The time of flight (TOF) spectrum is obtained as the difference of the times measured by the RPC2015DS and one of the RPC2012 counters. The system time resolution (the contributions of both RPC2015DS and one of the RPC2012 counter) is defined by the standard deviation σ of the Gauss function fitted to the TOF distribution.

With the aim to reach steady-state operating conditions for studying the working curve of the RPC2012 detectors, they were operated for about 8 hours with the same parameters: applied high voltage corresponding to 157 kV/cm electric field in the gas gap and 205 mV threshold set to the FEE. The detector performance was monitored over this period in terms of efficiency and time resolution and a clear conditioning effect was evidenced. Through this conditioning effect, the resistive electrodes reach a steady state potential and the microimpurities accidentally existing inside the detector are burned out. As it is shown in Fig. 3, the efficiency does not change during conditioning period while a slight improvement in the time resolution is observed.

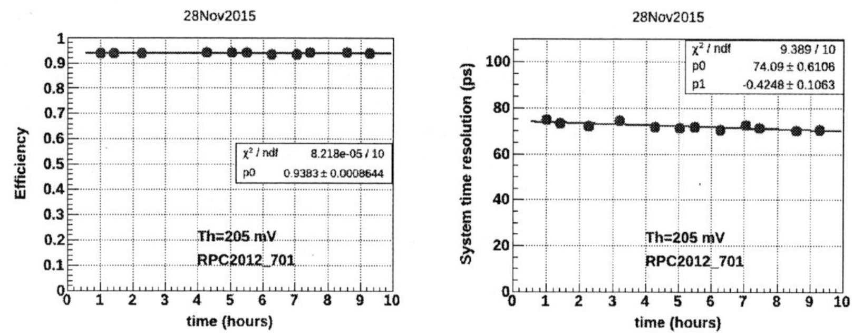

Figure 3. Efficiency - left and time resolution - right as a function of time for 157 kV/cm electric field and 205 mV FEE threshold. The statistical errors are within the marker size.

In the left side of Fig. 4 is presented the time difference spectrum between one of the four counters positioned in the upper, left side of the text box (called RPC2012_1) and RPC2015DS (see reference [14, 13]), with the same inner geometry and the same strip pitch. The 63 ps

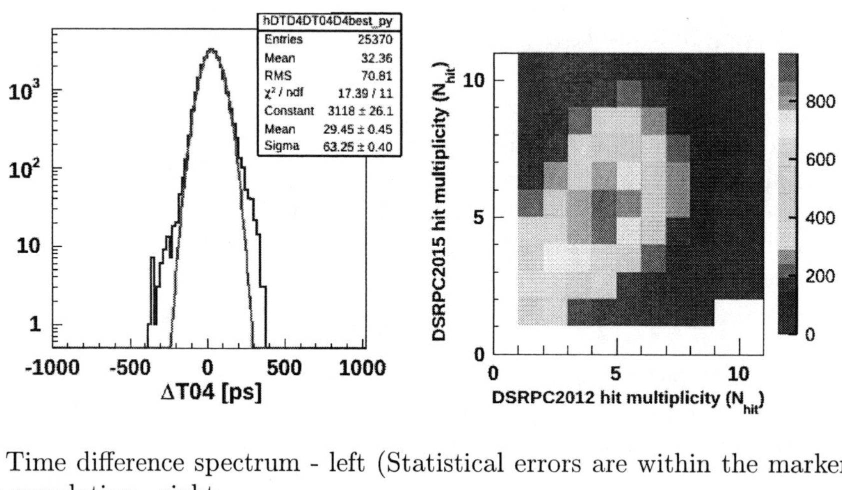

Figure 4. Time difference spectrum - left (Statistical errors are within the marker size). Hit multiplicity correlation - right.

standard deviation (σ) of the Gauss fit demonstrates the very good system time resolution. A single counter time resolution of 44 ps, including the electronic contribution is obtained supposing equal contributions of the two detectors. The comparison of the 63 ps standard deviation with the 71 ps RMS of the spectrum shows a non-significant contribution of the non-Gaussian tails, estimated to be of 2.8% in a $\pm(3\cdot\sigma)$. The correlation of the hit multiplicities in the two counters (Fig. 3 - right) shows that almost the same number of hits are simultaneously incident on both detectors. The detectors were operated at 2 x 5.5 kV voltage (157 kV/cm electric field).

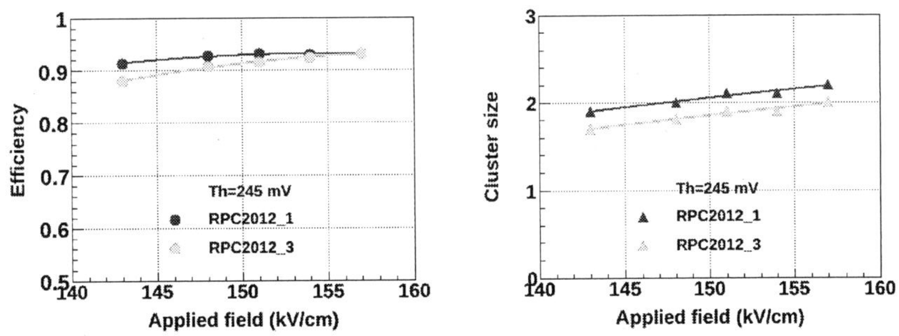

Figure 5. Efficiency - left (statistical errors are within the marker size) and cluster size - right as a function of high voltage for RPC2012_1 and RPC2012_3.

The expected behaviour of the efficiency and cluster size as a function of high voltage for RPC2012_3 and RPC2012_1 (upper - right side relative to the axis shown in Fig.2), for a PADI threshold of 245 mV can be followed in Fig.5. In the mentioned geometry of the experiment, a 93% efficiency for both counters was obtained. Due to their staggered positions, position cuts on both x and y directions were applied in the RPC2015DS, considered as reference counter, for efficiency estimation. However, due to a partial overlap of the active area of each of the four counters relative to the RPC2015DS, the obtained values are still affected by the hits lost at the edges of the overlapped area. The average cluster size was found to be 2.2 strips for RPC2012_1 and 2.0 strips for RPC2012_3 in the region of efficiency plateau. Average system time resolution of 73 ps for RPC2012_1 and of 67 ps for RPC2012_3 remains almost unchanged over the investigated high voltage range, as it is shown in Fig. 6. The detector performance stability

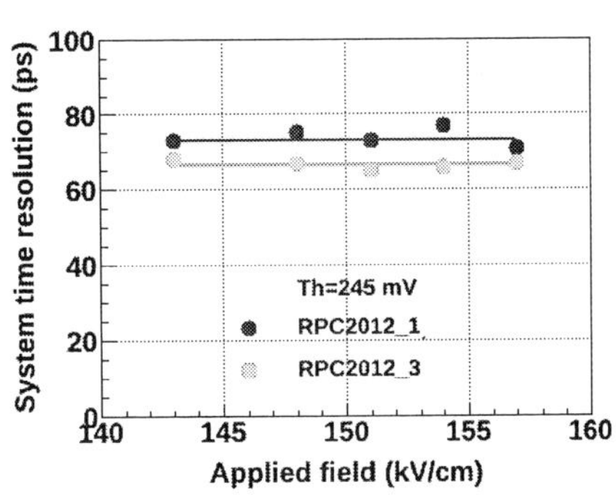

Figure 6. Time resolution as a function of high voltage for RPC2012_1 and RPC2012_3. Statistical errors are within the marker size.

was checked for a period of operation of about 15 hours with the same settings (157 kV/cm) for both DUT and REF counters.

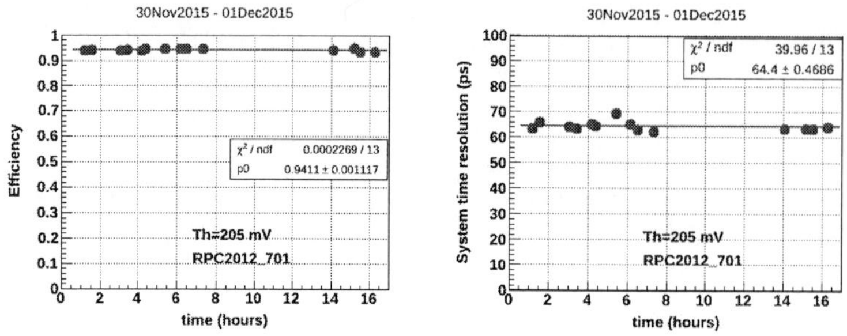

Figure 7. RPC2012_1 performance during 15 hours operation (the high voltage and threshold settings were the same for the entire period). Statistical errors are within the marker size.

The stability of the detector performance is demonstrated by the plots presented in Fig. 7 and Fig. 8. The average efficiency of 94% and a time resolution better than 70 ps remains constant over the entire measurement period of about 15 hours. This demonstrates that the performance of the detector remanains unchanged over the whole period in which the counters were exposed to the reaction products, about two weeks of in-beam test.

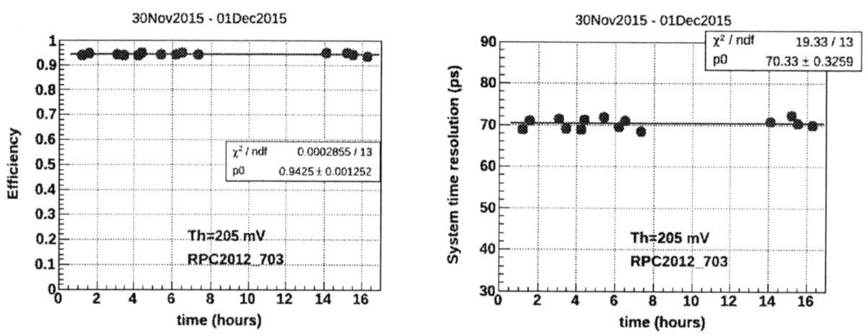

Figure 8. RPC2012_3 performance during 15 hours operation (the high voltage and threshold settings were the same for the entire period). Statistical errors are within the marker size.

5. Conclusions

The results presented in this contribution show that the MSMGRPC prototypes and the proposed basic architecture fulfill the performance requirements for the inner zone of the CBM-TOF wall. The proposed geometry will be implemented in the architecture of the modules designed for the region of low-polar angles of the CBM-TOF wall.

Acknowledgments

This work was supported by RO-FAIR - F04 and NASR/NUCLEU PN16420104 Projects.

References

[1] CBM Collaboration. Eur. Phys. J. A, 53:60, 2017
[2] CBM-TOF Collaboration, CBM-TOF TDR, October 2014
[3] E. Cerron Zeballos et al., Nucl. Instrum. Meth. A374, (1996), 132
[4] M. Petrovici et al., Nucl. Instrum. Meth. A487, (2002), 337
[5] M. Petrovici et al, 2012 Journal of Instrumentation, Volume 7, 2012 (2012 JINST 7 P11003).
[6] M. Petriş et al., CBM Progress Report 2012 (2013), p.68
[7] M. Petrş et al., Journal of Physics: Conference Series, 724, (2016), 012037
[8] F. Anghinolfi et al., Nucl.Instr.and Meth. A533(2004)183
[9] C. Simon et al., CBM Progress Report 2015, (2016), p.90
[10] M. Ciobanu et al., CBM Progress Report 2013, (2014), p.84

[11] C. Ugur et al., GSI Scientific Report 2014 (2015), p.212
[12] M. Traxler et al., GSI Scientific Report 2014 (2015), p.514
[13] V. Aprodu et al., CBM Progress Report 2015 (2016), p.98
[14] D. Bartos et al., arxiv[physics.ins-det] 1708.02707
[15] M. Petriş et al., CBM Progress Report 2016 (2017), 129
[16] M. Petriş et al., Journal of Instrumentation, Volume 11, September 2016 (2016 JINST 11 C09009)

IOP Conf. Series: Journal of Physics: Conf. Series **1023** (2018) 012008 doi:10.1088/1742-6596/1023/1/012008

Status of Simulations for the Cyclotron Laboratory at the Institute for Nuclear Research and Nuclear Energy

G Asova[a], **N Goutev**[a], **D Tonev**[a], **A Artinyan**[a]

Institute for Nuclear Research and Nuclear Energy, BAS, BG-1784 Sofia, Bulgaria

E-mail: gasova@inrne.bas.bg

Abstract. The Institute for Nuclear Research and Nuclear Energy is preparing to operate a high-power cyclotron for production of radioisotopes for nuclear medicine, research in radiochemistry, radiobiology, nuclear physics, solid state physics. The cyclotron is a TR24 produced by ASCI, Canada, capable to deliver proton beams in the energy range of 15 to 24 MeV with current as high as 400 μA. Multiple extraction lines can be fed. The primary goal of the project is the production of PET and SPECT isotopes as ^{18}F, ^{67,68}Ga, ^{99m}Tc, etc.

This contribution reports the status of the project. Design considerations for the cyclotron vault will be discussed for some of the target radioisotopes.

1. Introduction

The Institute for Nuclear Research and Nuclear Energy (INRNE) at the Bulgarian Academy of Sciences is building a facility dedicated to: production of well-established medical radioisotopes; research and development on emerging radioisotopes for nuclear medicine; fundamental and applied research with radiotracers in areas related to life sciences and industry [1]. The facility is based on a TR24 cyclotron capable to deliver proton beams with current up to 400 μA and energy as high as 24 MeV. The cyclotron has already been produced by ASCI and delivered at the INRNE with its ancillary equipment like RF and water-cooling systems, a single Y-beamline equipped with steering and focusing magnets, etc. The construction of the building housing the cyclotron vault and the radiochemistry laboratories still has to be done. Meanwhile efforts are dedicated to numerical studies on the possibilities to produce various isotopes for medicine and radiological characterization of the setup.

The major objective for the start-up is the widely used ^{18}F–FDG produced via ^{18}O (p, n) ^{18}F reaction in ^{18}O-enriched target volume and, therefore, the radiological characterization has to take into account production of neutrons and γ-rays. Part of the neutrons result from the nuclear reaction within the target volume, whereas another part from primary particles (protons) or secondaries (neutrons, γ-rays, etc.) interacting with cyclotron components, target, walls of the facility. γ-rays can be generated by ongoing ^{18}F–decay during irradiation or by neutron-induced reaction with various components. This study shows numerical estimations of some internal hazards coming along with the operation of the facility at maximal beam power. Radioactive materials generated in the ^{18}F–target and the vault will be discussed.

2. Considerations for radiological characterization

2.1. Methodology for the simulations

For the simulations shown here FLUKA [2, 3] has been used due to its capabilities to numerically estimate emission and transport of secondary particles and to evaluate the radioactive waste generated as a result of reactions related to the secondary emission sources.

2.2. Radiation source

To adjust the simulation settings a real-size target for ^{18}F was initially irradiated using various sets of beam parameters [4]. Despite the fact that the highest beam current and energy are not needed to be used for ^{18}F production, afterward such were chosen to irradiate the target due to the stringent conditions they would impose on the radiation environment [5]. The target was irradiated for six hours and the particles, emitted in a spherical volume as a result of the (p, x) reactions with any of the target components, have been collected to be used later as sources irradiating further components. Here x denotes neutrons and γ–rays as they are of practical interest when dealing with proton accelerators [5]. As shown with red bars in Fig. 1, the energy

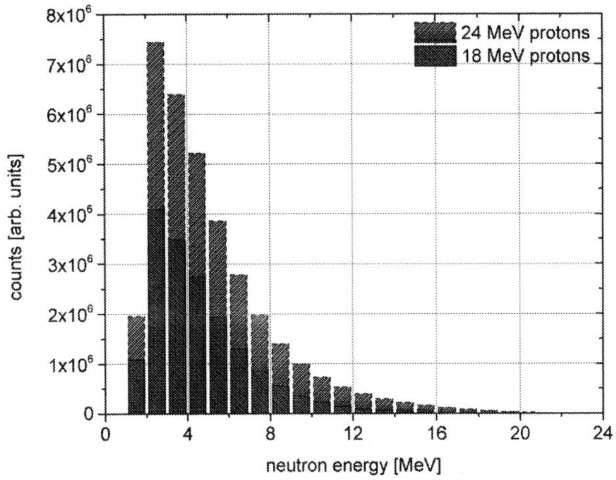

Figure 1. Spectrum of neutrons emitted from the liquid water volume at proton beam energies of 18 and 24 MeV and current of 400 μA. Here the proton beam meets a Havar foil separating the enriched water from the vacuum chamber at normal angle of incidence.

of the neutrons covers the full range of energy up to the energy of the primary particles corrected with the Q-value of the ^{18}O (p, n) ^{18}F reaction ($Q = -2.44$ MeV) and the energy lost on a Havar foil, separating the area of the liquid target from the vacuum chamber of the setup. Majority of the neutrons are low energy to fast ones. The same trend is seen also for energy of 18 MeV.

The simulated spectrum for 18 MeV and 100 μA results in fluence rate of about 2.5e11 n/cm^{2}s which can be compared to the measurement results shown in [6] where 2.4e10 n/cm^{2}s are given for 15 μA and 17 MeV. The difference of an order of magnitude might be due to the lower beam current used in the measured data.

Figure 2 shows the fluence of secondary neutrons overlaid on the target geometry.

2.3. Fields of secondary particles within the cyclotron vault

The vault is designed to house the cyclotron and four irradiation areas as only one of them will be used for ^{18}F production. There are two possible locations for the ^{18}F–target in the vault - close to the cyclotron but still externally to it and in a separated irradiation room. Figure 3 shows the distribution of neutrons when using both positions - the highest density shown in black is the exact center of mass of the source from Section 2.2. The source obtained in the step described above is positioned in either location of the target and the particles are transported further depending on their spatial coordinates and momentum vector components in the source

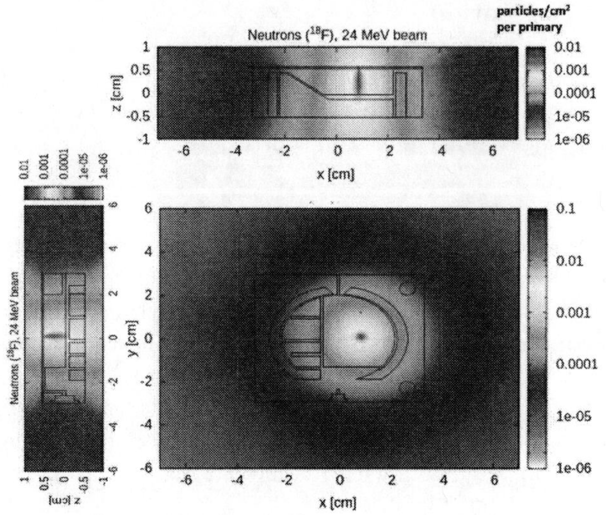

Figure 2. Fluence (particles per cm^2 per primary particle) of neutrons crossing the target body in the three planes. Proton beam with energy of 24MeV and current of 400 μA impinges into the (x, y) plane.

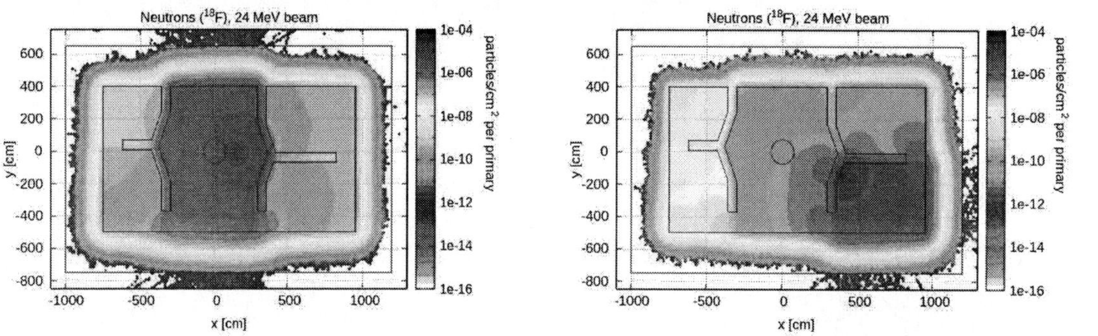

(a) Neutrons in the vaults when positioning the target within the vault inner area.

(b) Neutrons in the vaults when positioning the target outside the vault inner area in a dedicated irradiation room.

Figure 3. Top view of the neutron fields in the cyclotron vault for two different positions of the liquid target. The cyclotron is shown with a black circle in the center of the bunker. Thinner concrete walls separate the cyclotron area from the four irradiation areas.

phase space. The vault space is surrounded by 2.5 m thick walls made of high-density concrete as shown in Fig. 3. The target area is not shielded additionally. The neutron distribution homogeneity around the target is disturbed by cyclotron metal parts mostly in Fig. 3(a), inner and outer shield walls (Fig. 3(b)). As it might be assumed that the areas outside the vault will be used for chemical synthesis, target preparation, etc., it is desirable to minimize any leakages, thus, making the position of the target shown with Fig. 3(b) preferable if no further shielding is used. Additionally, the shielding can be improved as it is shown with Fig. 4 where a few centimeters thick marble-enriched concrete is used locally around the target. Further optimization of the thickness of this shielding and its chemical composition still has to be done.

3. Activation of the vault walls

The distribution shown in Fig. 2 can be used to study the activation of components of the target itself. It has already been shown [4, 7] that the nuclides generated in the Havar foil contain a

IOP Conf. Series: Journal of Physics: Conf. Series **1023** (2018) 012008 doi:10.1088/1742-6596/1023/1/012008

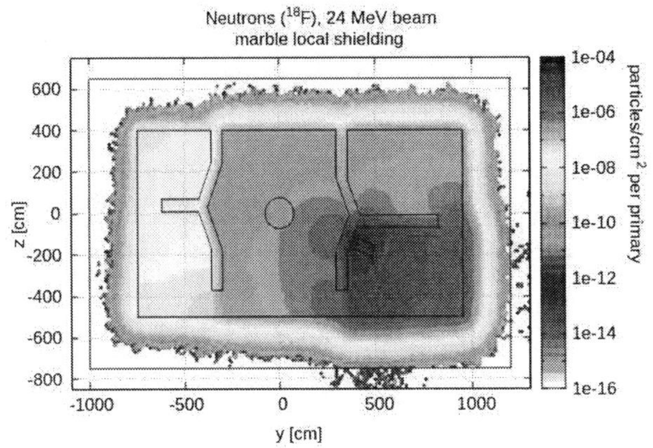

Figure 4. Top view of the neutron fields in the vault if the target is positioned as in Fig. 3(b) and shielded locally with marble-enriched concrete.

number of long-living ones (^{3}H, ^{58}Co). This study focuses on the activation of the vault using the distributions from Fig. 2 and Fig. 3. It is assumed that a sacrificial layer takes 20 cm from the wall thickness including also the walls, separating the cyclotron from the irradiation areas. The last ones have thickness of 60 cm. Table 1 shows some of the nuclides generated in the sacrificial layer after a month of operation with daily irradiation conditions as above - 6 hours at 24 MeV and 400 μA, immediately after End of Beam (EOB) and after 3 weeks of cooling time.

Table 1. Some of the nuclides generated in the outmost 20 cm of the walls of the cyclotron vault. The irradiating target is positioned as in Fig. 3(b).

Isotope	Activity at EOB [Bq]	Activity in 4 weeks [Bq]	Parent nucleus
^{59}Fe	2.1×10^{10}	1.4×10^{10}	^{58}Fe
^{56}Mn	1.8×10^{9}	-	^{56}Fe
^{55}Fe	1.8×10^{10}	1.7×10^{10}	^{55}Mn
^{54}Mn	4.3×10^{9}	4.1×10^{9}	^{54}Cr, ^{54}Fe
^{51}Cr	1.1×10^{9}	5.7×10^{8}	^{52}Cr
^{47}Ca	1.1×10^{9}	1.6×10^{7}	^{46}Ca
^{47}Sc	1.0×10^{9}	5.0×10^{7}	^{47}Ti
^{45}Ca	2.3×10^{10}	2.0×10^{10}	^{44}Ca
^{41}Ca	2.5×10^{6}	2.5×10^{6}	^{40}Ca
^{42}K	2.2×10^{11}	-	^{41}K
^{41}Ar	5.9×10^{6}	-	^{40}Ar
^{39}Ar	8.1×10^{7}	8.1×10^{7}	^{39}K
^{37}Ar	5.2×10^{11}	2.9×10^{11}	^{40}Ca
^{36}Cl	2.8×10^{4}	2.8×10^{4}	^{39}K
^{24}Na	1.7×10^{12}	-	^{27}Al
^{14}C	2.2×10^{3}	2.2×10^{3}	^{13}C
^{3}H	3.1×10^{6}	3.1×10^{6}	(p, ^{3}H)

^{41}Ar resulting from neutron capture on ^{40}Ar is also seen in the air of the vault in activities in the range of 1×10^{7} Bq - above the limit of 200 Bq/m^{3} with respect to the volume of the vault increasing the cooling time before maintenance access. In 18 hours the activity decreases to 3×10^{4} Bq with which the specific activity within the inner cyclotron room decreases below the exemption limit. Further estimations are needed considering that the usable lifetime of a cyclotron can be about 20 years. At the moment such estimations with the software package used can be done for operation time of one month.

4. Conclusion

The distribution of neutrons emitted as a result of primary reaction on ^{18}F target from a mid-energy high-intensity cyclotron was evaluated using Monte-Carlo simulations. Two different locations of the target were considered as the one positioned in a separate irradiation room showed to be preferable with respect to possible leakages of low-energy neutrons through the walls of the vault. It was also shown that local target shielding with marble-enriched concrete would lead to further improvements.
A number of toxic and long-living isotopes were seen in the sacrificial layer of the inner walls (^{55}Fe, ^{45}Ca, ^{41}Ca, ^{39}Ar). ^{41}Ar is also seen in the air limiting the time needed for maintenance access.

5. Future steps

Characterization of the vault depends on the targets being irradiated and therefore more isotopes will be studied. At the moment efforts are spent on the production routes of ^{67,68}Ge and ^{99m}Tc and they will also be used to assess the activation in the vault. Additionally the local target shielding needs to be optimized with respect to ease of usage (weight), composition, leackage and neutrons and γ-rays.

Acknowledgments

The research has been supported by the Bulgarian Science Fund under Contracts No. DN 08/6, 13.12.2016 and DM 18/2, 12.12.2017 and DFNP-17-165/03.08.2017.

References

[1] Tonev D, Goutev N, Georgiev L S 2015, Cyclotron laboratory of the Institute for Nuclear Research and Nuclear Energy *J. Phys.: Conf. Series* **724** 012049
[2] Böhlen T T, Cerutti F, Chin M P W, Fassò A, Ferrari A, Ortega P G, Mairani A, Sala P R, Smirnov G, Vlachoudis V 2014, The FLUKA code: Developments and challenges for high energy and medical applications *Nuclear Data Sheets* **120** 211 214
[3] Ferrari A, Sala P R, Fassò A, Ranft J 2005, FLUKA: A multi-particle transport code (program version 2005), **CERN-2005-010, INFN-TC-05-11, SLAC-R-773**
[4] Yavahchova M, Asova G, Goutev N, Tonev D 2016, FLUKA simulations on the thick target yield of 18F production by proton-induced reaction, Science and Technologies **VI** 3
[5] IAEA Technical Report Series No 283 1988, *Radiological Safety Aspects of the Operation of Proton Accelerators* (Vienna)
[6] Sadat-Eshkevar S et al. 2012 *Assessment of the staff absorbed dose related to cyclotron operation and service in the production of 18F radiopharmaceuticals Nukleonika* **57-3** 407-410
[7] Asova G, Artinyan A, Goutev N, Mineva M, Tonev D, Ivanova S, Zheleva N 2017 *Monte-Carlo simulations on the radiological characteristics of a small medical cyclotron C. R. Acad. Bulg. Sci* **70** 10

Effects of direct interactions on deuteron induced reactions[1]

M. Avrigeanu and V. Avrigeanu
Horia Hulubei National Institute for Physics and Nuclear Engineering, P.O. Box MG-6, 76900 Bucharest, Romania

E-mail: marilena.avrigeanu@nipne.ro

Abstract. A detailed analysis of the direct-interaction key role in the deuteron-induced reactions on target nuclei from Al to U has been carried out. Particular comments concern the main importance of the deuteron breakup mechanism in the deuteron interaction with heavy target nuclei at incident energies around Coulomb barrier.

1. Introduction

An update of the theoretical analysis of deuteron-nuclei interaction within an unitary and consistent account of the related reaction mechanisms is highly requested by on-going strategic research programs as the International Thermonuclear Experimental Reactor (ITER) [1], the International Fusion Material Irradiation Facility (IFMIF) [2], in connection with the ITER program, and the Neutron For Science (NFS) project at SPIRAL-2 facility [3]. The discrepancies between the existing deuteron experimental data and even the latest version of the evaluated data library TENDL-2015 [4] demand further measurements as well as improved model calculations so that the deuteron evaluated data libraries may approach the standard of the current neutron data libraries.

So far the Hauser-Feshbach statistical model has been involved as the main tool to calculate the deuteron reaction cross sections at low and medium incident energies, the compound-nucleus (CN) mechanism being considered to be dominant within this energy range. However, specific non-compound processes make the reactions induced by deuterons different from those induced by other incident particles [5, 6, 7, 8]. The deuteron interaction at incident energies below and around the Coulomb barrier proceeds largely through direct reaction (DR) mechanisms of stripping and pick-up, while the pre-equilibrium emission (PE) and evaporation from CN become important with the increase of the incident energy. Moreover, in addition to these well known reaction mechanisms, the specific deuteron breakup (BU) process plays an important role that increases the complexity of the deuteron interaction analysis in the whole incident energy range due to the large variety of reactions initiated by the BU nucleons at their turn.

2. Deuteron breakup

While the theoretical models for DR, PE, and CN are already settled, an increased attention should be paid to the description of the BU mechanism and its components, namely the

[1] International School on Nuclear Physics and Applications, September 10-16, 2017 Varna, Bulgaria

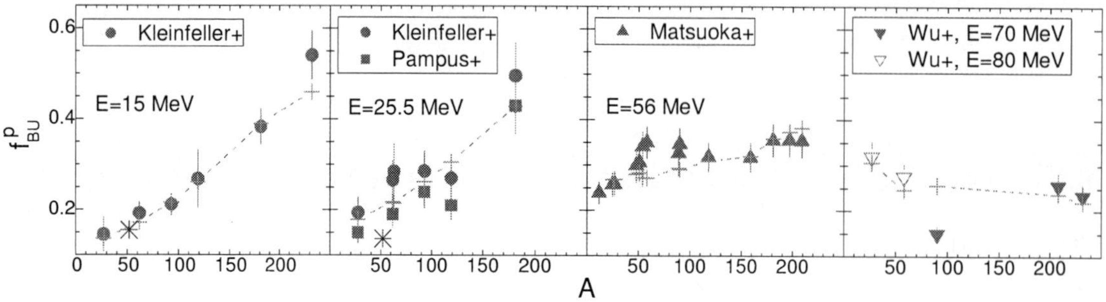

Figure 1. Comparison of the measured [13, 14, 15, 16] total BU proton-emission fractions with the empirical parametrization predictions (crosses) [11], connected by dashed lines for eye guiding, for target nuclei from ^{12}C up to ^{232}Th, at the deuteron energies of 15, 25.5, 56, 70, and 80 MeV.

elastic breakup (EB), in which the target nucleus stays in its ground state and both deuteron constituents fly apart, and the inelastic breakup or breakup fusion (BF), where one of breakup nucleons interacts non-elastically with the target nucleus. The compound nuclei in reactions induced by the BF nucleons differ by one unit of the atomic mass and maybe also atomic number than in deuteron-induced reactions, the partition of the BF cross section among various residual-nuclei population being triggered by the energy spectra of the breakup nucleons and the excitation functions of the reactions induced by these nucleons on the target nuclei [5, 6, 7, 8]. Overall, there are actually two opposite effects of the deuteron breakup on the deuteron activation cross sections that should be considered. Firstly, the deuteron total-reaction cross section σ_R, that is shared among different outgoing channels, is reduced by the value of the total BU cross section σ_{BU}. On the other hand, BU nucleons interactions with the target nucleus enhance the various reaction channels of the original deuteron interactions.

An empirical parametrization [11] of both the total breakup (EB+BF) and EB data has involved the assumption that the BF neutron-emission cross section σ^n_{BF} is the same as that for the BF proton-emission σ^p_{BF} (e.g., Ref. [12]), so that σ_{BU} is given by the sum $\sigma_{EB}+2\sigma^{n/p}_{BF}$. This parametrization has concerned the total breakup nucleon and EB fractions, i.e. $f^{n/p}_{BU} = \sigma^{n/p}_{BU}/\sigma_R$ and $f_{EB}=\sigma_{EB}/\sigma_R$, respectively. The dependence of these fractions on the deuteron incident energy E and the target-nucleus atomic Z and mass A numbers was obtained [11] through analysis of the experimental systematics of deuteron-induced reactions on target nuclei from ^{27}Al to ^{232}Th and incident energies up to 80 MeV for the former [13, 14, 15, 16] but within a more restricted up to 30 MeV [16] energy range for the later. In the absence of available experimental deuteron EB data at incident energies above 30 MeV, the correctness of the corresponding parametrization extrapolation has been checked by comparison of the related predictions [17] with results of the microscopic CDCC method [18].

The comparison of the measured total BU proton-emission fractions σ^p_{BU} at 15 [16], 25.5 [13, 16], 56 [15], and 70 and 80 MeV [14] with the parametrization prediction for deuterons incident on target nuclei from ^{12}C to ^{232}Th is shown in Fig. 1. The same scale is used for the f^p_{BU} values at all incident energies of the available experimental data, in order to make possible also an assessment of their energy dependence. At once with the variation with energy and the mass of the target nucleus, the f^p_{BU} values illustrate the importance of the BU process among the other reaction mechanisms related to the deuteron interaction. Among other features [11] this comparison shows the BU importance increasing with the target-nucleus mass, from ^{27}Al up to ^{232}Th, at the lower incident energies of 15-25.5 MeV. This increase is less significant at the energy of 56 MeV, and even reversed at 70-80 MeV.

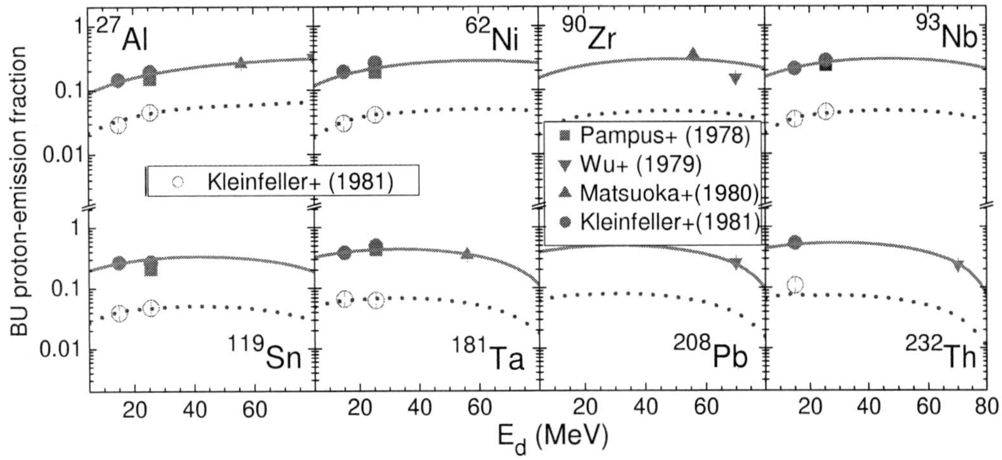

Figure 2. Comparison of experimental [13, 14, 15, 16] f^p_{BU} (solid symbols) and f_{EB} values (open circles) for deuterons incident on nuclei from Al to Th at energies up to 80 MeV. The parametrization predictions [11] are drown by solid (f^p_{BU}) and dotted (f_{EB}) curves.

Next, the comparison between total breakup proton-emission and elastic-breakup fractions for target nuclei from ^{27}Al to ^{232}Th shown in Fig. 2 emphasizes their large difference, pointing out the dominance of the BF component (f^p_{BF}=f^p_{BU}-f_{EB}) over the much weaker EB component.

In order to calculate the BF enhancement of the (d, xn) reaction cross sections, the BF proton-emission cross section σ^p_{BF} should be (i) multiplied by the ratios $\sigma_{(p,x)}/\sigma^p_R$, corresponding to the enhancing reaction, (ii) convoluted with the Gaussian line shape distribution of the BF–proton energy E_p for a given deuteron incident energy E_d, and followed by (iii) an integration over the BF proton energy. Consequently, the BF–enhancement cross section has the form [5, 6, 7]:

$$\sigma^{p,x}_{BF}(E_d) = \sigma^p_{BF}(E_d) \int_0^{E_d-B_d} dE_p \; \frac{\sigma_{(p,x)}(E_p)}{\sigma^p_R} \frac{1}{(2\pi)^{\frac{1}{2}} w} exp[-\frac{(E_p - E^0_p)^2}{2w^2}] \; , \tag{1}$$

where B_d is the deuteron binding energy, σ^p_R is the proton total reaction cross section, x stands for various γ, n, d, or α outgoing channels, while the Gaussian distribution parameters w and E^0_p are given by Kalbach [19].

To reduce as much as possible the supplementary uncertainties brought by additional theoretical calculations, the $\sigma_{(p,x)}$ values are taken from the experimental EXFOR library [20], while σ^p_R is provided by a proton optical model potential [21].

The BF enhancement could be quite important as it was shown [7] for the study of ^{231}Pa (d,3n)^{230}U reaction around the Coulomb barrier. In this case the absorbed BF proton contributes to enhancement of the ^{230}U activation cross section through ^{231}Pa(p,2n)^{230}U reaction. In order to calculate the corresponding BU enhancement, the convolution of the ratio of (p,2n) reaction cross section [20] and proton total-reaction cross section - that corresponds to the weight of the (p,2n) reaction induced by the BF protons on ^{231}Pa target nuclei - with the Gaussian distribution of the BU-proton energies corresponding to a given incident deuteron energy [19] was carried out as shown in Fig. 3(a) for three deuteron energies. The area of the related convolution results corresponds to the BF enhancement of (d,3n) reaction cross sections at each deuteron energy. The energy dependence of this BF enhancement to ^{231}Pa (d,3n)^{230}U activation cross section is shown in Fig. 3(b), where the corresponding total calculated activation of ^{230}U is finally compared with experimental data. As expected, the realistic treatment of the BF enhancement

IOP Conf. Series: Journal of Physics: Conf. Series **1023** (2018) 012009 doi:10.1088/1742-6596/1023/1/012009

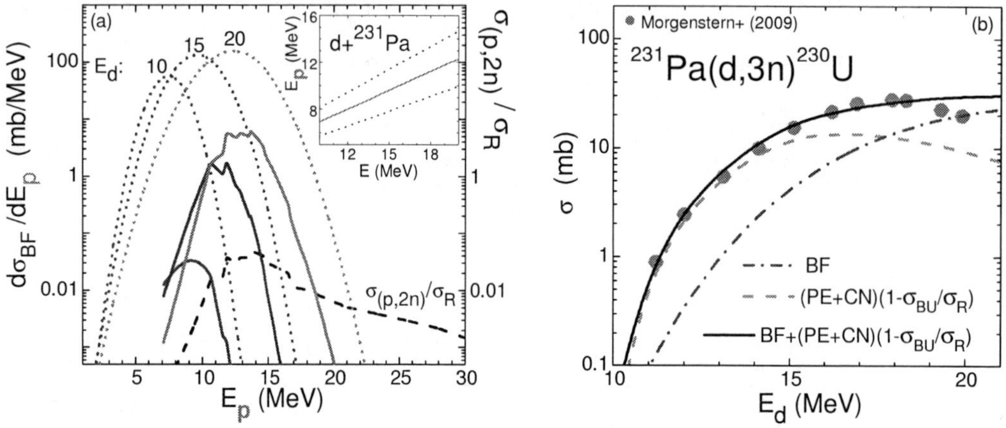

Figure 3. (a) BF enhancement (solid curves) to the ^{231}Pa(d,3n)^{230}U reaction obtained from the convolution of the cross sections ratio $\sigma_{(p,2n)}/\sigma_R^p$ for the target nucleus ^{231}Pa (dashed curve) with the Gaussian distribution (dotted curves) of breakup–proton energies for deuterons on ^{231}Pa at incident energies of 10, 15 and 20 MeV as noted on their top; (b) BF enhancement (dash-dotted curve) to the ^{231}Pa(d,3n)^{230}U reaction excitation function (solid curve) [7] in addition to PE+CN contribution corrected for initial flux leakage towards BU (dashed curve) (see text).

by taking into account the large widths of BU-proton energy distributions shown in the upper insertion of Fig. 3(a) led to a rather accurate description of data.

3. Transfer reactions

Apart from the BU contributions to deuteron interaction, an increased attention has to be devoted to the DR, in spite of related very poor attention or being even not accounted so far in deuteron activation analysis. The calculation of the DR stripping (d,p) and (d,n), and pick-up (d,t) and (d,α)) mechanism contributions has been performed using the distorted-wave Born approximation (DWBA) method within the FRESCO [22] computer code. The post/prior form distorted-wave transition amplitudes for $(d,n/p)$ stripping and respectively $(d,t/\alpha)$ pick-up reactions, and the finite-range interaction have been considered in this respect. The n-p effective interaction in deuteron [23] as well as d-n effective interaction in triton [24] were assumed to have a Gaussian shape, at the same time with a Woods-Saxon shape [25] of the d-d effective interaction in the α particle. The transferred nucleon and deuteron bound states were generated in a Woods-Saxon real potential [5, 6, 8] while the transfer of the deuteron cluster has been taken into account for the (d,α) pick-up cross section calculation. The populated discrete levels and the corresponding spectroscopic factors which have been available within the ENSDF library [26] were used for the DWBA calculations.

The suitable description of the experimental proton, neutron, triton, and alpha-particle angular distributions, for stripping and pick-up transitions, respectively, to states of the corresponding residual nuclei, has been standing for the validation of the spectroscopic information used and the finally calculated total stripping and pick-up reaction cross sections [5, 6, 9, 10]. Thus, the description of the experimental double-differential cross sections of the populations of low-lying levels in ^{239}U and ^{237}U through ^{238}U(d,p)^{239}U and ^{238}U(d,t)^{237}U stripping and pick-up reactions, respectively, shown in in Fig. 4(a,b), validates the correctness of the DR cross-section calculations. Consequently it has been proved the accuracy of the corresponding (d,p) and (d,t) total excitation functions in Fig. 4(c, bottom). Sum of the total (d,p), (d,t), and BU cross sections gives a lower limit of the DI contribution to the deuteron

IOP Conf. Series: Journal of Physics: Conf. Series **1023** (2018) 012009 doi:10.1088/1742-6596/1023/1/012009

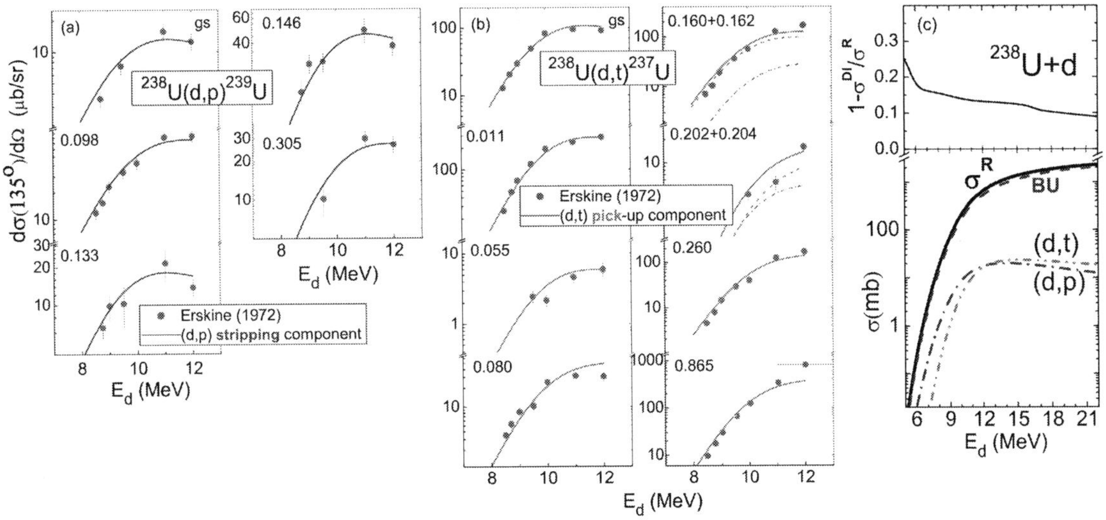

Figure 4. Comparison of calculated (solid curves) and measured [27] excitation functions for the populations at 135° of low-lying levels in (a) ^{239}U and (b) ^{237}U, through (d,p) and (d,t) direct reactions, respectively. (c) (bottom) BU (dashed curve), stripping (dash-dotted curve), and pick-up (dash-dot-dotted curve) cross sections for deuterons on ^{238}U, and the reduction factor of the deuteron flux going towards statistical processes (top) [9].

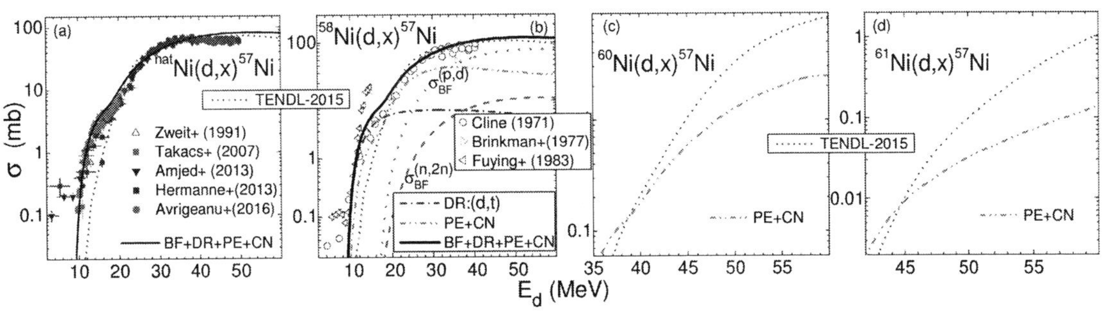

Figure 5. Comparison of measured [6, 20], TENDL-2015 [4] evaluated (short-dotted curves), and theoretically predicted (solid curves) cross sections for ^{nat}Ni$(d,x)^{57}$Ni reactions, with due consideration of contributions from the BF enhancement (dashed and dotted curves), pick-up DR (dash-dotted curve), and PE+CN (dash-dot-dotted curves) reaction mechanisms [6].

interaction with ^{238}U target nucleus, while the deuteron total-reaction cross section that remains to be available for the PE+CN mechanisms has to be corrected for the incident-flux leakage through the DI processes [5, 6, 9, 10] shown on top of Fig. 4(c). Results presented in Fig. 4(c) point out the DI (mainly breakup) dominant role in the deuteron interaction with ^{238}U around Coulomb barrier, which is a specific feature for the heavy target nuclei [13, 16].

The lack of due consideration of the DI mechanisms within theoretical frame of deuteron surrogate-reaction method should be considered the main reason for the failure of the $(d,p\gamma)$ surrogate–reaction validation tests comparing already well–known (n,γ) cross sections with those provided by deuteron surrogate reaction $(d,p\gamma)$ [28]. Our detailed analyzes [9, 10] of the nuclear reaction mechanisms involved in the $(d,p\gamma)$ surrogate reactions on ^{232}Th and ^{238}U target nuclei have pointed out the key role of the direct interactions, i.e. breakup, stripping and pick-up

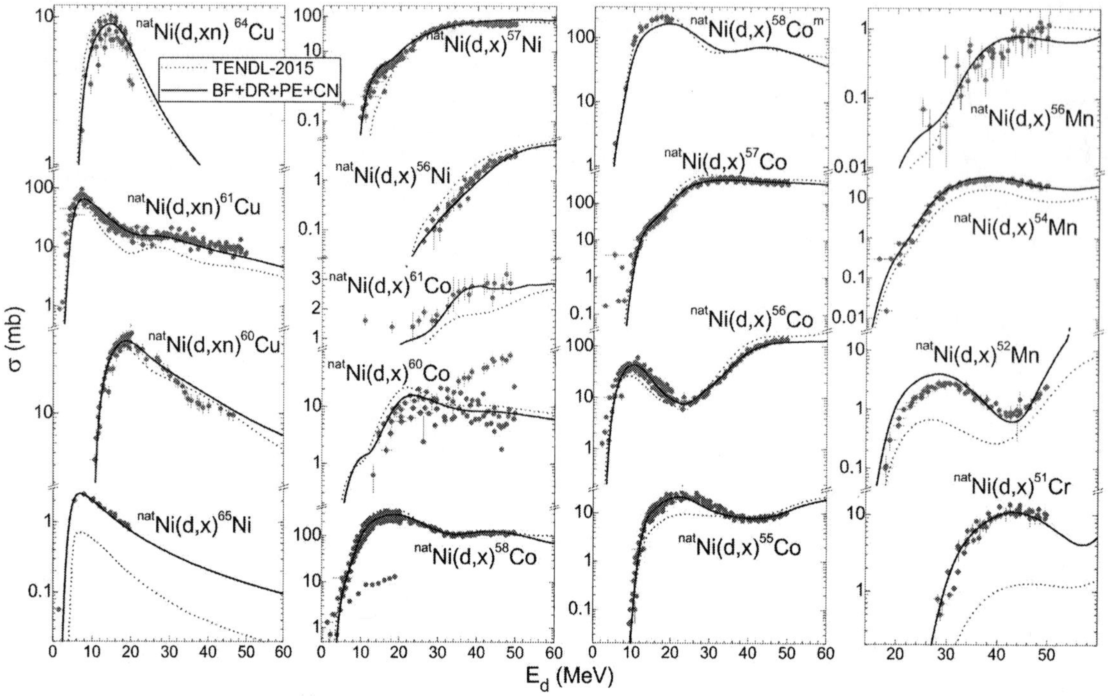

Figure 6. Comparison of measurements (solid circles) [6, 20], TENDL-2015 [4] evaluation (short-dotted curves), and model calculations (solid curves) of excitation functions for deuteron-induced reactions on ^{nat}Ni [6].

mechanisms, while only the CN mechanism was considered by the above-mentioned studies ([28] and Refs. therein).

A particular note of transfer reactions signature should concern the (d,t) pick-up reaction. In spite of being usually neglected in deuteron activation analysis, the (d,t) pick-up process is fully responsible for the lowest-energy side of the corresponding (d,x) excitation function, e.g., Fig. 5(a,b), namely at the energies between its threshold and those of the (d,dn) and (d,p2n) reactions that contribute to the population of the same residual nucleus [5, 6].

4. Statistical particle emission

The statistical PE+CN reaction mechanisms which complete the deuteron interaction analysis along an enlarged nuclear-interaction time scale, become important with the increase of the incident energy above the Coulomb barrier. The corresponding reaction cross sections are calculated using the TALYS [21] code, taking into account also the above-discussed BU, stripping, and pick–up results through a reduction factor of the OMP total-reaction cross section, given by the sum of the DI cross-sections. Another particular point of these calculations is the use of the same model parameters to account for different reaction mechanisms as, e.g., the same OMP parameters for calculation of transmission coefficients as well as PE transition rates (using the value 3 for the preeqmode TALYS keyword). Additional PE+CN calculations have been carried out with the code STAPRE-H [29] if other particular options of various input parameters could be useful (e.g., for gamma-ray strength functions or initial p-h configurations. The due consideration of all BU+DR+PE+CN is proved by the description of all measured data corresponding to deuteron interaction with a specific natural element target [5, 6], e.g., activation excitation functions for d+^{nat}Ni interaction process shown in Fig. 6 [6].

5. Conclusions

The present work has concerned a deeper analysis of the key role of DI, particularly of the breakup mechanism, in deuteron-induced reactions. The overall agreement between the measured data and model calculations supports the description of nuclear mechanisms taken into account for the deuteron-nucleus interaction, emphasizing the effects of direct interactions so far ignored in the evaluation procedures.

However, while the associated theoretical frames are already settled for stripping, pick-up, PE and CN mechanisms, an increased attention should be given to the breakup mechanism. Thus further work has to be done concerning its theoretical description including the inelastic component. The improvement of deuteron breakup description requires complementary experimental studies involving deuterons, protons and neutron induced reactions too.

Acknowledgments

This work has been partly supported by Autoritatea Nationala pentru Cercetare Stiintifica (PN-16420102) and partly carried out within the framework of the EUROfusion Consortium and has received funding from the Euratom research and training programme 2014-2018 under grant agreement No 633053. The views and opinions expressed herein do not necessarily reflect those of the European Commission.

References

[1] http://www.iter.org/proj
[2] http://www.ifmif.org/b/
[3] http://pro.ganil-spiral2.eu/spiral2/instrumentation/nfs
[4] A.J. Koning *et al.* Dec. 2015, *TENDL-2015*, http://www.talys.eu/TENDL2015/
[5] P. Bém *et al.* 2009 *Phys. Rev.* C **79**, 044610; E. Šimečková *et al.* 2011 *Phys. Rev.* C **84**, 014605; M. Avrigeanu *et al.* 2013 *Phys. Rev.* C **88**, 014612; M. Avrigeanu *et al.* 2014 *Phys. Rev.* C **89**, 044613;
[6] M. Avrigeanu *et al.* 2016 *Phys. Rev.* C **94**, 014606
[7] M. Avrigeanu, V. Avrigeanu, and A.J. Koning 2012 *Phys. Rev.* C **85**, 034603
[8] M. Avrigeanu and V. Avrigeanu 2015 *Phys. Rev.* C **92**, 021601
[9] M. Avrigeanu, V. Avrigeanu, and C. Costache 2017 *EPJ Web of Conferences* **146** 12020; M. Avrigeanu, V. Avrigeanu, and C. Costache 2017 arXiv:1703.05956v1 [nucl-th], 17 March.
[10] M. Avrigeanu and V. Avrigeanu 2016 *J. Phys: Conf. Ser.* **724**, 012003; 2015 CERN-Proceedings-2015-001, 203
[11] M. Avrigeanu *et al.* 2009 *Fusion Eng. Design* **84**, 418; M. Avrigeanu and V. Avrigeanu 2017 *Phys. Rev.* C **95**, 024607
[12] M.G. Mustafa, T. Tamura, and T. Udagawa 1987 *Phys. Rev.* C **35**, 2077
[13] J. Pampus *et al.* 1978 *Nucl. Phys.* A **311**, 141
[14] J.R. Wu, C.C. Chang, and H.D. Holmgren 1979 *Phys. Rev.* C **19**, 370
[15] N. Matsuoka *et al.* 1980 *Nucl. Phys.* A **345**, 1
[16] J. Kleinfeller *et al.* 1981 *Nucl. Phys.* A **370**, 205
[17] M. Avrigeanu and A. M. Moro 2010 *Phys. Rev.* C **82**, 037601
[18] M. Kamimura *et al.* 1986 *Prog. Theor. Phys.* **89**, Suppl. 1, 1; N. Austern *et al.* 1987 *Phys. Rep.* **154**, 125; A. Deltuva *et al.* 2007 *Phys. Rev.* C **76**, 064602; K. Ogata and K. Yoshida 2016 *Phys. Rev.* C **94**, 051603(R)
[19] C. Kalbach Walker 2003 *TUNL Progress Report* **XLII** *(2002-2003)*, p. 82-83
[20] Experimental Nuclear Reaction Data (EXFOR), www-nds.iaea.or.at/exfor
[21] A.J. Koning, S. Hilaire, and S. Goriely 2015 v. TALYS-1.8, http://www.talys.eu
[22] I.J. Thompson 1988 Comput. *Phys. Rep.* **7**, 167; 2011 v. FRES 2.9
[23] M. Kawai, M. Kamimura, and K. Takesako 1969 *Prog. Theor. Phys. Suppl.* **184**, 118
[24] P. Guazzoni *et al.* 2011 *Phys. Rev.* C **83**, 044614
[25] R.M. DelVecchio 1973 *Phys. Rev.* C **7**, 677
[26] Evaluated Nuclear Structure Data File (ENSDF), http://www.nndc.bnl.gov/ensdf/
[27] J.R. Erskine 1972 *Phys. Rev.* C **5**, 959
[28] J.N. Wilson et al., 2012 *Phys. Rev.* C **85**, 034607; Q. Ducasse et al. 2016 *Phys. Rev.* C **94**, 024614
[29] M. Avrigeanu and V. Avrigeanu 1995 *IPNE Report* NP-86-1995, Bucharest, and Refs. therein; http://www.oecdnea.org/tools/abstract/detail/iaea0971/

IOP Conf. Series: Journal of Physics: Conf. Series **1023** (2018) 012010 doi:10.1088/1742-6596/1023/1/012010

From *ab initio* structure predictions to reaction calculations via EFT

P. Capel[1,2,3], V. Durant[2,3], L. Huth[2,3], H.-W. Hammer[2,3], D. R. Phillips[4], and A. Schwenk[2,3,5]

E-mail: pierre.capel@ulb.ac.be,
durant@theorie.ikp.physik.tu-darmstadt.de,
lukashuth@theorie.ikp.physik.tu-darmstadt.de,
hans-werner.hammer@physik.tu-darmstadt.de,
phillid1@ohio.edu,
schwenk@physik.tu-darmstadt.de
[1]Physique Nucléaire et Physique Quantique (CP 229), Université libre de Bruxelles (ULB), B-1050 Brussels
[2]Institut für Kernphysik, Technische Universität Darmstadt, 64289 Darmstadt, Germany
[3]ExtreMe Matter Institute EMMI, GSI Helmholtzzentrum für Schwerionenforschung GmbH, 64291 Darmstadt, Germany
[4]Institute of Nuclear and Particle Physics and Department of Physics and Astronomy, Ohio University, Athens, OH 45701,USA
[5]Max-Planck-Institut für Kernphysik, Saupfercheckweg 1, 69117 Heidelberg, Germany

Abstract. We study how EFT can improve the description of reactions used to study the structure of exotic nuclei. Using Halo EFT helps constraining the potential that simulates the interaction between the projectile constituents. It enables us to include *ab initio* results within the reaction model and to clearly identify the nuclear properties probed by the reaction. New local chiral EFT nucleon-nucleon interactions can also be used to derive nucleus-nucleus optical potentials from first principles, which can provide reliable inputs in reaction modelling.

1. Introduction

Halo nuclei have been discovered thanks to the development of radioactive-ion beams in the mid-80s [1]. Compared to their isobars, they exhibit very large reaction cross sections on light targets that have been interpreted as the sign of a matter radius larger than the $A^{1/3}$ behaviour observed for stable and many other nuclei. This unusual size is qualitatively understood as resulting from a strongly clusterised structure: halo nuclei are seen as a compact core to which one or two neutrons are loosely bound. Thanks to this loose binding the valence neutrons exhibit a high probability of presence at a large distance from the other nucleons, forming a diffuse halo around the core. This exotic nuclear structure is encountered at the limit of stability, where the binding energy for one or two neutrons becomes very small. The nucleus ^{11}Be is an archetypical one-neutron halo nucleus, while ^{11}Li exhibits a two-neutron halo. Due to their exotic structure, halo nuclei are a challenge for nuclear-structure models. Very recently, Calci *et al.* have been able to compute ^{11}Be within a No-Core Shell Model with continuum calculation (NCSMC) [2], managing to describe the halo structure, the ^{10}Be-n continuum, and the well-known shell inversion between the $1/2^+$ ground state and the $1/2^-$ bound excited state.

IOP Conf. Series: Journal of Physics: Conf. Series **1023** (2018) 012010 doi:10.1088/1742-6596/1023/1/012010

Being located at the limit of stability, halo nuclei exhibit very short lifetimes, which makes it difficult to employ the usual spectroscopic techniques to study their structure. Therefore, they are usually studied indirectly through reactions like transfer (see the contribution of Yang in these proceedings [3]) or breakup [4]. In the latter reaction, the halo dissociates from the core during the collision with a target. To get valuable information from such measurements, an accurate reaction model coupled to a realistic description of the projectile and reliable inputs for the models are needed. In this contribution, we explore how effective field theory (EFT) can help us build a realistic few-body description of ^{11}Be from the *ab initio* calculation of Calci *et al.* [2] and how it could help us derive the optical potentials needed as inputs for the reaction model from first principles. The former task is performed within the EFT description of halo nuclei [5, 6]; for the latter, we test the double-folding of a local chiral EFT nucleon-nucleon interaction with the density of the projectile and the target [7].

After a quick presentation of the reaction model we consider in this study, we present in Sec. 3 how the *ab initio* results are described within the Halo EFT. In Sec. 4, we summarise the first results obtained using a double-folding procedure to derive optical potentials from first principles. Finally, we conclude in Sec. 5

2. Reaction model

We consider reactions in which the one-neutron halo nucleus ^{11}Be interacts with a target. The simplest way to describe such collisions is to consider a two-body model of the projectile, hence seen as an inert ^{10}Be core in its 0^+ ground state to which a neutron is loosely bound [8]. Such structure is described by the Hamiltonian

$$H_0 = T_r + V_{\rm cn}(\boldsymbol{r}), \tag{1}$$

where the effective potential $V_{\rm cn}$ simulates the interaction between the ^{10}Be core c and the valence neutron n. The particulars about this interaction will be detailed in Sec. 3.

The internal structure of the target T is neglected and its interaction with the projectile constituents is described by optical potentials V_{cT} and $V_{{\rm n}T}$. So far these potentials are taken from the literature, where they have been adjusted to reproduce elastic-scattering data. In Sec. 4, we will explore how they could be derived from first principles using a double-folding procedure.

Within this model, the study of the reaction reduces to solving the following three-body Schrödinger equation

$$[T_R + H_0 + V_{cT} + V_{{\rm n}T}]\, \Psi(\boldsymbol{r}, \boldsymbol{R}) = E_T\, \Psi(\boldsymbol{r}, \boldsymbol{R}) \tag{2}$$

with the initial condition that the incoming projectile is initially in its ground state ϕ_0:

$$\Psi(\boldsymbol{r}, \boldsymbol{R}) \underset{Z\to-\infty}{\longrightarrow} e^{iKZ}\phi_0(\boldsymbol{r}). \tag{3}$$

Various techniques exist to solve this problem, see Ref. [8] for a recent review. We will use the dynamical eikonal approximation (DEA) [9, 10], which is very efficient and precise at the beam energies considered here [11].

3. Including Halo EFT within reaction models

3.1. EFT description of ^{11}Be

Usually, the potential that simulates the c-n interaction within reaction models exhibits a Woods-Saxon form factor. The depth of that potential is then adjusted to reproduce the basic properties of the projectile, like its binding energy and the quantum numbers of its ground state. We suggest

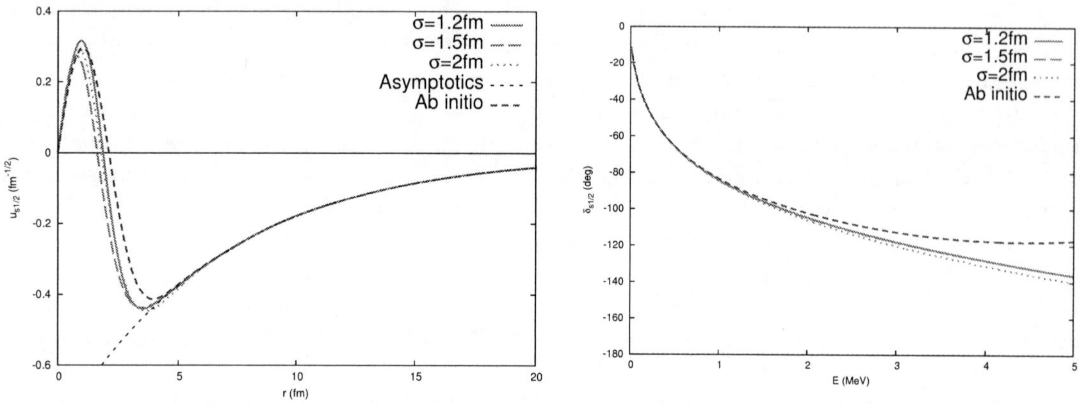

Figure 1. Left: $1s_{1/2}$ reduced radial wave function describing the $1/2^+$ ground state of ^{11}Be. Right: $s_{1/2}$ phase shift. Calculations performed within the Halo EFT description of ^{11}Be with different Gaussian width σ are confronted to the *ab initio* predictions [2].

to replace this phenomenological interaction by an effective interaction based on Halo EFT [5]. We use a Gaussian shape which, at next-to-leading order (NLO), can be taken to be [6]

$$V_{lj}(r) = V_{lj}^{0}\, e^{-\frac{r^2}{2\sigma^2}} + V_{lj}^{2}\, r^2 e^{-\frac{r^2}{2\sigma^2}}. \tag{4}$$

In each partial wave of orbital angular momentum l and total spin j, such a form gives us two parameters that can be adjusted to reproduce the energy of the projectile states and their asymptotic normalisation constant (ANC), for bound states, or their width, for resonant states. The low-energy spectrum of ^{11}Be is well known, however the ANC of its two bound states have not been directly measured. To help us in this task, we rely on the *ab initio* calculation of Calci *et al.* which is the most advanced calculation of ^{11}Be available so far [2].

The Gaussian width σ is a length scale that helps us evaluate the sensitivity of the reaction model to short-range physics. In the following, we consider $\sigma = 1.2$, 1.5, and 2 fm.

This two-body model of ^{11}Be assumes that its $1/2^+$ ground state corresponds to the single-particle state in which the valence neutron sits in the intruder $1s_{1/2}$ partial wave. The $1/2^-$ bound excited state corresponds to the regular $0p_{1/2}$ shell-model state. The reduced radial wave function is plotted in the left panel of Fig. 1 for the different Gaussian widths $\sigma = 1.2$ (red solid line), 1.5 (green dashed line), and 2 fm (magenta dotted line). These Halo EFT wave functions are compared to the *ab initio* result [2] (blue short-dashed line). All wave function exhibit the same asymptotic behaviour since our Gaussian potentials have been fitted to reproduce the binding energy and the ANC of the *ab initio* calculation. However, they significantly differ in the interior as expected from the difference in the short-range scale chosen for each potential.

The right panel of Fig. 1 displays the $s_{1/2}$ phase shift obtained by the three different Gaussian potentials. Since they have been adjusted to the same binding energy and ANC, they exhibit very similar phase shifts, especially at low energy, where their effective-range expansion is nearly identical [6]. These phase shifts compare very well with the *ab initio* prediction up to $E = 1$–1.5 MeV in the ^{10}Be-n continuum. Similar results are obtained in the $p_{1/2}$ partial wave. In the strict Halo EFT taken at NLO, the ^{10}Be-n interaction is switched off in all other partial waves, either because there is no low-energy state that generates a significant phase shift, or because their orbital angular momentum is too high.

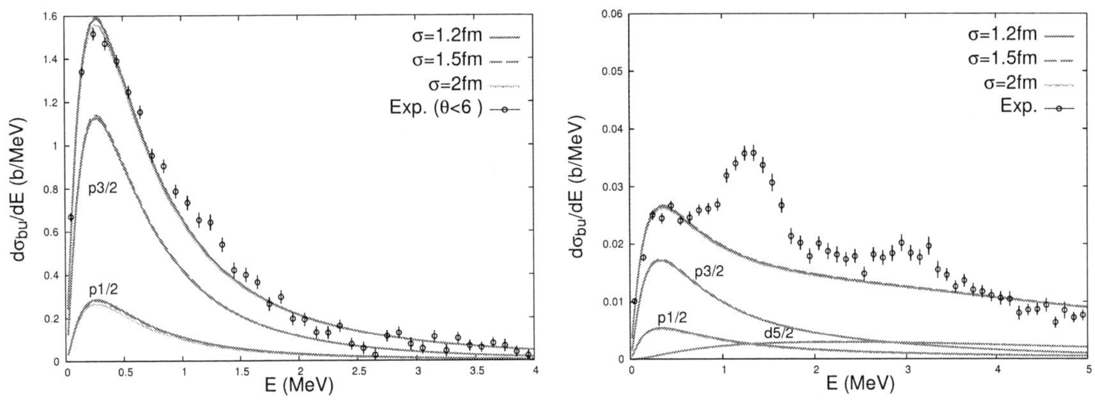

Figure 2. Energy distributions for the breakup of ^{11}Be on Pb at 69AMeV (left) and on C at 67AMeV (right). The data are from Ref. [4].

3.2. Breakup of ^{11}Be

Starting from the effective description of ^{11}Be derived in the previous section, we perform breakup calculations within the DEA using the numerical condition and inputs (viz. optical potentials) of Ref. [10]. We consider the experimental conditions of the RIKEN experiment [4]: a Pb target at 69AMeV and a C target at 67AMeV. The corresponding breakup cross section are plotted in Fig. 2 as a function of the ^{10}Be-n relative energy E after dissociation [6]. The contributions of the dominant partial waves are depicted as well.

It is interesting to note that in both cases, the computed cross sections are nearly identical for all three different descriptions of ^{11}Be. This confirms that—at least in the non-resonant continuum—breakup is a peripheral process, in the sense that it probes only the tail of the wave function [12]. Accordingly the three different descriptions, which produce identical ANCs for the ground state, lead to identical cross sections. Interestingly, this is true even on light targets, for which the reaction is dominated by the short-range nuclear interaction.

As expected from a collision on a heavy target, the breakup on Pb (Fig. 2 left) is mostly Coulombian. Therefore, the major partial waves populated in the continuum are the p waves. In this case, the agreement with the RIKEN data are excellent, without any fitting parameter. This confirms the value of the ground-state ANC predicted by the *ab initio* calculation.

The collision on C (Fig. 2 right) populates, in addition to the p waves, a significant amount of the d-wave continuum. In this case, the agreement with the data is less good. Although it produces the right order of magnitude and the general trends of the data, our calculations clearly miss the low-energy peak that corresponds to the $5/2^+$ state in the ^{10}Be-n continuum. That state is usually described as a single-particle $d_{5/2}$ resonance [13]. Unfortunately, adding that state in an extension of our Halo EFT description of ^{11}Be does not fully solve the problem [6]. This suggests that some significant degrees of freedom are missing in order to correctly describe that state and its influence on the reaction calculation. Within the Halo EFT expansion scheme, the first degree of freedom that has been overlooked in our model would most certainly be the first 2^+ excited state of the ^{10}Be core. Adding this state in an extension of the present model would most likely improve our comparison with the data. This is one of our goals for the future.

4. Deriving optical potential from first-principle chiral EFT NN interactions

4.1. Double-folding procedure

The core of halo nuclei is usually itself radioactive; it is therefore difficult to find the proper optical potential to describe its interaction with the target. To see if it can be derived from first principles, we study the development of an optical potential from chiral EFT nucleon-nucleon (NN) interactions using a double-folding procedure. To perform these calculations, we take advantage of the recent development of local chiral EFT NN interactions [14, 15].

The potential is built at the Hartree-Fock level, including both the direct and exchange terms [16]

$$V_F(\boldsymbol{r}, E_{\rm cm}) = V_{\rm D}(\boldsymbol{r}) + V_{\rm Ex}(\boldsymbol{r}, E_{\rm cm}) \tag{5}$$

The former is calculated by integrating the NN interaction over the neutron (n) and proton (p) density distributions $\rho_1^{\rm n,p}$ and $\rho_2^{\rm n,p}$ of the colliding nuclei,

$$V_{\rm D}(\boldsymbol{R}) = \sum_{i,j={\rm n,p}} \int\int \rho_1^i(\boldsymbol{r}_1)\, v_{\rm D}^{ij}(\boldsymbol{S})\, \rho_2^j(\boldsymbol{r}_2)\, d^3\boldsymbol{r}_1 d^3\boldsymbol{r}_2\,, \tag{6}$$

where $\boldsymbol{R}$ is the relative coordinate between the centre of mass of the nuclei, $\boldsymbol{r}_1$ and $\boldsymbol{r}_2$ are the coordinates from the centre of mass of each nucleus, $\boldsymbol{S} = \boldsymbol{R} - \boldsymbol{r}_1 + \boldsymbol{r}_2$, and the sum i, j is over neutrons and protons with their respective densities. The exchange term reads

$$V_{\rm Ex}(\boldsymbol{R}, E_{\rm cm}) = \sum_{i,j={\rm n,p}} \int\int \rho_1^i(\boldsymbol{r}_1, \boldsymbol{r}_1 + \boldsymbol{S})\, v_{\rm Ex}^{ij}(\boldsymbol{S}) \rho_2^j(\boldsymbol{r}_2, \boldsymbol{r}_2 - \boldsymbol{S}) \exp\left[\frac{i\boldsymbol{K}(\boldsymbol{R}) \cdot \boldsymbol{S}}{\mu/m_{\rm N}}\right] d^3\boldsymbol{r}_1 d^3\boldsymbol{r}_2\,, \tag{7}$$

where $\mu = m_{\rm N} A_1 A_2/(A_1 + A_2)$ is the reduced mass of the colliding nuclei (with $m_{\rm N}$ the nucleon mass) and the integral is over the density matrices $\rho^i(\boldsymbol{R}, \boldsymbol{R} \pm \boldsymbol{S})$ of the nuclei. The momentum for the NN relative motion $\boldsymbol{K}$ is related to $E_{\rm cm}$ and the double-folding potential [16, 7].

Following this method, we calculate double-folding potentials at LO, NLO and N^{2}LO, using different cutoffs R_0 [7]. Examples of the potentials we obtain are presented in the left panel of Fig. 3. We observe a systematic order-by-order behaviour expected in EFT. Moreover, we see that the potential exhibits a rather small dependence on the cutoff, indicating that it is not very sensitive to the short-range details of the NN interaction.

In addition to the real part provided by Eq. (5), optical potentials also include an imaginary part that simulates the absorption from the elastic channel due to other open channels, like inelastic scattering or transfer. In this first attempt, we follow the São Paulo group and assume the imaginary part to be proportional to the real part obtained by double folding [18]

$$U_F(r) = (1 + i\, N_W)\, V_F(r), \qquad \text{with } N_W = 0.6\text{–}0.8. \tag{8}$$

4.2. ^{16}O-^{16}O scattering and fusion

To evaluate the validity of our potentials, we perform calculations for ^{16}O-^{16}O elastic scattering at different energies and compare them to data [7]. This is illustrated in the right panel of Fig. 3, where the ratio of the elastic-scattering cross section is presented as a ratio to the Mott cross section at $E_{\rm lab} = 350$ MeV. Interestingly, we observe the very same systematic order-by-order behaviour expected in the EFT as in the double-folding calculation (see the left panel of Fig. 3). We also note that the sensitivity to the cutoff R_0 is rather small and becomes noticeable only at sufficiently large angles ($\theta > 20°$–$25°$). The agreement with the data is quite good, given that there is no parameter adjusted to fit the data. However, there remains some uncertainty to the choice of the scaling of the imaginary part N_W [7].

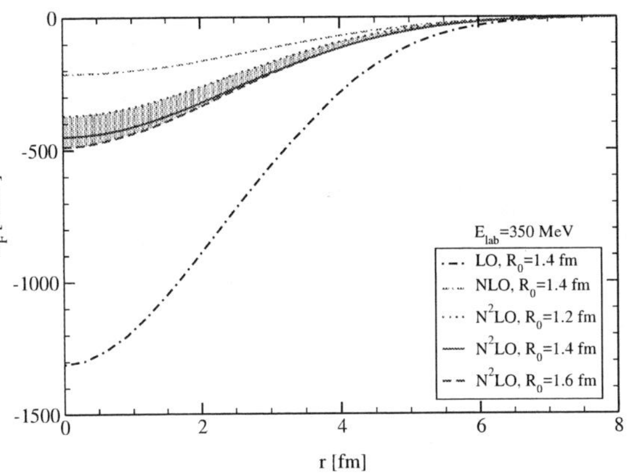

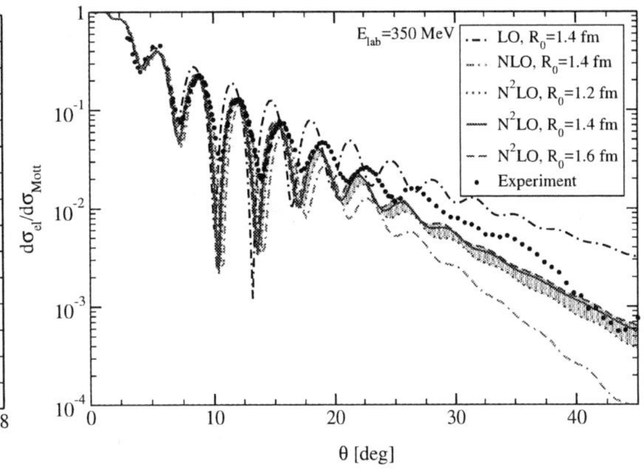

Figure 3. ^{16}O-^{16}O elastic scattering at $E_{\text{lab}} = 350$ MeV. Left: double-folding potential built from local chiral EFT NN interactions at LO, NLO and N^{2}LO [7]. Right: elastic-scattering cross section as a ratio to Mott cross section obtained with these potentials. Data are from Ref. [17].

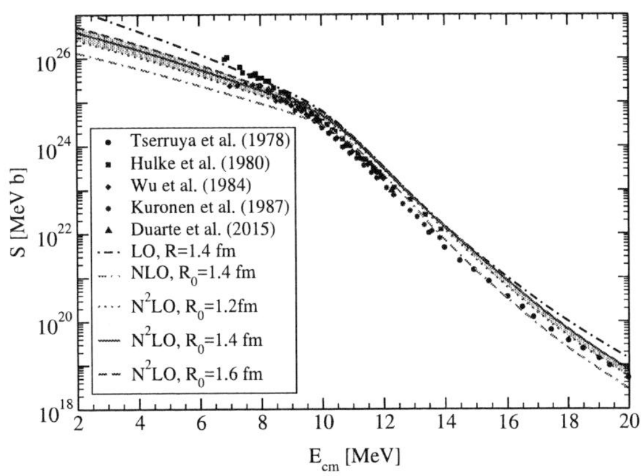

Figure 4. Astrophysical S factor for the ^{16}O-^{16}O fusion as a function of the centre-of-mass energy E_{cm} [7]. The data are from Refs. [19, 20, 21, 22, 23].

We can also check the reliability of our double-folding potential by computing the astrophysical S factor for ^{16}O-^{16}O fusion, which has been measured at various energies [19, 20, 21, 22, 23]. The results of this test are displayed in Fig. 4. As in the previous case, we observe the expected EFT order-by-order behaviour and note once more that the sensitivity of this observable to the details of the NN interaction remains small [7]. The agreement with the data are rather good, given that no parameter has been adjusted to fit the data.

5. Conclusion

In this work, we study the possibilities offered by EFT to improve the modelling of reactions involving halo nuclei. First, Halo EFT [5] provides an efficient tool to include the relevant degrees of freedom of the projectile nuclear structure within the few-body model of reactions. It also enables us to account for the outputs of *ab initio* nuclear-structure models within reaction calculations. The excellent agreement we have obtained for non-resonant breakup calculations using these outputs confirm their validity. The systematic study of these reactions with an EFT description of the projectile also shows which degrees of freedom are required in order to properly describe the projectile and obtain theoretical cross sections in agreement with experiments. In particular, our analysis shows that the ^{10}Be core excitation plays a significant role in the resonant breakup of ^{11}Be in the collision on a C target.

Second, we have shown how the recently derived local chiral EFT NN interactions [14, 15] can be used to build the nucleus-nucleus optical potentials required in the modelling of nuclear reactions. Using a double-folding technique, we have produced optical potentials that provide elastic-scattering cross sections and astrophysical S factors for fusion in good agreement with ^{16}O-^{16}O data [7]. These preliminary results are very encouraging. An important uncertainty remains about the choice of the imaginary part of the potential, and we plan to focus our future work in this direction.

These first attempts of combining EFT used in nuclear structure with an accurate few-body description of nuclear reactions are very promising. Our results show that it could lead to a significant improvement of the reaction modelling for exotic nuclei and, accordingly, to a better analysis of reaction measurements at radioactive-beam facilities.

Acknowledgments

This work was supported in part by the European Union's Horizon 2020 research and innovation programme under grant agreement No 654002 and from the European Research Council Grant No. 307986 STRONGINT, the GSI-TU Darmstadt cooperation. This research was also supported by the U.S. Department of Energy under grant DE-FG02-93ER40756, by the German Federal Ministry of Education and Research under contract 05P15RDFN1, and by the Deutsche Forschungsgemeinschaft within the Collaborative Research Center 1245 (SFB 1245).

References

[1] Tanihata I 1996 *J. Phys. G* **22** 157
[2] Calci A, Navrátil P, Roth R, Dohet-Eraly J, Quaglioni S and Hupin G 2016 *Phys. Rev. Lett.* **117** 242501
[3] Yang J and Capel P 2017 contribution to this workshop
[4] Fukuda N, Nakamura T, Aoi N, Imai N, Ishihara M, Kobayashi T, Iwasaki H, Kubo T, Mengoni A, Notani M, Otsu H, Sakurai H, Shimoura S, Teranishi T, Watanabe Y X and Yoneda K 2004 *Phys. Rev. C* **70** 054606
[5] Hammer H W, Ji C and Phillips D R 2017 *J. Phys. G* **44** 103002
[6] Capel P, Phillips D and Hammer H W 2017 in preparation
[7] Durant V, Capel P, Huth L, Balantekin A B and Schwenk A 2017 *ArXiv e-prints* (*Preprint* 1708.02527)
[8] Baye D and Capel P 2012 *Clusters in Nuclei, Vol. 2* vol 848 ed Beck C (Heidelberg: Springer)
[9] Baye D, Capel P and Goldstein G 2005 *Phys. Rev. Lett.* **95** 082502
[10] Goldstein G, Baye D and Capel P 2006 *Phys. Rev. C* **73** 024602
[11] Capel P, Esbensen H and Nunes F M 2012 *Phys. Rev. C* **85** 044604
[12] Capel P and Nunes F M 2007 *Phys. Rev. C* **75** 054609
[13] Capel P, Goldstein G and Baye D 2004 *Phys. Rev. C* **70** 064605
[14] Gezerlis A, Tews I, Epelbaum E, Gandolfi S, Hebeler K, Nogga A and Schwenk A 2013 *Phys. Rev. Lett.* **111** 032501
[15] Huth L, Tews I, Lynn J E and Schwenk A 2017 *ArXiv e-prints* (*Preprint* 1708.03194)
[16] Furumoto T, Horiuchi W, Takashina M, Yamamoto Y and Sakuragi Y 2012 *Phys. Rev. C* **85** 044607
[17] Bohlen H G, Stiliaris E, Gebauer B, von Oertzen W, Wilpert M, Wilpert T, Ostrowski A, Khoa D T, Demyanova A S and Ogloblin A A 1993 *Z. Phys. A* **346** 189
[18] Pereira D, Lubian J, Oliveira J R B, de Sousa D P and Chamon L C 2009 *Phys. Lett. B* **670** 330
[19] Tserruya I, Eisen Y, Pelte D, Gavron A, Oeschler H, Berndt D and Harney H L 1978 *Phys. Rev. C* **18** 1688
[20] Hulke G, Rolfs C and Trautvetter H P 1980 *Z. Phys. A* **297** 161
[21] Wu S C and Barnes C 1984 *Nucl. Phys. A* **422** 373
[22] Kuronen A, Keinonen J and Tikkanen P 1987 *Phys. Rev. C* **35** 591
[23] Duarte J G, Gasques L R, Oliveira J R B, Zagatto V A B, Chamon L C, Medina N, Added N, Seale W A, Alcántara-Núñez J A, Rossi Jr E S *et al.* 2015 *J. Phys. G* **42** 065102

IOP Conf. Series: Journal of Physics: Conf. Series **1023** (2018) 012011 doi:10.1088/1742-6596/1023/1/012011

Chaos in nuclei: Theory and experiment

L. Muñoz[1], R. A. Molina[2], J. M. G. Gómez[1]

[1] Grupo de Física Nuclear, Facultad de Ciencias Físicas, Universidad Complutense, E-28040 Madrid, Spain

[2] Instituto de Estructura de la Materia, IEM-CSIC, Serrano 123, Madrid, E-28006, Spain

E-mail: jmgomezg@ucm.es

Abstract.
During the last three decades the quest for chaos in nuclei has been quite intensive, both with theoretical calculations using nuclear models and with detailed analyses of experimental data. In this paper we outline the concept and characteristics of quantum chaos in two different approaches, the random matrix theory fluctuations and the time series fluctuations. Then we discuss the theoretical and experimental evidence of chaos in nuclei. Theoretical calculations, especially shell-model calculations, have shown a strongly chaotic behavior of bound states in regions of high level density. The analysis of experimental data has shown a strongly chaotic behavior of nuclear resonances just above the one-nucleon emission threshold. For bound states, combining experimental data of a large number of nuclei, a tendency towards chaotic motion is observed in spherical nuclei, while deformed nuclei exhibit a more regular behavior associated to the collective motion. On the other hand, it had never been possible to observe chaos in the experimental bound energy levels of any single nucleus. However, the complete experimental spectrum of the first 151 states up to excitation energies of 6.20 MeV in the ^{208}Pb nucleus have been recently identified and the analysis of its spectral fluctuations clearly shows the existence of chaotic motion.

1. Introduction

The atomic nucleus is generally considered a paradigmatic case of quantum chaos. Intuitively one can expect that fast moving nucleons interacting with the strong nuclear force and bound in the small nuclear volume should give rise to a chaotic motion. During the last three decades the quest for chaos in nuclei has been quite intensive, both with theoretical calculations using nuclear models and with detailed analyses of experimental data. Statistical spectroscopy studies in nuclei have been also motivated by a desire to understand the implications of chaotic behavior in many-body quantum systems [1].

In this paper we outline the concept and characteristics of quantum chaos in two different approaches, the Random Matrix Theory (RMT) fluctuations [2] and the time series fluctuations [3]. Although these two approaches are conceptually different, they lead to the same conclusions, namely, that both chaotic and regular quantum systems exhibit a universal but different behavior in the statistical properties of the spectrum. Furthermore it has been proven that, when the fluctuations of random matrix ensembles are studied with methods from time series analysis, all the ensembles exhibit $1/f$ noise, independently of the system symmetries [4].

Then we discuss the theoretical and experimental evidence of chaos in nuclei. Theoretical calculations, especially shell-model calculations, have shown a strongly chaotic behavior of bound

IOP Conf. Series: Journal of Physics: Conf. Series **1023** (2018) 012011 doi:10.1088/1742-6596/1023/1/012011

states at higher excitation energy, in regions of high level density [5]. In the low energy region most shell-model calculations show chaotic behavior as well, however, there are exceptions, for example, in neutron-rich Ca and Pb isotopes [6, 7]. The analysis of experimental data has shown a strongly chaotic behavior of nuclear resonances just above the one-nucleon emission threshold [8]. For bound states, combining experimental data of a large number of nuclei, a tendency towards chaotic motion is observed in spherical nuclei, while deformed nuclei exhibit a more regular behavior. For a comprehensive review of chaos in nuclei see for example Gómez *et al.* [1] and Weidenmüller and Mitchell [9]. On the other hand, it had never been possible to observe chaos in the experimental bound states of any single nucleus. However, very recently the complete experimental spectrum of 151 states up to excitation energies of 6.20 MeV in the ^{208}Pb nucleus have been identified at the Maier-Leibnitz-Laboratorium at Garching, Germany [10], and the analysis of its spectral fluctuations clearly shows the existence of chaotic motion in the nuclear bound states of ^{208}Pb [11, 12].

2. Quantum chaos and spectral fluctuations

2.1. Random Matrix Theory and quantum chaos

The concept of chaos in classical mechanics is well established, but the criteria that characterize classical chaos, like exponential divergence of trajectories, cannot be carried to quantum mechanics. From the study of quantum billiards with chaotic classical analogues, Bohigas, Giannoni and Schmidt (BGS) reached the following conjecture: the spectral fluctuation properties of simple quantum systems known to be ergodic in the classical limit, are universal and follow very closely those of the Gaussian ensembles of random matrices [2]. The BGS conjecture has been proven in the semiclassical limit [13]. The essential feature of chaotic energy spectra in quantum systems is the existence of level repulsion and correlations. On the contrary, classically integrable systems give rise to uncorrelated adjacent energy levels, that are well described by Poisson statistics [14].

The most commonly used statistics to characterize spectral fluctuations are the nearest neighbor spacing (NNS) distribution $P(s)$ and the spectral rigidity $\Delta_3(L)$ of Dyson and Mehta [15]. For the sequence of unfolded levels $\epsilon_i, i = 1, ..., N$, the nearest level spacing s_i is defined by $s_i = \epsilon_{i+1} - \epsilon_i$ and the average values are $\langle s \rangle = 1$, $\langle \epsilon_n \rangle = n$.

For the uncorrelated levels or Poisson case, $P_P(s) = \exp(-s)$, $P_P(0) = 1$, and $\langle \Delta_3(L) \rangle = L/15$. For the Gaussian orthogonal ensemble (GOE), $P(s)$ follows very closely the Wigner surmise $P_W(s) = (\pi/2)s \exp(-\pi s^2/4)$ and the $\langle \Delta_3(L) \rangle$ grows logarithmically instead of linearly.

2.2. Time series approach to quantum chaos

One of the main tools in the study of time series fluctuations is the power spectrum of the signal. For many different phenomena in nature, physics and social sciences, the power spectrum of a signal is characterized by a power law. To study correlations in the fluctuations of quantum spectra, we can consider the energy spectrum as a discrete signal, and the sequence of energy levels as a time series, where the level order index plays the role of a discrete time [3]. The power spectrum $S(k)$ of a discrete spectrum with $N + 1$ levels is calculated as

$$\delta_n = \sum_{i=1}^{n} (s_i - \langle s \rangle), \quad \hat{\delta}_k = \frac{1}{N} \sum_n \delta_n \exp\left(\frac{-2\pi i k n}{N}\right), \quad S(k) = \left|\hat{\delta}_k\right|^2. \tag{1}$$

The study of quantum billiards, nuclei and random matrices, suggested the following conjecture: the energy spectra of chaotic quantum systems are characterized by $1/f$ noise, while the energy spectra of regular quantum systems are characterized by $1/f^2$ noise [3]. This conjecture relates chaotic quantum systems with other complex systems in nature that are also characterized by $1/f$ noise [16].

IOP Conf. Series: Journal of Physics: Conf. Series **1023** (2018) 012011 doi:10.1088/1742-6596/1023/1/012011

3. Quantum chaos in nuclei

3.1. Chaos in experimental energy spectra

For a quantum system like the atomic nucleus, which has no classical limit, the term chaos started to be used when Haq, Pandey and Bohigas [8] analyzed the spectral fluctuations of experimental neutron and proton $J^\pi = 1/2^+$ resonances just above the one-nucleon emission threshold and showed that they agree very well with those of the Gaussian Orthogonal Ensemble (GOE) of Random Matrix Theory (RMT). Thus, in this sense, it is clear that nuclei are very chaotic in the energy region just above the one-nucleon emission threshold.

But for bound states, the situation is not so clear, because a good analysis of fluctuations in experimental energy spectra requires the knowledge of sufficiently long, pure and complete sequences, i.e. with the same $J^\pi T$ values and without missing levels or $J^\pi T$ misassignments. But this ideal situation is rarely found in nuclei. For very light nuclei the number of bound levels is not sufficient for statistical purposes. For medium and heavy nuclei the identified levels are limited to the ground state region, because at higher energy the level density becomes very high and the experimental identification of J^π values becomes generally impossible. In order to improve statistics, Shriner *et al.* combined level spacings from different nuclei into a single set to analyze the behavior of the $P(s)$ distribution [17]. Separating the data in six different mass regions a clear trend from GOE to Poisson is observed as the nuclear mass increases. Generally spherical nuclei are closer to GOE and deformed nuclei are closer to Poisson.

3.2. Chaos in shell-model calculations

In order to get a deeper understanding of what happens in the low-energy region we can use the shell model with configuration mixing. Realistic shell-model calculations in *sd* shell nuclei show a strong chaotic behavior [5]. As an illustrative example of the RMT approach, Fig. 1 shows the shell-model $P(s)$ and $\langle\Delta_3(L)\rangle$ values for the $J^\pi = 2^+$, $T = 0$ states of ^{28}Si obtained by Zelevinski *et al.* [5]. Clearly the agreement with GOE is excellent. As an example of the time series approach, Fig. 2 shows the $1/f$ noise behavior in the average power spectrum of a combination of sequences of states with different values of J^π in ^{34}Na and ^{24}Mg.

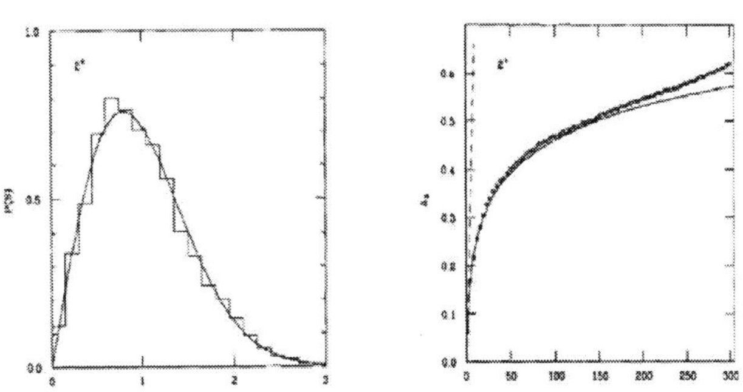

Figure 1. Spectral statistics for the $J^\pi = 2^+$, $T = 0$ states of ^{28}Si. *Left panel* $P(s)$ distribution. *Right panel* $\langle\Delta_3(L)\rangle$ [5].

In *pf* shell nuclei the dimensionalities of the shell-model valence space are much larger than in the *sd* shell and it becomes possible to study with realistic shell-model calculations how chaos depends on isospin and excitation energy [6]. In nuclei with both protons and neutrons in the valence space, the shell-model results show spectral fluctuations close to GOE. However, when all valence particles are neutrons, the spectral fluctuations deviate from chaotic behavior. These results can be interpreted in terms of the strength of the nuclear residual interaction. The mean

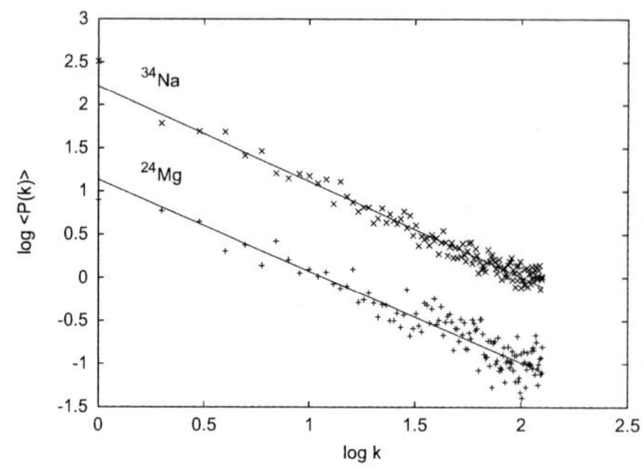

Figure 2. Average power spectrum of levels of the nuclei ^{34}Na and ^{24}Mg [3].

field generates regular motion and the residual interaction produces strong correlations that lead to chaotic motion. But, since the nn residual interaction is much weaker than the pn interaction, valence neutrons only may not be enough to strongly perturb the regular mean-field motion. Fig. 3 shows the isospin dependence of $\langle\Delta_3(L)\rangle$ for $J^\pi = 0^+$, $T = T_z$ levels in $A = 46$ isobars. In ^{46}Ca, there are only valence neutrons and $\langle\Delta_3(L)\rangle$ quickly deviates from the GOE curve for $L > 8$. In ^{46}Sc there are one proton and five neutrons outside the core and the chaotic behavior is very clear up to $L \approx 26$. In ^{46}Ti, the number of pn interactions is larger and there are no significant deviations from GOE.

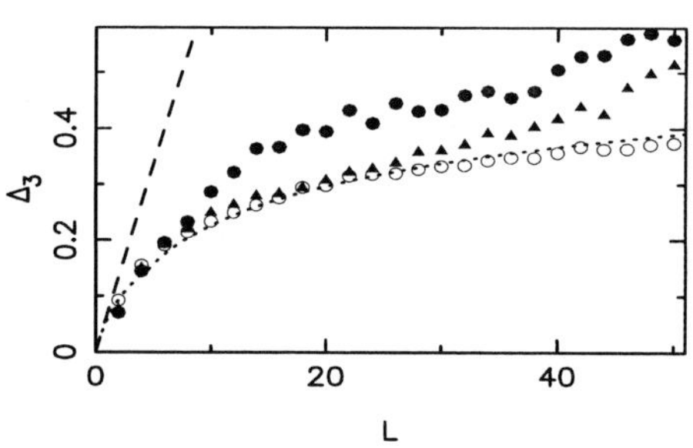

Figure 3. $\langle\Delta_3(L)\rangle$ for $J^\pi = 0^+$, $T = T_z$ levels in $A = 46$ isobars, ^{46}Ca (full dots), ^{46}Sc (triangles), ^{46}Ti (empty dots). The dotted and dashed curves represent the GOE and Poisson values, respectively [6].

3.3. *Spectral statistics in the experimental ^{208}Pb bound states.*

Experimental nuclear spectra become increasingly plagued with unidentified states, missing levels and some misassignments as excitation energy increases. But we consider that the recent accurate data on ^{208}Pb [10] enable a meaningful statistical analysis of level fluctuations in a nucleus, with pure, complete, and reasonably long sequences.

Fluctuations are the departure of the actual level density from a local uniform density. Therefore it is essential to eliminate the smooth part of the exponential increase of the nuclear density, mapping the actual spectrum onto a quasiuniform spectrum with mean spacing $\langle s\rangle = 1$. This step, called unfolding, is delicate and of utmost importance, because some of the unfolding procedures used in the literature can lead to completely wrong results on the behavior of level fluctuations [18]. In this work we have used the constant temperature formula [17]. A separate unfolding has been performed for all J^π sequences with a minimum of 5 known consecutive states. The longest sequence corresponds to the $J^\pi = 3^-$ states, with 19 consecutive levels. Gathering the unfolded spacings for all J^π into a single set, there are 115 spacings.

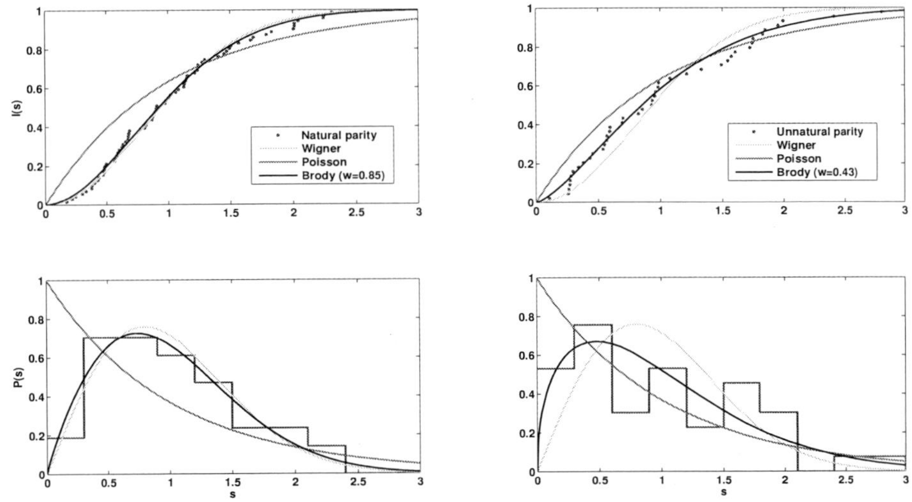

Figure 4. Comparison of spectral fluctuation statistics for natural parity states (left panels) and unnatural parity states (right panels) in ^{208}Pb. Panels (a) and (b) show the cumulative nearest neighbor spacing (cumulative NNS) distribution for experimental levels (dots). Panels (c) and (d) show the NNS distribution $P(s)$ for experimental levels (histogram). In all panels the green (light gray) line is the Wigner surmise, the red dashed line is the Poisson distribution, and the black solid line is the best fit Brody distribution.

A simple way to interpolate between the Poisson limit $P_P(s)$ and the Wigner surmise $P_W(s)$ is provided by the Brody distribution,

$$P_B(s,\omega) = (\omega+1)a_\omega s^\omega \exp(-a_\omega s^{\omega+1}), \;\; a_\omega = \left[\Gamma\left(\frac{\omega+2}{\omega+1}\right)\right]^{\omega+1}, \tag{2}$$

where Γ is the gamma function. The Brody parameter ω is given by the best fit to the histogram of $P(s)$. When the number of spacings is not very large, the $P(s)$ distribution may depend on the choice of histogram binning size. It is then preferable to fit the cumulative distribution $I(s) = \int_0^s dx P(x)$. For the full set of 115 spacings the best Brody fit to the experimental $I(s)$ is obtained for $\omega = 0.68$. This value is certainly much closer to the Wigner limit ($\omega = 1$) than to the Poisson limit ($\omega = 0$) but it is still rather far from full chaotic behavior.

3.4. Comparison of short-range spectral fluctuations in natural and unnatural parity states
Heusler *et al.* have shown [10] that the experimental unnatural parity levels in ^{208}Pb ($J^\pi = 0^-, 1^+, 2^-, \ldots, 11^+$) agree better than the natural parity states ($J^\pi = 0^+, 1^-, 2^+, \ldots, 12^+$) with the so-called "extended schematic shell model", which is a simplified shell-model consisting mainly of 1p-1h mean-field configurations . The excitation energies of 70 states with unnatural parity agree within about 0.2 MeV with the extended schematic shell model. By contrast, the excitation energies of about 20 natural parity states are more than 0.5 MeV lower than the model prediction. They conclude theerefore that the residual interaction is much stronger for natural than for unnatural parity states.

IOP Conf. Series: Journal of Physics: Conf. Series **1023** (2018) 012011 doi:10.1088/1742-6596/1023/1/012011

Table 1. Number of spacings, Brody parameter ω and rms deviation from Wigner and Poisson distributions for different parity sets in the experimental states of ^{208}Pb at $E_x < 6.20$ MeV.

Parity	all	natural	unnatural
Number of spacings	115	71	44
Brody ω	0.68	0.85	0.43
$(\mathrm{RMSD})_W$	0.040	0.025	0.077
$(\mathrm{RMSD})_P$	0.115	0.129	0.088

To check if this effect can be observed in the fluctuation measures, we have analyzed separately the NNS distribution of experimental natural and unnatural parity states. Table 1 shows the number of unfolded spacings of each type at $E_x < 6.20$ MeV in ^{208}Pb and the corresponding values of the Brody parameter obtained from the fit of the cumulative distribution, as well as the root mean square deviation (RMSD) of the experimental $I(s)$ from the Wigner and Poisson limits. Fig. 4 shows the distributions $I(s)$ and $P(s)$ for natural and unnatural parity states, compared to Wigner, Poisson and the best fit of the Brody distribution.

For natural parity states the behavior is definitely chaotic, with $\omega = 0.85$. The Brody and Wigner curves for $P(s)$ nearly coincide, and the dots representing the cumulative distribution clearly follow the Wigner distribution. To our knowledge, this is the closest GOE behavior ever observed in experimental nuclear bound states. It is worth to comment that the Wigner surmise is only a good analytical approximation to the GOE distribution, and that the best fit of $P_B(s, \omega)$ to the exact $P(s)$ distribution for GOE is obtained for $\omega = 0.957$, not for $\omega = 1$ [19].

By contrast, for unnatural parity states the cumulative distribution is intermediate between the two extremes, somewhat closer to Poisson, with $\omega = 0.43$. There are only 44 spacings and the histogram of $P(s)$ oscillates a lot. But notice especially the different behavior of $P(s)$ for small spacings in natural and unnatural parity states. Level repulsion is seen to be much stronger for natural parity states. Strong level repulsion is characteristic of chaotic (Wigner-like) spectra, whereas for the regular motion $P(s)$ is maximal for small spacings.

We should keep in mind that the experimental J^π sequences in nuclei are generally quite short and that any missing levels or misassignments always bias the statistical measures towards Poisson. In the case of ^{208}Pb all the states with $E_x < 6.20$ MeV are now well identified and we are confident that the NNS distribution is not biased by missing levels or misassignments. The results reflect the degree of chaos caused by the residual interaction, and this is clearly seen comparing the NNS distributions of natural and unnatural parity states.

4. Conclusions

Many shell-model calculations of bound states in spherical nuclei have clearly shown the existence of chaotic dynamics produced by the strong residual NN interaction which destroys the mean-field regular motion. The study of spectral fluctuations in experimental bound states is generally limited by the short length of $J^\pi T$ sequences, uncertainties in some quantum number assignments and possible missing levels. Nevertheless, combining data from many different nuclei along the nuclear chart, a clear trend from GOE to Poisson is observed as the nuclear mass increases. However, a fully chaotic spectrum has never been observed in these kind of analysis. This fact can be understood because at lower excitation energies many states are dominated by a few shell-model configurations. On the other hand, generally spherical nuclei are closer to GOE and deformed nuclei are closer to Poisson. This makes sense because the lower energy spectrum of deformed nuclei comes mainly from collective rotational motion, which is a regular motion.

The recent identification with spin and parity assignment of all the 151 states at $E_x < 6.20$

MeV in ^{208}Pb by Heusler *et al.* [10] has provided exceptionally long J^π sequences of consecutive states free of missing levels and misassignments, enabling us to perform a reliable analysis of spectral fluctuations in this nucleus. Comparison of the experimental spectrum with extended 1p-1h schematic shell-model calculations clearly indicate that the residual interaction is much stronger for natural than for unnatural parity states [10]. A recent analysis of spectral fluctuations has shown [11] that natural parity states exhibit results close to GOE and the unnatural parity states are far from GOE behavior. Thus these results indicate that chaotic and non-chaotic states coexist in the energy region from the ground state up to 6.20 MeV excitation energy (the neutron threshold in ^{208}Pb is $S_n = 7.368$ MeV). Furthermore, our analysis of the experimental spectrum has confirmed, to our knowledge for the first time, a well-known shell-model prediction, namely that chaos in nuclei arises when the residual interaction is strong enough to destroy the ordered motion of nucleons in the nuclear mean field.

Finally, let us point out that another analysis of spectral fluctuations of the same experimental states of ^{208}Pb has been published [12]. They use two chaoticity measures different from ours and do not observe a significant difference between experimental natural and unnatural parity states. Possibly this difference with our results is due to the different unfolding procedures used. The influence of the unfolding method can be important for such a small number of levels as the unnatural parity states of ^{208}Pb. On the other hand, the spectral fluctuations have been analyzed by the same authors for two different shell-model calculations in ^{208}Pb. It should be noticed that the shell-model J^π sequences are much larger than the experimental ones and therefore the unfolding procedure is more reliable. The shell-model values of the chaoticity parameters are different for natural and unnatural parity, the natural parity states being more chaotic, in agreement with our results.

5. Acknowledgment

This research was conducted with support of MINECO/FEDER Grant Nos. FIS2015-63770-P and FPA2015-65035-P.

References

[1] J.M.G. Gómez, K. Kar, V.K.B. Kota, R.A. Molina, A. Relaño, and J. Retamosa, Phys. Rep. **499**, 103 (2011).
[2] O. Bohigas, M.J. Giannoni, C. Schmit, Phys. Rev. Lett. **52**, 1 (1984).
[3] A. Relaño, J.M.G. Gómez, R.A. Molina, J. Retamosa, E. Faleiro, Phys. Rev. Lett. **89**, 244102 (2002).
[4] E. Faleiro, J.M.G. Gómez, R.A. Molina, L. Muñoz, A. Relaño, J. Retamosa, Phys. Rev. Lett. **93**, 244101 (2004).
[5] V. Zelevinsky, B.A. Brown, N. Frazier, M. Horoi, Phys. Rep. **276**, 85 (1996).
[6] R.A. Molina, J.M.G. Gómez, J. Retamosa, Phys. Rev. C **63**, 014311 (2000).
[7] R.A. Molina, Eur. Phys. J. A **28**, 125 (2006).
[8] R.U. Haq, A. Pandey, O. Bohigas, Phys. Rev. Lett. **48**, 1086 (1982).
[9] H.A. Weidenmüller, G.E. Mitchell, Rev. Mod. Phys. **81**, 539 (2009).
[10] A. Heusler, R. V. Jolos, T. Faestermann, R. Hertenberger, H.-F. Wirth, and P. von Brentano, Phys. Rev. C **93**, 054321 (2016).
[11] L. Muñoz, R.A. Molina, J.M.G. Gómez, A. Heusler, Phys. Rev. C **95** , 014317 (2017).
[12] B. Dietz, A. Heusler, K.H. maier, A. Richter, B.A. Brown, Phys. Rev. Lett. **118**, 012501 (2017).
[13] S. Müller, S. Heusler, P. Braun, F. Haake, A. Altland, Phys. Rev. Lett. **93**, 014103 (2004).
[14] M.V. Berry, M. Tabor, Proc. R. Soc. Lond. **A 356**, 375 (1977).
[15] F.J. Dyson, M.L. Mehta, J. Math. Phys. **4**, 701 (1963).
[16] L.M Ward, P.E. Greenwood, Scholarpedia **2**, 1537 (2007).
[17] J.F. Shriner Jr., G.E. Mitchell, T. von Egidy, Z. Phys. A – Hadrons and Nuclei **338**, 309 (1991).
[18] J.M.G. Gómez, R.A. Molina, A. Relaño, J. Retamosa, Phys. Rev. E **66**, 03
[19] T.A. Brody, J. Flores, J.B. French, P.A. Mello, A. Pandey, S.S.M. Wong, Rev. Modern Phys. **53**, 385 (1981).

Structure of mirror pairs ^{8}Li-^{8}B and ^{9}Be-^{9}B in Fermionic Molecular Dynamics

K. R. Henninger
iThemba Laboratory for Accelerator Based Sciences, Old Faure road, Western Cape, South Africa

E-mail: katharinehenninger@gmail.com

Abstract. Structure of light weakly-bound nuclei like ^{8}B or ^{9}Be is of astrophysical interest. Microscopic models of their structure face challenges, since these nuclei feature clusters; difficult to access using a basis of spatially-isotropic single-particle states. The Fermionic Molecular Dynamics (FMD) model offers a basis general enough to describe both compact (shell model like) states and clustering. We discuss FMD calculations for the mirror pairs ^{8}B-^{8}Li and ^{9}Be-^{9}B. These are interesting because one nucleus of each pair is weakly-bound or unbound. It is shown that the FMD calculations reproduce energy levels and radii fairly, and that subtraction of the Coulomb energy calculated from the FMD states restores degeneracy of the mirror levels within at most $\sim$ 100 keV.

1. Introduction

Weakly-bound nuclei have low-lying states proximal to thresholds for decay by particle emission. This means that, according to the Ikeda hypothesis [1], the asymptotic part of the wave function for these states will be well-described by clustering configurations.

Mirror nuclei are systems of nucleons that would be degenerate in energy were it not for isospin-symmetry breaking in the NN interaction: effectively, without the Coulomb interaction (the primary breaker of isospin symmetry). The Coulomb interaction and the uniformity of the charge distribution in the nucleus are connected [2]. Thus, structural differences between mirror nuclei (such as different clustering) can be considered as mainly driven by the Coulomb interaction, and show up in the location of decay thresholds relative to different states. For example, in ^{8}B, the ^{7}Be+p threshold is ≈ 0.3 MeV above the ground state, but the corresponding ^{7}Li+n threshold is ≈ 3 MeV above the ^{8}Li ground state. By the Ikeda hypothesis, this suggests different asymptotic structure of these levels, *i.e.* different clustering. If we can account for these structural differences when calculating Coulomb energy, does subtracting it restore degeneracy of mirror energy levels? In this paper, we try to answer this question for the mirror nuclei ^{8}Li-^{8}B and ^{9}Be-^{9}B; pairs which are of interest to astrophysics (*e.g.* [3, 4]).

Clusters are difficult to model when beginning from spatially-isotropic single-particle distributions [5]. In this study we use the Fermionic Molecular Dynamics (FMD) model [6]. FMD offers a Hilbert space in which clustering may either be introduced explicitly, or arise naturally in the basis states, allowing one to access both compact and clustering configurations consistently [5]. Another highlight of FMD is its treatment of the proton-neutron mass difference: while many models neglect this, it is taken into account in FMD [5], making comparison of energy-levels in mirror nuclei possible. The model is outlined in Section 2.

2. Fermionic Molecular Dynamics

Starting with protons and neutrons as basic constituents of a nuclear system, we choose intrinsic many-body configurations $|Q\rangle$ that are Slater determinants of single-particle states $|q\rangle$, or [7]:

$$|Q\rangle = \hat{\mathcal{A}}\left\{|q_1\rangle \otimes ... \otimes |q_A\rangle\right\} \tag{1}$$

where $\hat{\mathcal{A}}$ is the antisymmetrization operator and $Q = \{q_1, ..., q_A\}$ denotes the set of single-particle parameters. The single-particle states corresponding to the protons and neutrons are:

$$|q\rangle = \sum_i |a_i, \vec{b}_i\rangle \otimes |\chi_i^{\uparrow}, \chi_i^{\downarrow}\rangle \otimes |\xi\rangle c_i, \tag{2}$$

Parameters $\vec{b}$ relate to the mean position and mean momentum and and a to the width of the wave-packets [8] respectively. The ket $|\chi^{\uparrow}, \chi^{\downarrow}\rangle$ allows all possible orientations of the nucleon spin. The isospin is defined in $|\xi\rangle$. In co-ordinate space one has Gaussian wave-packets:

$$\langle\vec{x}|a, \vec{b}\rangle = \exp\left\{-\frac{(\vec{x}-\vec{b})^2}{2a}\right\}. \tag{3}$$

Slater determinants $|Q\rangle$ like that in Eq.(1) are in general not eigenstates of the angular momentum or parity operators, so these need to be projected out via projection operators [8]:

$$\hat{P}^{\pi} = \frac{1}{2}(1+\pi\hat{\Pi}) \tag{4a}$$

$$\hat{P}^{J}_{MK} = \frac{2J+1}{8\pi^2}\int d\Omega\, D^{J}_{MK}(\Omega)^{*}\hat{R}(\Omega), \tag{4b}$$

where $D^J_{MK}(\alpha, \beta, \gamma)$ is the Wigner D-matrix, $\hat{R}(\alpha,\beta,\gamma) = e^{-i\alpha\hat{J}_z}e^{-i\beta\hat{J}_y}e^{-i\gamma\hat{J}_z}$ is the rotation operator and $\hat{\Pi}$ is the parity operator. The $\Omega = \{\alpha, \beta, \gamma\}$ symbolizes the three Euler angles. The projected states are such that the result is a many-body state with quantum numbers J, K, M. By diagonalizing the Hamiltonian in the set of K-projected states,we obtain the FMD many-body basis states

$$|Q; J^{\pi}M\alpha\rangle = \sum_K |Q; J^{\pi}MK\rangle c_K^{J^{\pi}\alpha}. \tag{5}$$

Additional basis states can be generated by performing minimization subject to constraints. These constraints could be on *e.g.* radii, moments or deformation parameters [5]. One may also create explicit cluster states: States which are antisymmetrized product states of Slater determinants derived via the above-described minimization scheme, *i.e.*

$$|Q_{c1+c2}\rangle = \hat{\mathcal{A}}\left\{|Q_{c1}\rangle \otimes |Q_{c2}\rangle\right\}. \tag{6}$$

The Hamiltonian is then diagonalized in this basis set, and one obtains the eigenstates. As already mentioned, the FMD takes account of the proton-neutron mass difference, setting the mass of a proton to 938.27231 MeV and the mass of a neutron to 939.56563 MeV.

IOP Conf. Series: Journal of Physics: Conf. Series **1023** (2018) 012012 doi:10.1088/1742-6596/1023/1/012012

2.1. Coulomb energy

In the FMD, matrix elements for a two-body $\frac{1}{r}$ interaction like the Coulomb interaction are [5]:

$$\langle a_\kappa \vec{b}_\kappa, a_\lambda \vec{b}_\lambda | \frac{1}{r} | a_\mu \vec{b}_\mu, \alpha_\nu \vec{b}_\nu \rangle = R_{\kappa\mu} R_{\lambda\nu} \frac{1}{\sqrt{\rho^2_{\kappa\lambda\mu\nu}}} \mathrm{erf} \left\{ \sqrt{\frac{\rho^2_{\kappa\lambda\mu\nu}}{2\alpha_{\kappa\lambda\mu\nu}}} \right\}. \tag{7}$$

where all the terms are as defined in [5], and "erf" is the Error Function.

This determination of Coulomb energy allows it to be extracted from the structural information in our calculations and does not rely on (extreme) approximations to the nucleon distribution, as does the well-known uniform sphere approximation used in the Semi-Empirical Mass Formula (SEMF, or Bethe-Weizsäcker formula):

$$E_{SEMF} = a_C \frac{Z(Z-1)}{A^{1/3}}, \tag{8}$$

where a_C is 0.691 MeV [9]. Comparing this value to one obtained in Eq.(7) can be interesting, as how much the two deviate is a good indicator of how "non uniform" or "non-spherical" a charge distribution is.

3. Calculations

The basis sets used for these calculations comprised FMD VAP states with constraints on the matter radius, projected on the angular momenta and parties of the ground state and first two excited states of the nuclei of interest. The states to be included in each basis were then chosen by an optimization process [10]. To these states were added explicit cluster configurations corresponding to the lowest-lying thresholds (listed in Table 1). In each cluster configuration, the core consisted of a single VAP state projected onto the angular momentum and parity of the core's ground state.

To create equivalent model spaces for the mirror nuclei, it was decided to use the basis states of the mirror, with protons and neutrons exchanged. This gives a total basis size of up ≈40 VAP states and 11 explicit cluster configurations in each basis set.

Table 1. Summary of the cluster configurations included in the basis sets for each nucleus.

Nucleus	Clustering
^{8}Li	^{7}Li+n
^{8}B	^{7}Be+p
^{9}Be	^{8}Be+n
^{9}B	^{8}Be+p

The Argonne *v18* interaction [11] was used, transformed by the Unitary Correlation Operator Method (UCOM) [12]. The transformation enables the inclusion of short-range central and tensor correlations. Three-body and higher-order terms are not included.

4. Results

Calculated level-schemes are provided in Fig. 1, where they are compared to the experimental levels-schemes. In all cases, one sees under-binding in our calculations, which is primarily a feature of the interaction used [13]. The important calculated quantity is thus binding relative to the calculated thresholds, which compares fairly with experiment. Thresholds were calculated within an FMD model space of comparative size to that of the calculations, and also using the same interaction.

In Fig. 1, one may also note that the higher-lying levels in all four systems are somewhat more poorly-reproduced than the lower-lying levels (*e.g.* note the 0^+ states of ^{8}Li and ^{8}B). This is because the bases described above (Section 3) are not adequately large to accurately reproduce the high-lying states ($\approx$ 10 MeV) in any of these nuclei, as these states may be broad resonances or strongly mixed with the continuum [14], and are thus unlikely to be well-described in a model space that can accurately describe the lower-lying states.

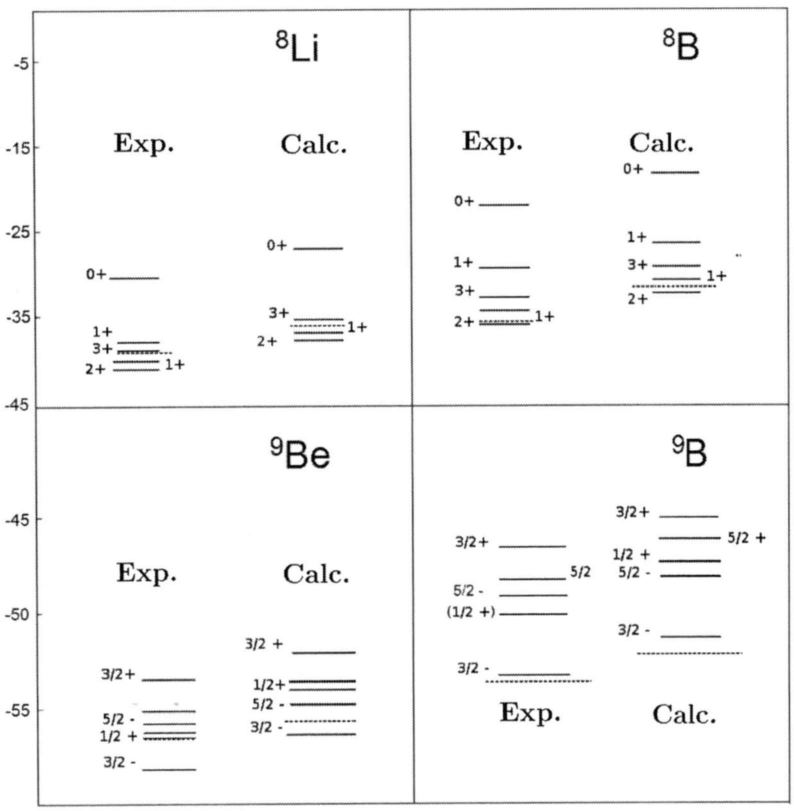

Figure 1. Experimental and calculated energy levels (labeled Exp. and Calc. respectively). Dotted lines indicate the lowest-lying thresholds. Note reproduction of binding energy relative to threshold. Calculations are labeled for the respective nuclei. For A=9 mirror nuclei (bottom panels), note that the $5/2^-$ and $1/2^+$ levels are reversed in order in our calculations. The $1/2^+$ state is very broad ($\approx$ 0.3 MeV broad in ^{9}Be, and $\approx$ 3 MeV broad in ^{9}B [14]), explaining the better reproduction of the $5/2^-$ state in this model-space.

Good indicators of whether structure is being effectively reproduced by the calculations are the calculated matter- and charge-radii. Table 2 compares the matter- and charge-radii (where known) with experiment. Note the difference between the calculated matter- and charge radii

IOP Conf. Series: Journal of Physics: Conf. Series **1023** (2018) 012012 doi:10.1088/1742-6596/1023/1/012012

of ^{8}B, indicative of the fact that we reproduce a proton halo in our calculations. This indicates that the calculations capture the structural features of ^{8}B, at least in the ground state.

Table 2. Calculated matter- and charge-radii, compared to experiment. Note the difference between the matter- and charge radii of ^{8}B, indicative of a proton halo.

Nucleus	Matter radius [fm]		Charge radius [fm]	
	Exp. [a]	Calc. C	Exp.[b]	Calc. C
^{8}Li	2.37(2)	2.243	2.29(8)	2.211
^{8}B	2.38(4)	2.282	-	2.537
^{9}Be	2.38(1)	2.338	2.519(12)	2.417
^{9}B	-	2.358	-	2.556

[a] All experimental matter radii from [14].

[b] Charge radius of ^{8}Li from [15], and for ^{9}Be from [16].

In Figs. 2, the effect of subtraction of the Coulomb energies is demonstrated graphically. As described in the caption, plots illustrate the effect of subtracting the Coulomb energy calculated firstly using the Semi-Empirical Mass Formula, and then that calculated from the FMD states as described in Sect. 2.1. It can clearly be seen that subtraction of the "FMD Coulomb energy" comes close to restoring the degeneracy of the analogous energy levels, both in the case of the $A = 8$ and $A = 9$ systems.

To be more precise, the difference between calculated energies of the analog states in the $A = 8$ system is of the order of 100 keV (see Table 3), while for the $A = 9$ system, it is of the order of tens of keV. A tentative explanation of why the method seems more effective for the $A = 9$ systems is that both ^{9}Be and ^{9}B are probably modeled equally well (or badly!) in model-spaces of the size used (both have quite complicated three-body structure even in the ground state [14], meaning that they are both hard to access in a limited model space), while for the $A = 8$ group, it may be argued that our approximation is worse for ^{8}B than for ^{8}Li (since ^{8}Li has a "simpler" structure (see *e.g.* [14]), leading to a greater discrepancy in the results due to a failure to capture some structural information in ^{8}B.

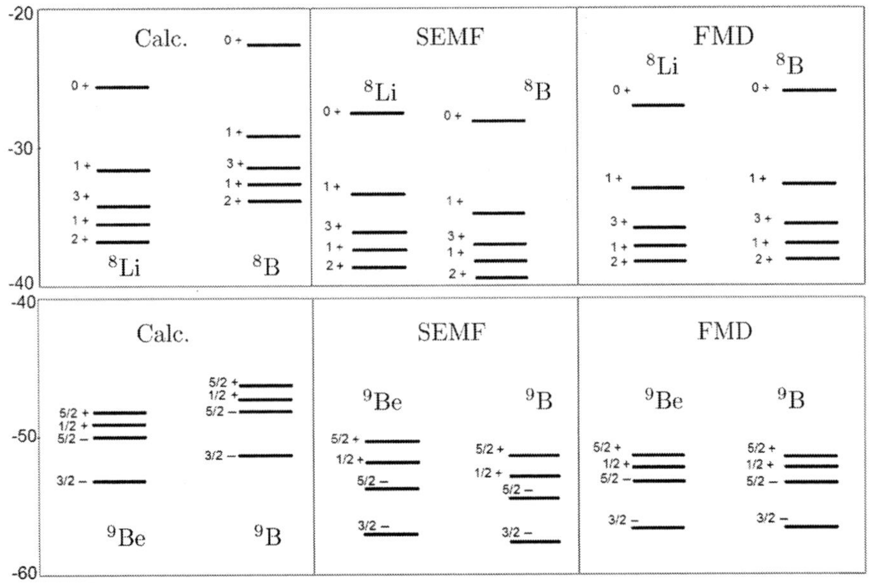

Figure 2. Calculated energy levels for the A=8 and $A = 9$ mirror systems (labeled Calc.), shown before and after subtraction of the SEMF Coulomb energy and the FMD Coulomb energy (in panels labeled SEMF and FMD, respectively). Levels are labeled for the nuclei for which they are calculated. Note how close to degenerate the levels become after subtraction of the FMD Coulomb energy.

Level	Energy less FMD Coulomb [MeV]	Energy less SEMF Coulomb [MeV]	Diff. FMD (Abs.) [MeV]	Diff. SEMF (Abs.) [MeV]
^{8}Li 2+	-38.050	-38.376		
^{8}B 2+	-37.907	-39.700	0.143	1.324
^{8}Li 1+	-36.412	-36.754		
^{8}B 1+	-36.249	-38.130	0.163	1.376
^{9}Be 3/2 -	-56.661	-57.351		
^{9}B 3/2 -	-56.603	-58.147	0.058	0.796
^{9}Be 1/2 +	-52.265	-53.195		
^{9}B 1/2 +	-52.219	-54.209	0.046	1.014
^{9}Be 5/2 -	-53.382	-54.120		
^{9}B 5/2 -	-53.322	-54.949	0.060	0.829
^{9}Be 5/2 +	-51.403	-52.355		
^{9}B 5/2 +	-51.351	-53.331	0.052	0.976

Table 3. The differences in calculated energy of analogous energy levels before and after subtraction of the FMD Coulomb energy (column 2) and the SEMF Coulomb energy (column 3). Note that the degeneracy is restored by $\sim$ 10 keV for the $A = 9$ systems and by $\sim$ 100 keV for the $A = 8$ systems.

5. Summary and Conclusion

The FMD calculations of energy and matter- and charge-radii for the ^{8}Li-^{8}B and ^{9}Be-^{9}B systems compare, for the most part, favorably with experiment. The subtraction of a Coulomb energy "based on structure", or one which takes into account the asymptotic structure of these nuclei, has been shown to restore degeneracy of the mirror levels with at most $\sim$ 100 keV.

As the FMD does take into account the proton-neutron mass difference, the results presented here indicate that the major differences between the energies of the mirror nuclei are brought about by structural considerations arising through the Coulomb interaction. This conclusion is probably unsurprising, but is interesting for drip-line physics where it is important to know to what extent the proton:neutron ratio plays a role, through what mechanism it plays a role, and how to account for it.

As an extension to this study, it must also be noted that the UCOM-transformed AV18 interaction used is truncated at two-body level. The missing three-body strength is accounted for by using a coefficient of 2 for the spin-orbit force. Investigations of the effect of three-body terms would be worth carrying out.

6. References

[1] Ikeda K, Takigawa N and Horuichi H 1968 Prog. Theor. Phys. *(Japan)* **Supplement** 464
[2] Sagawa H 2007 J. Phys. **G349** 949
[3] Reeves H and Meyer J P 1978 Astrophys. J **226** 613
[4] Parker P D 1966 Phys. Rev. **150** 851
[5] Neff T and Feldmeier H 2010 *Cluster Structure in the Fermionic Molecular Dynamics Approach Cluster Structure of Atomic Nuclei* ed Brenner M (Research Signpost publishing) pp 67–94
[6] Feldmeier H 1990 Nucl. Phys. **A515** 147
[7] Feldmeier H and Schnack J 1997 Prog. Part. Nucl. Phys. **39** 343
[8] Neff T and Feldmeier H 2008 Eur. Phys. J. **156** 69
[9] Wong S S M 2002 *Introductory Nuclear Physics* (New Jersey: Pearson Prentice-Hall publishers) pp 11-13, 261-274 and 372-375.
[10] Henninger K R 2016 J. Phys. Conf. Ser. **724** 012019
[11] Wiringa R B, Stokes B G J and Schiavilla R 1995 Phys. Rev ***C*** **51** 38
[12] Roth R, Hergert H, Papakonstantinou P, Neff T and Feldmeier H 2005 *Phys. Rev.* ***C*** **72** 034002
[13] Dickhoff W H 2016 IOP: Conf. Ser. **702** 012013
[14] Tilley D R, Kelley H, Godwin J L, Millener D J, Purcell J E, Sheu C G and Weller H R 2004 (revised 2014) Nucl. Phys. **A745** 155
[15] Ewald G, Nörtershäuser W, Dax A, Götte S, Kirchner R, Kluge H J, Kühl T, Sanchez R, Wojtaszek A, Bushaw B A, Drake G W F, Yan Z C and Zimmermann C 2004 Phys. Rev. Lett. **93** 113002
[16] Nörtershäuser W, Tiedemann D, Žáková M, Andjelkovic Z, Blaum K, Bissell M L, Cazan R, Drake G W F, Geppert C, Kowalska M, Krämer J, Krieger A, Neugart R, Sánchez R, Schmidt-Kaler F, Yan Z C, Yordanov D T and Zimmermann C 2009 *Phys. Rev. Lett.* **102**(6) 062503 URL `https://link.aps.org/doi/10.1103/PhysRevLett.102.062503`

Structure of the low-lying positive parity states in the proton-neutron symplectic model

H. G. Ganev

Joint Institute for Nuclear Research, Dubna, Russia and
Institute of Nuclear Research and Nuclear Energy, Bulgarian Academy of Sciences, Sofia, Bulgaria

Abstract. The proton-neutron symplectic model with $Sp(12, R)$ dynamical symmetry is applied for the simultaneous description of the microscopic structure of the low-lying states of the ground state, γ and β bands in ^{166}Er. For this purpose, the model Hamiltonian is diagonalized in the space of stretched states by exploiting the $SU_p(3) \otimes SU_n(3)$ symmetry-adapted basis. The theoretical predictions are compared with experiment and some other microscopic collective models, like the one-component $Sp(6, R)$ symplectic and pseudo-$SU(3)$ models. A good description of the energy levels of the three bands under consideration, as well as the enhanced intraband $B(E2)$ transition strengths between the states of the ground and γ bands is obtained without the use of effective charges. The results show the presence of a good $SU(3)$ dynamical symmetry. It is also shown that, in contrast to the $Sp(6, R)$ case, the lowest excited bands, e.g., the β and γ bands, naturally appear together with the ground state band within a single $Sp(12, R)$ irreducible representation.

1. Introduction

Experimental spectra in heavy nuclei show the emergence of simple collective patterns represented primarily by the nuclear collective rotation. The microscopic shell-model structure of these low-lying rotational states is still a challenge for the microscopic many-particle nuclear theory. This is particularly so because the model space dimensionalities rule out the use of standard shell-model theory. As a consequence, different algebraic models which capitalize on symmetries, exact or approximate, have been developed to reduce the model space in manageable size.

In the present work, the recently proposed [1, 2] proton-neutron symplectic model (PNSM) with $Sp(12, R)$ dynamical symmetry is applied for the simultaneous description of the microscopic structure of the low-lying states of the ground state, γ and β bands in ^{166}Er. For this purpose, the model Hamiltonian is diagonalized in the space of stretched states by exploiting the $SU_p(3) \otimes SU_n(3)$ symmetry-adapted basis. The theoretical predictions are compared with experiment and some other microscopic collective models, like the one-component $Sp(6, R)$ symplectic [3] and pseudo-$SU(3)$ [4, 5, 6] models. A good description of the energy levels of the three bands under consideration, as well as the enhanced intraband $B(E2)$ transition strengths between the states of the ground and γ bands is obtained without the use of effective charges. The results reveal almost the pure $SU(3)$ structure, traditionally assumed for well-deformed heavy mass nuclei in the macroscopic theory of nuclear rotation, like, e.g., the Interacting Boson Model [7].

2. The Proton-Neutron Symplectic Model

Collective observables of the proton-neutron symplectic model, which span the $Sp(12, R)$ algebra, are given by the following one-body operators [1]:

$$Q_{ij}(\alpha,\beta) = \sum_{s=1}^{m} x_{is}(\alpha)x_{js}(\beta), \tag{1}$$

$$S_{ij}(\alpha,\beta) = \sum_{s=1}^{m} \Big(x_{is}(\alpha)p_{js}(\beta) + p_{is}(\alpha)x_{js}(\beta) \Big), \tag{2}$$

$$L_{ij}(\alpha,\beta) = \sum_{s=1}^{m} \Big(x_{is}(\alpha)p_{js}(\beta) - x_{js}(\beta)p_{is}(\alpha) \Big), \tag{3}$$

$$T_{ij}(\alpha,\beta) = \sum_{s=1}^{m} p_{is}(\alpha)p_{js}(\beta), \tag{4}$$

where $i, j = 1, 2, 3;\ \alpha, \beta = p, n$ and $s = 1, \ldots, m = A - 1$. In Eqs.(1)–(4), $x_{is}(\alpha)$ and $p_{is}(\alpha)$ denote the coordinates and corresponding momenta of the translationally-invariant Jacobi vectors of the m-quasiparticle two-component nuclear system and A is the number of protons and neutrons.

In terms of the harmonic oscillator creation and annihilation operators

$$b^{\dagger}_{i\alpha,s} = \sqrt{\frac{m_\alpha \omega}{2\hbar}} \Big(x_{is}(\alpha) - \frac{i}{m_\alpha \omega} p_{is}(\alpha) \Big), \quad b_{i\alpha,s} = \sqrt{\frac{m_\alpha \omega}{2\hbar}} \Big(x_{is}(\alpha) + \frac{i}{m_\alpha \omega} p_{is}(\alpha) \Big), \tag{5}$$

the many-particle realization of the $Sp(12, R)$ Lie algebra is given by [2]

$$F_{ij}(\alpha,\beta) = \sum_{s=1}^{m} b^{\dagger}_{i\alpha,s} b^{\dagger}_{j\beta,s},\ G_{ij}(\alpha,\beta) = \sum_{s=1}^{m} b_{i\alpha,s} b_{j\beta,s},\ A_{ij}(\alpha,\beta) = \frac{1}{2}\sum_{s=1}^{m} (b^{\dagger}_{i\alpha,s} b_{j\beta,s} + b_{j\beta,s} b^{\dagger}_{i\alpha,s}). \tag{6}$$

An $Sp(12, R)$ unitary irreducible representation is characterized by the $U(6)$ quantum numbers $\sigma = [\sigma_1, \ldots, \sigma_6]$ of its lowest-weight state $|\sigma\rangle$, i.e. $|\sigma\rangle$ satisfies

$$G_{ab}|\sigma\rangle = 0; \quad A_{ab}|\sigma\rangle = 0, \quad a < b; \quad A_{aa}|\sigma\rangle = \Big(\sigma_a + \frac{m}{2}\Big)|\sigma\rangle \tag{7}$$

for the indices $a \equiv i\alpha$ and $b \equiv j\beta$ taking the values $1, \ldots, 6$. If we introduce the $U(6)$ tensor product operators $P^{(n)}(F) = [F \times \ldots \times F]^{(n)}$, where $n = [n_1, \ldots, n_6]$ is a partition with even integer parts, then by an $U(6)$ coupling of these tensor products to the lowest-weight $U(6)$ state $|\sigma\rangle$, one constructs the whole basis of states for an $Sp(12, R)$ irrep

$$|\Psi(\sigma n \rho E \eta)\rangle = [P^{(n)}(F) \times |\sigma\rangle]^{\rho E}_{\eta}, \tag{8}$$

where $E = [E_1, \ldots, E_6]$ indicates the $U(6)$ quantum numbers of the coupled state, η labels a basis of states for the coupled $U(6)$ irrep E and ρ is a multiplicity index. In this way we obtain a basis of $Sp(12, R)$ states that reduces the subgroup chain $Sp(12, R) \supset U(6)$. To fix the basis η one has to consider further the reduction of the $U(6)$ to the 3-dimensional rotational group $SO(3)$. Thus, in order to completely classify the basis states, we use the following reduction chain [2]:

$$\begin{array}{ccccccccccc} Sp(12,R) & \supset & U(6) & \supset & SU_p(3) \otimes SU_n(3) & \supset & SU(3) & \supset & SO(3) & \supset & SO(2), \\ \sigma & & n\rho \ E & \gamma & (\lambda_p,\mu_p) \ (\lambda_n,\mu_n) & \varrho & (\lambda,\mu) & K & L & & M \end{array} \tag{9}$$

IOP Conf. Series: Journal of Physics: Conf. Series **1023** (2018) 012013 doi:10.1088/1742-6596/1023/1/012013

which defines a shell-model coupling scheme. The chain (9) corresponds to the following choice of the index $\eta = \gamma(\lambda_p, \mu_p)(\lambda_n, \mu_n)\varrho(\lambda, \mu)KLM$, labeling the basis states (8) of an $Sp(12, R)$ irrep. Each $Sp(12, R)$ irreducible representation is determined by a symplectic bandhead or an intrinsic $U(6)$ space, which in turn is fixed by the underlying proton-neutron shell-model structure. So, the theory becomes completely compatible with the Pauli principle.

3. Application

The first application of the one-component $Sp(6, R)$ symplectic model, using a phenomenological Hamiltonian consisting of harmonic oscillator and collective potential which is expressed as polynomial up to fourth order in the mass quadrupole moment, to ^{20}Ne was given in [8]. The results showed an excessive collectivity compared to the experimental data. A more realistic Hamiltonian with paring symplectic symmetry-breaking interaction was used subsequently for the same nucleus with obtaining a better agreement [9].

The contracted version of the symplectic model has been applied for the description of the ground state band (^{20}Ne) and ground state and γ bands (^{22}Ne, ^{24}Mg) within a single generic, triaxial, symplectic irreducible representation [10]. The contracted version was also extended and applied to some heavy mass even-even nuclei from the rare-earth and actinide regions for the description of the ground state and γ bands [11, 12] by using the underlying pseudo-$SU(3)$ scheme.

The first few excited states of the ground state rotational bands up to $L = 6$ in ^{126}Ba, ^{154}Sm, ^{164}Yb and ^{166}Er have been described in the stretched approximation of the symplectic model in Refs.[13, 14]. The latter is defined as the set of $SU(3)$ states $(\lambda_0 + 2n, \mu_0)$ [15], where (λ_0, μ_0) is the leading irreducible representation for the combined proton-neutron nuclear system and $n = 0, 1, 2, 3, \ldots$. The calculations within the framework of stretched approximation is often called $Sp(2, R)$ submodel of the $Sp(6, R)$ because the set of these basis states can be generated only by the raising symplectic generators adding oscillator quanta only along the z-axis which are generators of the subgroup $Sp(2, R) \subset Sp(6, R)$. The energy levels and $E2$ transitions between the states of the ground state band in ^{166}Er up to $L = 16$ were determined using the full $Sp(6, R)$ model with a Davidson interaction [16].

All calculations mentioned above proved the earlier observation of Arickx and collaborators for the case of light nuclei [17, 18, 19] that the dominant contributions to the wave functions are presented by the stretched $SU(3)$ states. Moreover, a comparison of the $Sp(2, R)$ and $Sp(6, R)$ models by their application to light and heavy nuclei up to spins $L = 6$ has been performed in Ref.[20]. It was confirmed there that not only the stretched $SU(3)$ states give dominant contribution but also that there is no practical difference from the full $Sp(6, R)$ calculations. The stretched $SU(3)$ states within the framework of the one-component $Sp(6, R)$ symplectic model and its contracted version usually contribute between 80% and 90% to the ground state band wave functions. For example, 90% of the ^{20}Ne ground state comes from the $(8, 0)$, $(10, 0)$ and $(12, 0)$ stretched states [8]. Similarly, the stretched states give rise up to 93, 7% to the ground state in ^{238}U using the contracted symplectec model (CSM) [11]. The same picture was obtained in the recent applications of the symplectic $Sp(6, R)$ scheme with algebraic and schematic many-particle interactions to light and intermediate-mass nuclei [21, 22, 23]. Hence, the restriction of the full symplectic basis to the subset of stretched states turns out to be a valuable approximation for the symplectic model calculations in the heavy nuclei and is the first step toward the comprehensive microscopic description of the low-lying quadrupole collective dynamics. In the present work, we restrict ourselves to this approximation only.

In our application, we use the following model Hamiltonian

$$H = N\hbar\omega - \frac{1}{2}\chi\left[Q_p \cdot Q_n - (Q_p \cdot Q_n)_{TE}\right] - \xi C_2[SU(3)] + aL^2 + bK^2, \qquad (10)$$

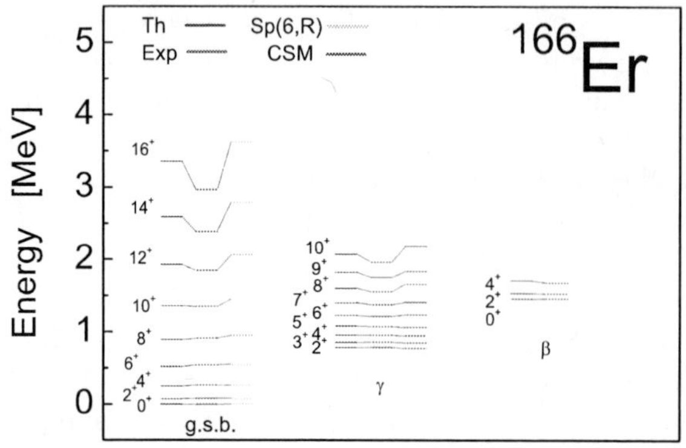

Figure 1. (Color online) Comparison of the theoretical and experimental energy levels for the ground, γ ($K^+ = 2^+$) and β ($K^+ = 0^+$) bands in ^{166}Er. The theoretical results of the $Sp(6, R)$ symplectic model [16] and the (pseudo)contracted symplectic model [12] for the ground band and γ band, respectively, are also shown.

where $N = N_p + N_n$ and $Q_\alpha \equiv Q(\alpha, \alpha)$ with $\alpha = p, n$ are given by Eq.(1). A similar Hamiltonian has been used in the pseudo-$SU(3)$ scheme calculations within the framework of the contracted symplectic model [11, 12]. The trace-equivalent part $(Q_p \cdot Q_n)_{TE}$ [24, 25, 26] is subtracted from the collective potential in order to preserve the mean-field shell structure under the action of the proton-neutron quadrupole-quadrupole interaction [11, 12, 10]. The $SU(3)$ second-order Casimir operator $C_2[SU(3)]$ splits energetically different $SU(3)$ multiplets and in this way determines the bandhead energies of excited bands with respect to the ground state band. Finally, the last two terms in (10), which represent a residual rotor part, allows the experimentally observed moment of inertia and band splitting features to be reproduced without altering the wave functions. The Hamiltonian (10) preserves the symplectic symmetry, thus having $Sp(12, R)$ as its dynamical symmetry. The full dynamics for it therefore occurs within a single irreducible representation of $Sp(12, R)$.

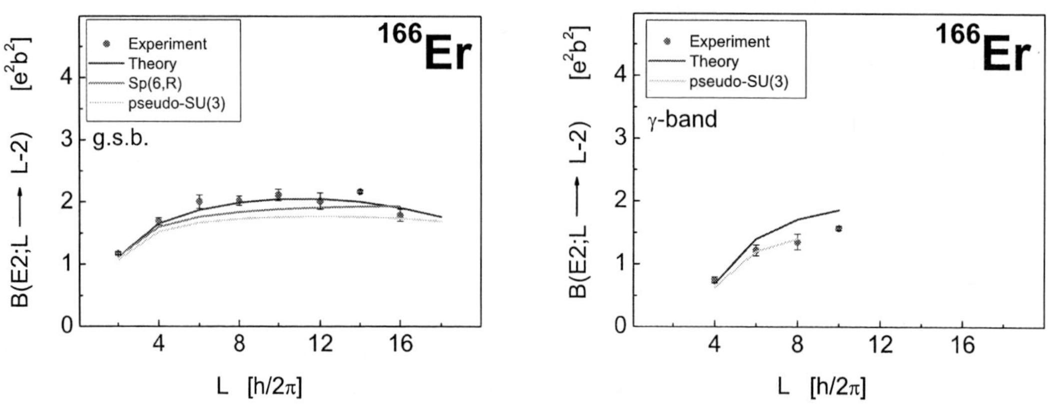

Figure 2. (Color online) Calculated and experimental intraband $B(E2)$ values between the states of the ground and γ ($K^+ = 2^+$) bands in ^{166}Er. For comparison, the theoretical results of the $Sp(6, R)$ symplectic model [16] and the pseudo-$SU(3)$ model [6] for the ground band and γ band, respectively, are also shown. In the pseudo-$SU(3)$ model calculations [6], the used value for the effective charge is $e_{eff} = 1.5e$

The first point in the practical application of the theory for description of the low-lying collective states in strongly deformed nuclei is the determination of the relevant irreducible representation of $Sp(12, R)$. Different approaches exist to determine the symplectic irrep by fixing the shell-model structure of the ground state using isotropic or anisotropic harmonic oscillator with or without spin-orbit interaction. It is well known that, for heavy mass nuclei from the rare-earth and actinide regions, the latter is strong and destroys the oscillator structure. Due to this, we use the pseudo-$SU(3)$ scheme to determine the relevant irreducible representation of $Sp(12, R)$. The shell-model considerations based on the pseudo-$SU(3)$ thus give the symplectic irrep $\langle\sigma\rangle = \langle 82+\frac{165}{2}, 52+\frac{165}{2}, 44+\frac{165}{2}, 44+\frac{165}{2}, 44+\frac{165}{2}, 44+\frac{165}{2}\rangle$ for ^{166}Er, which is fixed by the direct product proton-neutron shell-model structure of the ground state $(10, 4) \otimes (20, 4)$. The symplectic bandhead structure so obtained contains a plethora $of SU(3)$ multiplets, namely:

$$\begin{aligned}(10,4)\otimes(20,4) \rightarrow\ & (30,8),\ (31,6),(32,4),\ldots \\ & (28,9),\ (29,7),(30,5),\ldots \\ & (26,10),(27,8),(28,6),\ldots \\ & \ \ \vdots \qquad\quad \vdots \qquad\quad \vdots \end{aligned} \tag{11}$$

which are appropriate for the simultaneous shell-model description of the lowest collective bands, observed in the spectrum of ^{166}Er.

After obtaining the appropriate symplectic irrep, the model Hamiltonian (10) is further used to determine the microscopic structure of the low-lying collective states in ^{166}Er. For this purpose, we perform its diagonalization in the space of stretched states $(\lambda_0 + 2n, \mu_0)$ using the $SU_p(3) \otimes SU_n(3)$ symmetry-adapted basis. The results for the low-lying energy levels of the ground state, γ $(K^+ = 2^+)$ and β $(K^+ = 0^+)$ bands, compared with experiment [27], are shown in Fig.1. The major shell separation energy $\hbar\omega$ is determined by the standard formula $41A^{-1/3}$ MeV. The adopted values for the model parameters (in MeV) are as follows: $\chi = 0.0029$, $\xi = 0.0259$, $a = 0.010$ and $b = 0.187$. For comparison, the (two-parameter) $Sp(6, R)$ [16] and (four-parameter) (pseudo)contracted symplectic model [12] calculations for the ground state and γ bands, respectively, are also shown. The results of the CSM for the γ band are for ^{168}Er, but the pseudo-$SU(3)$ scheme gives the same leading irreducible representation $(30, 8)$ of the combined proton-neutron system for both ^{166}Er and ^{168}Er nuclei. So, the theory predicts similar collective properties and the usage of the γ band energies for ^{168}Er is a reasonable choice. From Fig.1 we see almost the same results for the ground state band between the PNSM and the full $Sp(6, R)$ calculations from one side, and for the γ band between the PNSM and the full CSM calculations from the other one. We recall that within the framework of the contracted symplectic model, the γ band together with the ground band belong to the single triaxial $Sp(6, R)$ irrep $(30, 8)$ and no low-lying β band exists. A low-lying β band can be obtained within the framework of one-component symplectic model by including symplectic symmetry-breaking terms, like,e.g., the spin-orbit or pairing interactions (see, e.g., Ref.[15]). The symplectic irrep of the $Sp(6, R)$ model calculations used in Ref.[16] is the axially symmetric one $(78, 0)$, so only the ground state band appears. From Fig.1 one also sees that, in contrast to the CSM, no staggering for the γ band levels is obtained in the PNSM, as observed in experiment.

We also computed the reduced intraband $E2$ electromagnetic transition strengths between the states of the ground and γ bands

$$B(E2; L_i \rightarrow L_f) = \frac{2L_f + 1}{2L_i + 1}\left(\frac{5}{16\pi}\right)\left(\frac{eZ}{A-1}\right)^2 |\langle f||Q(p,p)||i\rangle|^2, \tag{12}$$

Note that in the definition of the operator $Q(p, p)$ (c.f. Eq.(1)), the summation is over the $(A - 1)$ Jacobi quasiparticles. Thus, in order to obtain the charge quadrupole operator, the

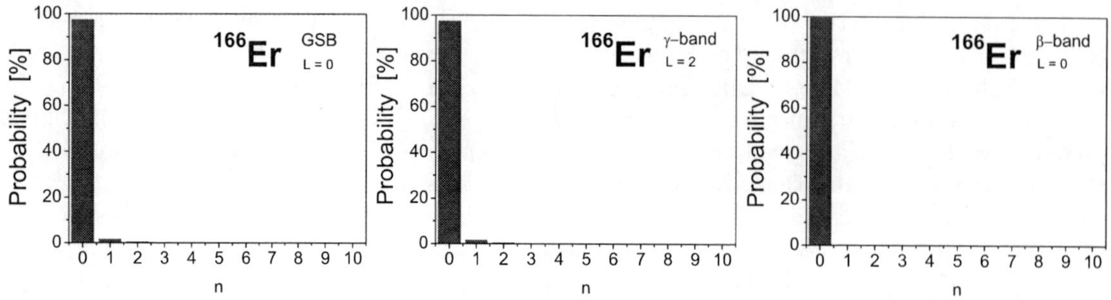

Figure 3. (Color online) Calculated probability distributions for the wave functions of the 0^+ state of the ground, 2^+ state of the γ and 0^+ state of the β bands as a function of n.

$Q(p,p)$ operator is multiplied by the factor $Z/(A-1)$. The results are shown in Fig.2 and compared with experiment [27] and the theoretical predictions of the $Sp(6,R)$ symplectic model [16] and the pseudo-$SU(3)$ model [6] for the ground state band and γ band, respectively. Again, we see that the PNSM, the full $Sp(6,R)$ and the pseudo-$SU(3)$ calculations reproduce equally well the general trend. Within the symplectic schemes which take the core polarization effects by construction no effective charge is used, i.e. $e=1$. In the pseudo-$SU(3)$ model calculations [6], the used value for the effective charge is $e_{eff}=1.5e$.

Concerning the interband transitions, the calculations within the stretched approximation give zero for $\beta\to g$ and $\beta\to\gamma$, which is a reasonable approximation to the small experimentally observed transition strengths, e.g., $0^+_\beta\to 2^+_g$ (2.7 w.u.) and $0^+_\beta\to 2^+_\gamma$ (2.4 w.u.). For $\gamma\to g$, we have non-zero values since both the ground and γ bands belong to the same set of stretched $SU(3)$ irreps $(30+2n,8)$. For instance, for $2^+_\gamma\to 0^+_g$ we obtain 18.45 w.u. compared with the experimental value 5.7 w.u. It is clear that the extension of the model space beyond the stretched approximation will produce a mixing of the two sets of $SU(3)$ irreps corresponding to the ground (γ) and β bands which will result in non-zero interband transitions between them.

In Fig.3, the squares of the amplitudes (probabilities) of the $SU(3)$ components from each $n\hbar\omega$ shell are plotted for the 0^+ state of the ground, 2^+ state of the γ and 0^+ state of the β bands as a function of n. From the latter the structure of the eigenstates obtained in the diagonalization of the model Hamiltonian becomes evident. In particular, we see that the structure of the ground and γ bands is predominantly determined by a single $SU(3)$ irrep which exhausts up to $\approx 97.5\%$ of the structure. For the β band we see that the probability of the 0^+ state is fully exhausted ($\approx 100\%$) by the $SU(3)$ irrep $(32,4)$. The results obtained for this particular nucleus, exploiting the stretched approximation in the $SU_p(3)\otimes SU_n(3)$ symmetry-adapted basis, thus clearly show the presence of a good $SU(3)$ dynamical symmetry.

4. Conclusions

In the present paper, the proton-neutron symplectic model with $Sp(12,R)$ dynamical symmetry is applied for the simultaneous description of the microscopic structure of the low-lying states of the ground state, γ and β bands in ^{166}Er. For this purpose, the model Hamiltonian is diagonalized in the space of stretched states by exploiting the $SU_p(3)\otimes SU_n(3)$ symmetry-adapted basis. The model Hamiltonian, which exhibits a symplectic dynamical symmetry, is restricted to the spherical harmonic oscillator shell-model part and the full major shell-mixing proton-neutron quadrupole-quadrupole interaction, plus an $SU(3)$ scalar term which take into account the band-head energies and a residual rotational part which allows the inertial and band splitting features of the observed spectrum to be reproduced without altering the wave functions. The theoretical predictions are compared with experiment and some other microscopic collective models, like the

IOP Conf. Series: Journal of Physics: Conf. Series **1023** (2018) 012013 doi:10.1088/1742-6596/1023/1/012013

one-component $Sp(6, R)$ symplectic and pseudo-$SU(3)$ models. A good description of the energy levels of the three bands under consideration, as well as the enhanced intraband $B(E2)$ transition strengths between the states of the ground and γ bands is obtained without the use of effective charges. The results show the presence of a good $SU(3)$ dynamical symmetry. It is also shown that, in contrast to the $Sp(6, R)$ case, the lowest excited bands, e.g., the β and γ bands, naturally appear together with the ground state band within a single $Sp(12, R)$ irreducible representation.

[1] H. G. Ganev, *Eur. Phys. J. A* **50**, 183 (2014).
[2] H. G. Ganev, Eur. Phys. J. A **51**, 84 (2015).
[3] D. J. Rowe and G. Rosensteel, Phys. Rev. Lett. **38**, 10 (1977).
[4] R. D. Ratna Raju, J. P. Draayer, and K. T. Hecht, Nucl. Phys. **A 202**, 433 (1973).
[5] J. P. Draayer and K. J. Weeks, Phys. Rev. Lett. **51**, 1422 (1983).
[6] J. P. Draayer and K. J. Weeks, Ann. Phys. **156**, 41 (1984).
[7] F. Iachello and A. Arima, *The Interacting Boson Model* (Cambridge University Press, Cambridge, 1987).
[8] G. Rosensteel and D. J. Rowe, Ann. Phys. **126**, 343 (1980).
[9] J. P. Drayeer, K. J. Weeks, and G. Rosensteel, Nucl. Phys. **A 413**, 215 (1984).
[10] O. Castanos and J. P. Draayer, Nucl. Phys. **A 491**, 349 (1989).
[11] O. Castanos, P. O. Hess, J. P. Draayer, and P. Rochford, Nucl. Phys. **A 524**, 469 (1991).
[12] D. Troltenier, J. P. Draayer, P. O. Hess, and O. Castanos, Nucl. Phys. **A 576**, 351 (1994).
[13] J. Carvalho, P. Park, D. J. Rowe, and G. Rosensteel, Phys. Lett. **B 119**, 249 (1982).
[14] P. Park, J. Carvalho, M. Vassanji, D. J. Rowe, and G. Rosensteel, Nucl. Phys. **A 414**, 93 (1984).
[15] D. J. Rowe, Prog. Part. Nucl. Phys. **37**, 4711 (1985).
[16] C. Bahri and D. Rowe, Nucl. Phys. **A 662**, 125 (2000).
[17] F. Arickx, P. Van Leuven, and M. Bouten, Nucl. Phys. **A 252**, 416 (1975).
[18] F. Arickx, Nucl. Phys. **A 268**, 347 (1976).
[19] F. Arickx, J. Broeckhove, and E. Deumens, Nucl. Phys. **A 318**, 269 (1979).
[20] P. O. Hess, Foundations of Physics, vol. **27**, 1061 (1997).
[21] T. Dytrych *et al.*, Phys. Rev. Lett. **111**, 252501 (2013).
[22] K. D. Launey, A. C. Dreyfuss, R. B. Baker, J. P. Draayer, and T. Dytrych, J. Phys.: Conf. Ser. **597**, 012054 (2015).
[23] G. K. Tobin et al., Phys. Rev. **C 89**, 034312 (2014).
[24] J. P. Draayer and G. Rosensteel, Phys. Lett. **B 124**, 281 (1983).
[25] J. P. Draayer and G. Rosensteel, Phys. Lett. **B 125**, 237 (1983).
[26] G. Rosensteel and J. P. Draayer, Nucl. Phys. **A 436**, 445 (1985).
[27] National Nuclear Data Center (NNDC), http://www.nndc.bnl.gov/

Symmetry Energy and Its Components in Finite Nuclei

A N Antonov[1], M K Gaidarov[1], D N Kadrev[1], P Sarriguren[2], E Moya de Guerra[3]

[1]Institute for Nuclear Research and Nuclear Energy, Bulgarian Academy of Sciences, Sofia 1784, Bulgaria
[2]Instituto de Estructura de la Materia, IEM-CSIC, Serrano 123, E-28006 Madrid, Spain
[3]Departamento de Fisica Atomica, Molecular y Nuclear, Facultad de Ciencias Fisicas, Universidad Complutense de Madrid, E-28040 Madrid, Spain

E-mail: gaidarov@inrne.bas.bg

Abstract. We derive the volume and surface components of the nuclear symmetry energy (NSE) and their ratio within the coherent density fluctuation model. The estimations use the results of the model for the NSE in finite nuclei based on the Brueckner and Skyrme energy-density functionals for nuclear matter. The obtained values of the volume and surface contributions to the NSE and their ratio for the Ni, Sn, and Pb isotopic chains are compared with estimations of other approaches which have used available experimental data on binding energies, neutron-skin thicknesses, and excitation energies to isobaric analog states (IAS). Apart from the density dependence investigated in our previous works, we study also the temperature dependence of the symmetry energy in finite nuclei in the framework of the local density approximation combining it with the self-consistent Skyrme-HFB method using the cylindrical transformed deformed harmonic-oscillator basis. The results for the thermal evolution of the NSE in the interval T=0–4 MeV show that its values decrease with temperature. The investigations of the T-dependence of the neutron and proton root-mean-square radii and the corresponding neutron skin thickness point out that the effect of temperature leads mainly to a substantial increase of the neutron radii and skins, especially in nuclei which are more rich of neutrons.

1. Introduction

The nuclear symmetry energy, as a fundamental quantity in nuclear physics and astrophysics, represents a measure of the energy gain in converting isospin asymmetric nuclear matter (ANM) to a symmetric system. Its value depends on the density ρ and temperature T. A natural and important way to learn more about the NSE is the transition from ANM to finite nuclei. Experimentally, the NSE is not a directly measurable quantity and is extracted indirectly from observables that are related to it (see, e.g., the review [1]). The need of information for the symmetry energy in finite nuclei (including the one theoretically obtained) is a major issue because it allows one to constrain the bulk and surface properties of the nuclear energy-density functionals (EDFs) quite effectively. More information on the nuclear symmetry energy is still required for understanding the structures of nuclei far away from the β-stability line, heavy-ion collisions, supernova explosions, and neutron star properties.

In Ref. [2] it has been demonstrated by Danielewicz that the ratio of the volume to surface symmetry energy is closely related to the neutron-skin thickness (see also Refs. [3, 4]). Discussions on the correlation between the bulk and surface symmetry energy are given also e.g., in Refs. [5, 6, 7, 8]. It has been shown in [9] by Lee and Mekjian by calculations of the thermal nuclear properties that the surface symmetry-energy term is more sensitive to the temperature than the volume energy term. The thermal behavior of the symmetry energy has a role in changing the location of the nuclear drip lines as nuclei warm up. Also, since the density derivative of the symmetry coefficient reflects the pressure difference on the neutrons and protons and is thus one of the determinants in fixing the neutron skin of nuclei, the nature and stability of phases within a warm neutron star, its crustal composition or its thickness [10] would be strongly influenced by the temperature dependence of the symmetry energy.

In the present work (see also Ref. [11]) we investigate the volume and surface contributions to the NSE within the coherent density fluctuation model (CDFM) [12, 13]. We use our results for NSE obtained using Brueckner EDF [14, 15] in Refs. [16, 17, 18], as well as the considerations of this subject mentioned above (e.g., [2, 3, 4, 6, 7, 8]). The present calculations are performed using both Brueckner and, in addition, Skyrme energy-density functionals with different Skyrme forces (e.g., Ref. [19]). The results are compared with those of other theoretical methods and with corresponding experimental data obtained from analyses of different nuclear quantities, such as binding energies, neutron-skin thicknesses, excitation energies to IAS and others. In this work (see also Ref. [20]) we focus on the determination of the symmetry energy coefficient depending on ρ and T, for which we have explored the local density approximation [21, 22, 23, 24] with some modifications. The temperature-dependent densities of these nuclei are calculated within a self-consistent Skyrme-HFB method using the cylindrical transformed deformed harmonic-oscillator basis (HFBTHO densities) [25, 26]. The kinetic energy density is calculated by the HFBTHO code, as well as, for a comparison, by the Thomas-Fermi (TF) expression up to T^2 term [9]. In addition, in Ref. [20] some results are presented for the ^{208}Pb nucleus with densities obtained within the extended TF method [27, 28] and the rigorous density functional approach [29]. The effect of temperature on the rms radii of protons and neutrons and the formation of neutron skin in hot nuclei is also analyzed and discussed.

2. Theoretical framework

An important result that expresses the ratio of the volume (a_A^V) to the surface (a_A^S) energy coefficients by means of the shape of the symmetry energy dependence on density $s(\rho)$ is given (in the local density approximation to the symmetry energy), e.g. in Refs. [2, 4, 7]:

$$\frac{a_A^V}{a_A^S} = \frac{3}{r_0} \int dr \frac{\rho(r)}{\rho_0} \left\{ \frac{s(\rho_0)}{s[\rho(r)]} - 1 \right\}. \tag{1}$$

In Eq. (1) $\rho(r)$ is the half-infinite nuclear matter density, ρ_0 is the nuclear matter equilibrium density, and r_0 is the radius of the nuclear volume per nucleon that can be obtained from

$$\frac{4\pi r_0^3}{3} = \frac{1}{\rho_0}. \tag{2}$$

For density-independent symmetry energy $s(\rho) = s(\rho_0) = a_A^V$ and, then it follows from Eq. (1) that the ratio $a_A^V/a_A^S = 0$ [4]. The density $\rho(r)$ in Eq. (1) is uniform in two Cartesian directions and generally nonuniform in the third, usually chosen to be z [6]. The integral in Eq. (1) is across the nuclear surface involving the shape of the density dependence [3]. In the Danielewicz's approximation only the symmetry energy of a finite nucleus has a mass dependence, while a_A^V, a_A^S, and their ratio a_A^V/a_A^S are A-independent quantities. The values of a_A^V and a_A^S differ for various Skyrme interactions in wide intervals (see Table I of Ref. [6]). At the same time, as shown in [2], a combination of empirical data on skin sizes and masses of nuclei constrains the volume symmetry parameter to $27 \leq a_A^V \leq 31$ MeV and the ratio a_A^V/a_A^S to $2.0 \leq a_A^V/a_A^S \leq 2.8$.

In the present paper and in [11] we develop, using as a base the Danielewicz's model [Eq. (1)], another approach to calculate the ratio a_A^V/a_A^S, as well as a_A^V and a_A^S within the CDFM. Our motivation is that numerous analyses of the volume and surface components of the NSE using a wide range of data on the binding energies, neutron-skin thicknesses and excitation energies to IAS give estimations of these quantities as functions of the mass number A that change in some intervals for different regions of nuclei. For instance, the reported values of a_A^V and a_A^S are consistent with each other in a wide mass region ($30 \leq A \leq 240$). In the CDFM we take nuclear matter values of the parameters to deduce their values in finite nuclei (using the self-consistently calculated nuclear density) which become dependent on the considered nucleus. For this purpose, we start from Eq. (1) but in it we replace the density $\rho(r)$ for the half-infinite nuclear matter in the integrand by the density distribution of finite nucleus (Eq. (18) of Ref. [11]). Then, we obtain approximately an expression that allows us to calculate the ratio a_A^V/a_A^S. It has the form:

$$\kappa \equiv \frac{a_A^V}{a_A^S} = \frac{3}{r_0\rho_0}\int_0^\infty dx|\mathcal{F}(x)|^2 x\rho_0(x)\left\{\frac{s(\rho_0)}{s[\rho_0(x)]} - 1\right\}. \tag{3}$$

In Eq. (3) $s(\rho_0) = s^{ANM}(\rho_0)$ and the quantity $s[\rho_0(x)] = s^{ANM}[\rho_0(x)]$ is the NSE within the chosen approach for the EDF. The weight function $|\mathcal{F}(x)|^2$ in Eq. (3) can be expressed by the density distribution $\rho(r)$ and in the case of the monotonically decreasing local density ($d\rho/dr \leq 0$) can be obtained using a known density (from experiments or from theoretical models) for a given nucleus:

$$|\mathcal{F}(x)|^2 = -\frac{1}{\rho_0(x)}\left.\frac{d\rho(r)}{dr}\right|_{r=x} \tag{4}$$

with the normalization $\int_0^\infty dx|\mathcal{F}(x)|^2 = 1$. Having calculated within the CDFM the values of s and κ, the expressions from which we can estimate the values of a_A^V and a_A^S separately can be obtained (see Ref. [11]):

$$a_A^V = s(1 + A^{-1/3}\kappa), \tag{5}$$

$$a_A^S = \frac{s}{\kappa}(1 + A^{-1/3}\kappa). \tag{6}$$

For finite systems, different definitions of the symmetry energy coefficient and its temperature dependence are considered in the literature. In the present paper we develop an approach to calculate the symmetry energy coefficient for a specific nucleus starting with the LDA expression given in [21, 22]:

$$e_{sym}(A,T) = \frac{1}{I^2A}\int \rho(r)e_{sym}[\rho(r),T]\delta^2(r)d^3r. \tag{7}$$

In Eq. (7) $I = (N-Z)/A$, $e_{sym}[\rho(r),T]$ is the symmetry energy coefficient at temperature T of infinite nuclear matter at the value of the total local density $\rho(r) = \rho_n(r) + \rho_p(r)$, $\delta(r) = [\rho_n(r) - \rho_p(r)]/\rho(r)$ is the ratio between the isovector and the isoscalar parts of $\rho(r)$, with $\rho_n(r)$ and $\rho_p(r)$ being the neutron and proton local densities. The symmetry energy coefficient $e_{sym}(\rho,T)$ can be evaluated in different ways. Following Refs. [21, 24], we adopt in this work the definition

$$e_{sym}(\rho,T) = \frac{e(\rho,\delta,T) - e(\rho,\delta=0,T)}{\delta^2}, \tag{8}$$

where $e(\rho,\delta,T)$ is the energy per nucleon in an asymmetric infinite matter, while $e(\rho,\delta=0,T)$ is that one of symmetric nuclear matter. These quantities are expressed by $e = \mathcal{E}(r,T)/\rho$, where $\mathcal{E}(r,T)$ is the total energy density of the system (for the Skyrme EDF in our case). The dependence on temperature of $\mathcal{E}(r,T)$ and nucleon effective mass comes from the T-dependence of the densities and kinetic energy densities (see Ref. [20]).

In our study aiming to investigate the temperature dependence of e_{sym} within a given isotopic chain, we introduce other definitions of $e_{sym}(A,T)$ in LDA that, in our opinion, would be more appropriate in this case. They concern the problem of calculating the term $e(\rho,\delta=0,T)$ of Eq. (8) for symmetric nuclear matter. In our LDA approach the latter is simulated by considering the $N=Z=A/2$ nucleus, but we analyze two possibilities. First, on the basis of Eqs. (7) and (8) with $e=\mathcal{E}(r)/\rho$, we present the integrand of the right-hand side of the following expression for $I^2 e_{sym}(A,T)$ as a difference of two terms with transparent physical meaning:

$$I^2 e_{sym}(A,T) = \int d\vec{r}\left[\frac{\mathcal{E}(\rho_A(r),\delta,T)}{A} - \frac{\mathcal{E}(\rho_{A1}(r),\delta=0,T)}{A1}\right], \tag{9}$$

in which the first one corresponds to the energy per volume and particle of nuclear matter $\mathcal{E}(\rho_A(r),\delta,T)/A$ with a density $\rho_A(r)$ equal to that of the considered nucleus with A nucleons, Z protons and N neutrons from the given isotopic chain. The second term $\mathcal{E}(\rho_{A1}(r),\delta=0,T)/A1$ is the analogous for the isotope with $A1=2Z$ ($N1=Z=A1/2$). For example, for the Ni isotopic chain the nucleus $A1$ is the double-closed shell nucleus ^{56}Ni ($Z=N1=28$), while for the Sn isotopic chain the nucleus $A1$ is the double-closed shell nucleus ^{100}Sn ($Z=N1=50$) and both ^{56}Ni and ^{100}Sn isotopes play a role of reference nuclei. Our second new definition of $e_{sym}(A,T)$ is based on an expression given in Ref. [24] for finite nuclei. The latter allows us, using the LDA, to present $e_{sym}(A,T)$ in the form:

$$I^2 e_{sym}(A,T) = \int \frac{d\vec{r}}{A}\left[\mathcal{E}(\rho_A(r),\delta,T) - \mathcal{E}(\rho_{\bar{A}}(r),N=\bar{A}/2,Z=\bar{A}/2,\delta=0,T)\right], \tag{10}$$

in which the mass number $\bar{A}=A$ is the same, but with different nucleon content, $A(Z,N)$ and $\bar{A}(Z=\bar{A}/2,N=\bar{A}/2)$. This consideration requires the even-even nucleus with $N=Z=\bar{A}/2$ to be bound.

3. Results and discussion

In the case of the Brueckner EDF the results of the calculations using Eq. (3) of the ratio κ as a function of the mass number A for the isotopic chain of Sn with different forces are given in Fig. 1. By means of Eqs. (5) and (6) and the values of the NSE obtained in our works [16, 17], we calculated the coefficients a_A^V and a_A^S. Their values as functions of A for the same isotopic chain are presented in Fig. 2. It can be seen from Fig. 1 (also for Ni and Pb isotopic chains [20]) that our results for the values of the ratio κ are within the range $2.10 \leq \kappa \leq 2.90$. This range of values is similar to the estimations of the ratio κ extracted from nuclear properties and presented in Ref. [7]:

$$2.6 \leq \kappa \leq 3.0 \tag{11}$$

from IAS and skins [3],

$$2.0 \leq \kappa \leq 2.8 \tag{12}$$

from masses and skins [2], and

$$1.6 \leq \kappa \leq 2.0 \tag{13}$$

from the analyses in Ref. [7] of masses and skins. As can be seen the ranges (11) and (12) are in a good agreement with our results. In addition, there exists a "kink" in the curve of $\kappa \equiv a_A^V/a_A^S$ as a function of A for the double-magic ^{132}Sn nucleus that can be seen in Fig. 1. It is seen from Fig. 2 (also for Ni and Pb isotopic chains [20]) that our CDFM results obtained with Brueckner EDF for a_A^V are between 41.5 and 43 MeV, while for a_A^S they are between 14 and 20 MeV. We would like to note that "kinks" appear also in the cases of a_A^V and a_A^S as functions of the mass number A. In Fig. 2(a) a "kink" appears for $a_A^V(A)$ not only for the double-magic ^{132}Sn, but also for the semi-magic ^{140}Sn nucleus. The latter is related to the closed $2f_{7/2}$ subshell for neutrons. Reiss *et al.* discussed in Ref. [30] that the region around $N = 90$ for neutron-rich tin isotopes is an interesting one because the shell structure is somewhat fluctuating. In addition, in Ref. [31] $N = 90$ was predicted to be submagic with Gogny D1S and D1M interactions at ^{140}Sn because the $2f_{7/2}$ orbit is fully occupied, but not with M3Y-P6 and P7 semi-realistic NN interactions.

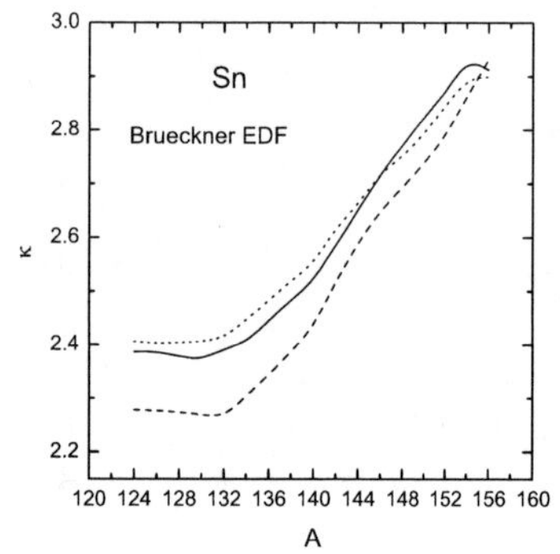

Figure 1. The ratio $\kappa = a_A^V/a_A^S$ as a function of A for the isotopic chain of Sn. Solid line: SLy4, dashed line: SGII, dotted line: Sk3 force.

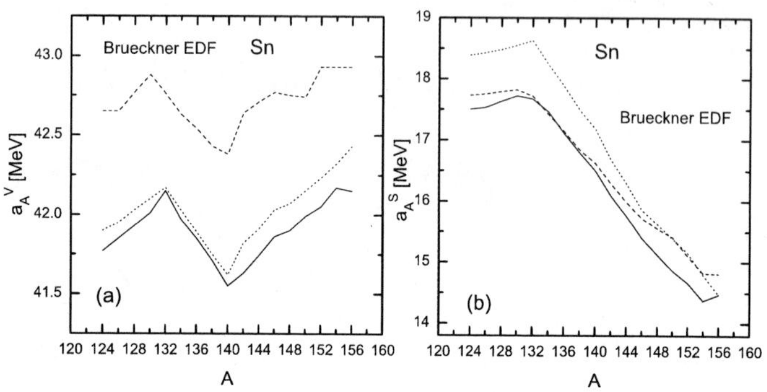

Figure 2. The values of a_A^V (a) and a_A^S (b) as functions of A for the isotopic chain of Sn. The SLy4 (solid line), SGII (dashed line), and Sk3 (dotted line) forces have been used in the HF+BCS calculations of the densities.

The mass dependence of $e_{sym}(A)$ is displayed in Fig. 3 for Ni, Sn, and Pb isotopic chains for SLy4 and SkM* interactions at three temperatures, T=0, 2, and 4 MeV. The results are obtained by using HFBTHO densities and the TF method up to T^2 term for the kinetic energy density (more details can be found in [20]). From one hand, one can see that the values of e_{sym} calculated with SLy4 overestimate those obtained with SkM* force. From another side, the difference between both sets of values decreases going to higher temperatures, in a way that it is small at the transition from T=0 to T=2 MeV and a "gap" appears between the results corresponding to T=2 and T=4 MeV. For Pb isotopic chain even a "crossover" of curves that correspond to temperatures T=0 and T=2 MeV and both parametrizations is observed in Fig. 3(c). We also would like to note the existence of a kink in the values of $e_{sym}(A)$ at zero temperature at the double-magic ^{78}Ni and ^{132}Sn nuclei (see Figs. 3(a) and 3(b)) as well as the lack of kinks in the Pb isotopic chain [Fig. 3(c)]. These results confirm our previous observations when studying the density dependence of the symmetry energy for Ni, Sn, and Pb isotopes [16, 17]. We also note that in the cases of $e_{sym}(A)$ for Ni and Sn isotopic chains the kinks exist for $T = 0$ MeV, but not for $T = 2$ and $T = 4$ MeV. The reason is the well-known fact that the shell effects can be expected up to $T \leq 2$ MeV.

4. Conclusions

The results of the present work can be summarized as follows:

i) We develop, using as a base the Danielewicz's model [Eq. (1)], another approach within the CDFM to calculate the ratio a_A^V/a_A^S between the volume and surface components of the symmetry energy s, as well as a_A^V and a_A^S separately, for finite nuclei. We obtain within the CDFM the expression for $\kappa \equiv a_A^V/a_A^S$ [Eq. (3)] that allows us to calculate this ratio using the ingredients of the model, the weight function $|\mathcal{F}(x)|^2$ and the nuclear matter symmetry energy $s^{ANM}[\rho_0(x)]$ from two energy-density functionals, Brueckner and Skyrme ones. The first one of

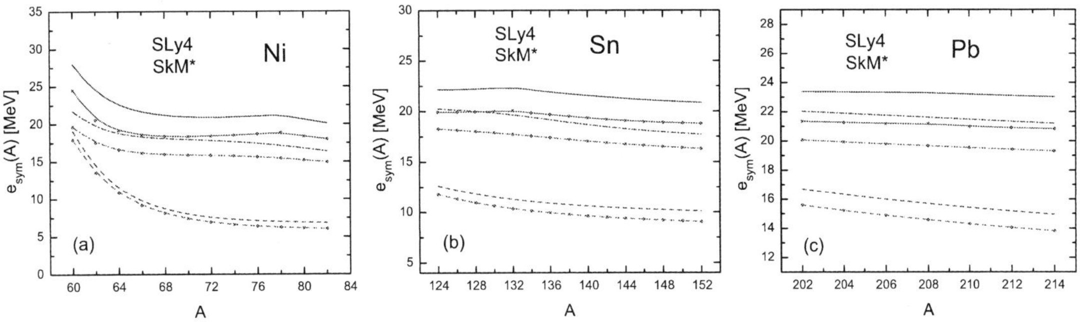

Figure 3. The mass dependence of the symmetry energy coefficient e_{sym} for Ni (a), Sn (b), and Pb (c) isotopic chains at temperatures $T = 0$ MeV (solid line), $T = 2$ MeV (dash-dotted line), and $T = 4$ MeV (dashed line) calculated with SLy4 (blue lines) and SkM* (green lines with points) Skyrme interactions.

them was used to calculate the NSE in our previous works [16, 17, 18]. In the CDFM we take nuclear matter values of the components of NSE to deduce their values in finite nuclei. Thus, our approach is different from the Danielewicz's formalism. Being motivated by the available empirical data that show A-dependence of a_A^V, a_A^S, and their ratio, we obtained in our approach a possibility to find a (weak) A-dependence of the theoretical results for these quantities within the CDFM.

ii) Studying firstly the isotopic sensitivity of a_A^V, a_A^S, and their ratio in the case of using the Brueckner EDF we observe peculiarities ("kinks") of these quantities as functions of the mass number A in the cases of the double-magic ^{78}Ni and ^{132}Sn isotopes for $\kappa \equiv a_A^V/a_A^S$, a_A^V, and a_A^S, as well as a "kink" of a_A^V for ^{140}Sn. The latter is related to the closed $2f_{7/2}$ subshell for neutrons. The origin of the "kinks" is in the different behavior of the density distributions $\rho(r)$ for the isotopes, because the derivative of $\rho(r)$ determines the weight function $|\mathcal{F}(x)|^2$ [Eq. (4)] that takes part in the expression for the ratio $\kappa \equiv a_A^V/a_A^S$ [Eq. (3)]. Similarly to the case when Brueckner EDF is used, in the case of the Skyrme EDF one can see also a "kink" in the behavior of κ for the chain of Ni at $A = 78$ and for the Sn chain at $A = 132$, as well as a lack of "kinks" for the Pb case. A "kink" in the Ni chain at $A = 78$ can be seen also in the A-dependence of a_A^S, as well as of a_A^S at $A = 132$ in the case of the Sn chain.

iii) A theoretical approach to the nuclear many-body problem has been used to study the temperature dependence of the symmetry energy coefficient in finite nuclei and other properties, such as the T-dependent nucleon densities and related rms radii, as well as the possibility of formation of neutron skins. The properties of hot nuclei were modelled in a temperature range T=0–4 MeV. In general, we have found that the density distributions decrease with the temperature in the center of the nucleus. Following the trend of the corresponding proton and neutron rms radii, the neutron-skin thickness grows significantly with the increase of T within a given isotopic chain.

iv) Our investigations of the T-dependent symmetry coefficients $e_{sym}(A,T)$ for finite nuclei (in particular, cases of Ni, Sn, and Pb isotopic chains) within the LDA with some modifications face the problem for the choice of density distributions and the kinetic energy densities. In our work both quantities are calculated through the HFBTHO code that solves the nuclear Skyrme-HFB problem by using the cylindrical transformed deformed harmonic-oscillator basis [26]. We have explored the LDA expression [Eq. (7)] for the symmetry energy $e_{sym}(A,T)$. Aiming to study the T-dependence of e_{sym} within a given isotopic chain, we introduced two new definitions of $e_{sym}(A,T)$ [Eqs. (9) and (10)] within the LDA, as an attempt to analyze in a more appropriate way the symmetry energy coefficient of finite nuclei within a given chain. In general, the results of e_{sym} calculated for various isotopes in the present work are in good agreement with theoretical predictions for some specific nuclei reported by other authors.

Having in mind the dependence of $e_{sym}(A,T)$ on its various definitions we note that more refined future investigations, for instance, of the temperature dependence of both volume and surface components of the symmetry energy coefficient [24], would provide better description of hot nuclei and could minimize the ambiguities due to the use of different definitions for the symmetry energy coefficient of finite nuclei.

5. Acknowledgments

Three of the authors (M.K.G., A.N.A., and D.N.K) are grateful for support of the Bulgarian Science Fund under Contract No. DFNI-T02/19. D.N.K. thanks for the partial support from Contract No. DFNI-E02/6 of the Bulgarian Science Fund. E.M.G. and P.S. acknowledge support from MINECO (Spain) under Contract FIS2014–51971–P.

References

[1] Shetty D V and Yennello S J 2010 *Pramana* **75** 259
[2] Danielewicz P 2003 *Nucl. Phys.* A **727** 233
[3] Danielewicz P 2004 arXiv: 0411115 [nucl-th]
[4] Danielewicz P 2006 arXiv: 0607030 [nucl-th]
[5] Steiner A W, Prakash M, Lattimer J M and Ellis P J 2005 *Phys. Rep.* **411** 325
[6] Danielewicz P and Lee J 2009 *Nucl. Phys.* A **818** 36
[7] Dieperink A E L and Van Isacker P 2007 *Eur. Phys. J.* A **32** 11
[8] Nikolov N, Schunck N, Nazarewicz W, Bender M and Pei J 2011 *Phys. Rev.* C **83** 034305
[9] Lee S J and Mekjian A Z 2010 *Phys. Rev.* C **82** 064319
[10] Steiner A W 2008 *Phys. Rev.* C **77** 035805
[11] Antonov A N, Gaidarov M K, Sarriguren P and Moya de Guerra E 2016 *Phys. Rev.* C **94** 014319
[12] Antonov A N, Nikolaev V A and Petkov I Zh 1979 *Bulg. J. Phys.* **6** 151; 1980 *Z. Phys.* A **297** 257; *ibid* 1982 **304** 239; 1985 *Nuovo Cimento* A **86** 23; Antonov A N *et al.*, *ibid* 1989 **102** 1701; Antonov A N, Kadrev D N and Hodgson P E 1994 *Phys. Rev.* C **50** 164
[13] Antonov A N, Hodgson P E and Petkov I Zh 1988 *Nucleon Momentum and Density Distributions in Nuclei* (Clarendon Press, Oxford); 1993 *Nucleon Correlations in Nuclei* (Springer-Verlag, Berlin-Heidelberg-New York)
[14] Brueckner K A, Buchler J R, Jorna S and Lombard R J 1968 *Phys. Rev.* **171** 1188
[15] Brueckner K A, Buchler J R, Clark R C and Lombard R J 1969 *Phys. Rev.* **181** 1543
[16] Gaidarov M K, Antonov A N, Sarriguren P and Moya de Guerra E 2011 *Phys. Rev.* C **84** 034316
[17] Gaidarov M K, Antonov A N, Sarriguren P and Moya de Guerra E 2012 *Phys. Rev.* C **85** 064319
[18] Gaidarov M K, Sarriguren P, Antonov A N and Moya de Guerra E 2014 *Phys. Rev.* C **89** 064301
[19] Chabanat E, Bonche P, Haensel P, Meyer J and Schaeffer R 1998 *Nucl. Phys.* A **635** 231; Beiner M, Flocard H, Van Giai N and Quentin P 1975 *Nucl. Phys.* A **238** 29; Van Giai N and Sagawa H 1981 *Phys. Lett.* B **106** 379
[20] Antonov A N, Kadrev D N, Gaidarov M K, Sarriguren P and Moya de Guerra E 2017 *Phys. Rev.* C **95** 024314
[21] Agrawal B K, De J N, Samaddar S K, Centelles M and Viñas X 2014 *Eur. Phys. J.* A **50** 19
[22] Samaddar S K, De J N, Viñas X and Centelles M 2007 *Phys. Rev.* C **76** 041602(R)
[23] Samaddar S K, De J N, Viñas X and Centelles M 2008 *Phys. Rev.* C **78** 034607
[24] De J N and Samaddar S K 2012 *Phys. Rev.* C **85** 024310
[25] Stoitsov M, Schunck N, Kortelainen M, Michel N, Nam H, Olsen E, Sarich J and Wild S 2013 *Computer Physics Communications* **184** 1592
[26] Stoitsov M V, Dobaczewski J, Nazarewicz W and Ring P 2005 *Comput. Phys. Comm.* **167** 43
[27] Brack M 1984 *Phys. Rev. Lett.* **53** 119
[28] Brack M, Guet C and Håkansson H-B 1985 *Phys. Rep.* **123** 275
[29] Stoitsov M V, Petkov I Z and Kryachko E S 1987 *C. R. Bulg. Acad. Sci.* **40** 45; Stoitsov M V 1987 *Nuovo Cimento* **98A** 725
[30] Reiß C, Bender M and Reinhard P.-G. 1999 *Eur. Phys. J.* A **6** 157
[31] Nakada H and Sugiura K 2014 *Prog. Theor. Exp. Phys.* **2014** 033D02

Large scale shell model calculations for the yrast line of ^{138}Xe

S.S. Dimitrova[1], **N. Lo Iudice** [2,3]

[1]Institute of Nuclear Research and Nuclear Energy, Bulgarian Academy of Sciences, 1784 Sofia, Bulgaria
[2]Dipartimento di Fisica, Università di Napoli Federico II, Monte S. Angelo, via Cintia, I-80126 Napoli, Italy
[3]Istituto Nazionale di Fisica Nucleare, Sezione di Napoli, Monte S. Angelo, via Cintia, I-80126 Napoli, Italy

E-mail: `sevdim@inrne.bas.bg`

Abstract. We have adopted an importance sampling iterative matrix diagonalization algorithm to compute a large scale shell model calculation of the yrast spectrum of ^{138}Xe up to high spin thereby extending a previous calculation confined to low-lying angular momenta. An effective nucleon–nucleon interaction derived from the CD-Bonn nucleon-nucleon potential is used to compute energies, $E2$ transition probabilities, and occupation numbers. A satisfactory agreement with the experimental data is reached.

1. Introduction

The region around the doubly magic ^{132}Sn represents a precious laboratory for investigating how collectivity versus shell effects evolve as nuclei depart from shell closure and/or move away from the stability valley.

The experiments performed on several chains of nuclei including Te, Xe, Ba, and Ce isotopes [1] have been supported by theoretical investigations carried out within the quasiparticle-phonon model (QPM) (see [2] for references), large scale shell model [3, 4, 5, 6, 7], and in the nucleon pair approximation [8].

All theoretical studies were focused mainly on the quadrupole collectivity of the lowest isoscalar and mixed symmetry 2^+ states and therefore treated only low-lying levels of low spin.

We intend to explore the possibility of extending the description of the spectroscopic properties of the nuclei in this region by computing the levels of higher spins. Our attention will be concentrated first on neutron-rich even-even Xe isotopes.

These isotopes were already studied within a shell model approach endowed with an importance sampling [9] using the CD-Bonn two-body potential in a large configuration space [7]. The calculation was confined to the levels of spin up to $J^\pi = 6^+$.

Here, we adopt the same approach to generate the yrast line of ^{138}Xe up to $J^\pi = 12^+$. After a brief description of the method we analyze the results by relating them to the available experimental data.

IOP Conf. Series: Journal of Physics: Conf. Series **1023** (2018) 012015 doi:10.1088/1742-6596/1023/1/012015

2. The method

The large-scale shell model under consideration is described in [9, 10]. Here we will outline just the main steps of the algorithm.

Let us consider a symmetric matrix representing a self-adjoint operator $\hat{A}$ in an orthonormal basis $\{| 1\rangle, | 2\rangle, \ldots, | N\rangle\}$. In order to determine the lowest m eigenvectors of the matrix we have to follow an initialization loop and a subsequent set of refinement loops.

*Step 1.*The initialization begins by diagonalizing the matrix (a_{ij}) (i,j $= 1,$n), where n fulfills the relation $(m < n \ll N)$. We choose the lowest m eigenvalues λ_i and eigenvectors $| \phi_i\rangle$, and constructing the new matrix of dimensions $(m + n')$

$$\alpha = \begin{pmatrix} \lambda & b_j \\ b_j & a' \end{pmatrix} \quad , \tag{1}$$

where $\{\lambda\}$ is a diagonal block composed of the eigenvalues $(\lambda_i^n, i = 1, m)$, a' is the new submatrix $a'_{jj'} = \langle j \mid \hat{A} \mid j'\rangle$ with $(j, j' = n+1, n')$, and $\{b_j\}$ is an off-diagonal block connecting λ to a and is composed of the elements $b_{ij} = \langle \phi_i^n \mid \hat{A} \mid j \rangle$ where $(i = 1, m;\ j = n + 1, n')$.

We add the lowest m eigenvalues $\lambda_i^{n'}$ together with the corresponding eigenvectors $| \phi_i^{n'}\rangle$ to a new subset of orthonormal states $| j\rangle$ to build a new matrix and proceed as we did for α. This initialization loop ends when the whole configuration space is exhausted. As a result, a zero order approximation to the lowest eigenvalues and eigenvectors is obtained

$$E_i^{(0)} \equiv \lambda_i^N, \quad | \psi_i^{(0)}\rangle \equiv| \phi_i^N\rangle = \sum_{j=1}^{N} c_j^{(N)} \mid j\rangle \ , \quad \{i = 1, m\}. \tag{2}$$

Step 2. The solutions of the eigenvalue Eqs. (2) are used as an entry to the first refinement loop, which goes through the same steps as described above with one difference. One should just solve an eigenvalue problem of general form since the vectors $| \phi\rangle$ and $| j\rangle$ are no longer orthogonal. It has been shown in [9] that the eigenvalues $E^{(n)}$ and eigenvectors $| \psi^{(n)}\rangle$ obtained after the n-th loop converge to the solution of the exact diagonalization of $\langle\hat{A}\rangle$.

Step 3. The implementation of the method requires an adequate sampling criterion for reducing the size of the configuration space. Bearing in mind that the algorithm provides accurate solutions already after the initialization loop, one can sample the configuration space as follows:

- Diagonalize the submatrix $\{a_{ij}\}$ $(i, j = 1, m)$ and obtain its eigenvalues λ_i;
- For $j = m + 1, \ldots, N$, diagonalize the $m + 1$-dimensional matrix

$$\alpha = \begin{pmatrix} \Lambda_m & \vec{b}_j \\ \vec{b}_j^T & a_{jj} \end{pmatrix}, \tag{3}$$

  where $\vec{b}_j = \{b_{1j}, b_{2j}, \cdot, b_{mj}\}$.

- Accept the new state only if

$$\sum_{i=1,m} \mid \lambda'_i - \lambda_i \mid > \epsilon \quad , \tag{4}$$

  otherwise ignore the state and continue the sampling process with a new vector j. In the relation above ϵ is a small parameter which allows to control the accuracy of the truncation. In the actual calculations we use an upgraded important sampling procedure [11, 12].

The main advantage of the method is that the two key elements – the diagonalization procedure and the importance sampling– are closely related. The algorithm provides robust and always ghost–free solutions and the accuracy of the truncation procedure is fully under control.

3. Calculations

We will describe ^{138}Xe as ^{132}Sn core plus four valence protons in the $(1g_{7/2}, d_{5/2}, 2d_{3/2}, 1h_{11/2}, 3s_{1/2})$ model space and two valence neutrons occupying the $(2f_{7/2}, 3p_{3/2}, 1h_{9/2}, 3p_{1/2}, 2f_{5/2}, 1i_{13/2})$ levels. The effective two–body potential is a renormalized G matrix [13] derived from the CD–Bonn potential [14]. The single-particle energies are the same as in [7]. They are listed in Table 1.

Table 1. Single particle energies in MeV.

Protons		Neutrons	
$1g_{7/2}$	0.00	$2f_{7/2}$	0.00
$2d_{5/2}$	0.96	$3p_{3/2}$	0.85
$2d_{3/2}$	2.71	$1h_{9/2}$	1.56
$1h_{11/2}$	2.80	$3p_{1/2}$	1.66
$3s_{1/2}$	3.50	$2f_{5/2}$	2.00
		$1i_{13/2}$	2.11

In the previous calculation [7] scaling factors for the pairing-like components of the two–body potential were adopted. For the optimal reproduction of the experimental values of the excitation levels up to $J^\pi = 12^+$ for ^{138}Xe [15] we need to scale just the proton–proton $J^\pi = 0^+$ components by the factor 0.85.

We perform the shell model calculations in the m–scheme. It is useful to replace the standard two–body Hamiltonian by the modified one

$$H = H + \alpha[\hat{\mathbf{J}}^2 - J(J+1)] \quad , \tag{5}$$

where α is a positive constant. Due to the additional term, the states with total spin different from J are pushed up in energy for a sufficiently large α. Thus, the diagonalization yields only the low-lying states of a given spin J.

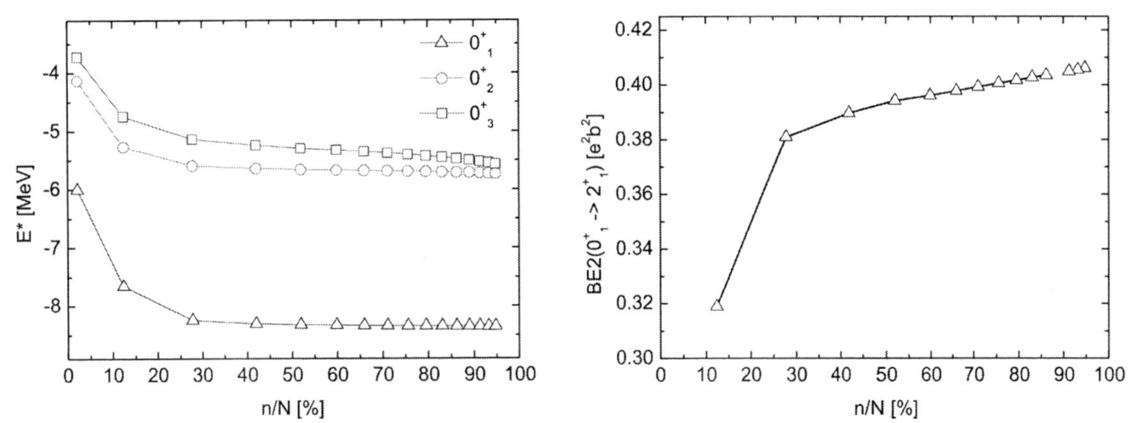

Figure 1. Convergence of the energy of the lowest 0^+ states in MeV and the BE2$(0^+_1 \to 2^+_1)[e^2fm^4]$ value in ^{138}Xe.

Although the shell model problem for ^{138}Xe does not need a severe truncation of the configuration space for any J value, the importance sampling procedure described in Sect. (2)

will be absolutely necessary for the heavier xenon isotopes. Thus we will use the importance sampling to restrict the size of the Hamiltonian and to study the convergence of the method.

First we determine a series of small positive values $\{\epsilon_1 > \epsilon_2 > ... > \epsilon_m\}$ and used them as a sampling criterion in Eq. (4). To each value of ϵ corresponds an unique number of configuration states which determine the Hamiltonian matrix to be diagonalize.

In Fig.1 we present the behavior of the energies of the lowest $J^\pi = 0^+$ states (left panel) and the value of the B($E2; 0^+_1 \to 2^+_1$) (right panel) as functions of the size of the Hamiltonian matrix. It is worthwhile to mention that, during the sampling procedure, all possible configurations are explored. The eigenvalues of the lowest $J^\pi = 0^+$ states and the B(E2) value have a steep exponential trend and, then, tend to their asymptotic values. The energies, however, converge to the exact eigenvalues with just a small fraction of basis states. The convergence of the $E2$ transition rate is considerably slower. The reason for this is that the sampling is based mostly on the energies.

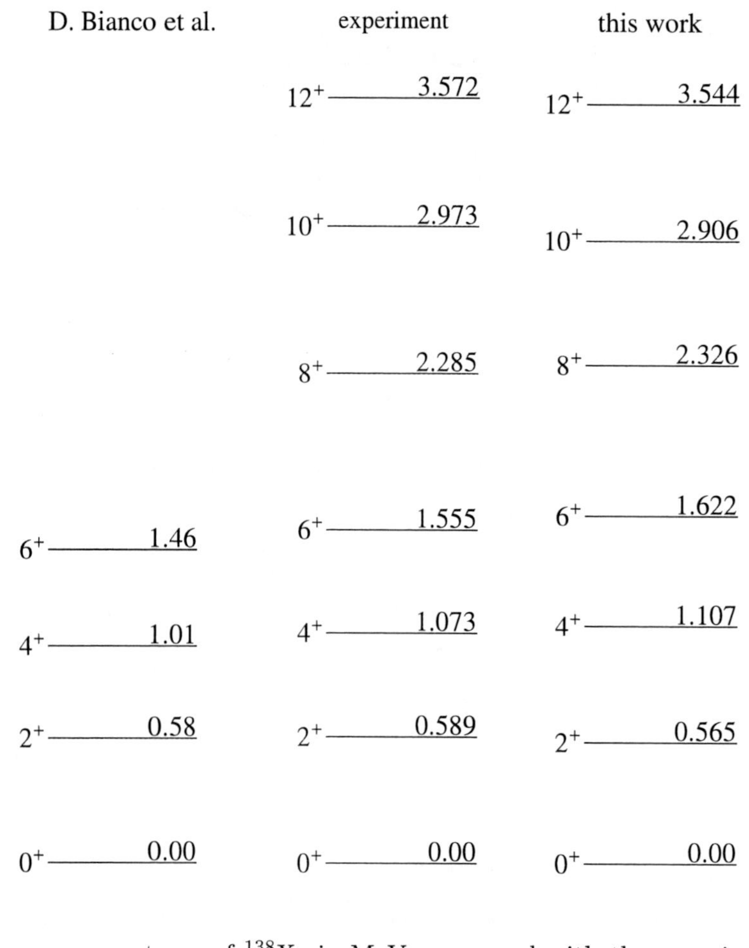

Figure 2. Low energy spectrum of ^{138}Xe in MeV, compared with the experimental data [15] and the previous calculation [7].

4. Results

The energy spectrum of ^{138}Xe is presented in Fig.2. A very good agreement with the experimental data [15] is achieved for the low lying excited states as well as for levels up to

$J^\pi = 12^+$.

Table 2 shows that the calculation reproduces also the experimental $E2$ transition rate [16] improving slightly the previous result [7]. In both cases the adopted effective charges were $e_\pi = 1.6e$ and $e_\nu = 0.7e$.

Table 2. Experimental and theoretical BE2($0_1^+ \rightarrow 2_1^+$)[e^2b^2].

Exp [16]	Bianco [7]	this work
0.40 (10)	0.379	0.406

A further insight into the shell structure of the nucleus is provided by the proton and neutron occupation numbers listed in Table 3. One should notice that our values deviate significantly from the independent particle model rule ($2(j+1)$ for the hole and 0 for the particle states).

There is a large depletion of the proton Fermi sea starting from 0.52 for the 0^+ ground state and decreasing to 0.45 for the first 12^+ state. This means that the excited states are highly correlated. For neutrons, the depletion is considerably smaller. There is also no dependence on the spin of the excited states as for protons.

5. Conclusions

The calculation has shown that the importance sampling algorithm can be extended successfully to high spin states at least in ^{138}Xe. It is, in fact, able to reproduce the yrast line up to $J^\pi = 12^+$ as well as the available experimental $E2$ transition rate The present work is the first step of a project intended to cover low and high spin spectra of the nuclei in the region around ^{132}Sn.

Table 3. Occupation numbers of the single particle levels for ^{138}Xe.

	0_1^+	2_1^+	4_1^+	6_1^+	8_1^+	10_1^+	12_1^+
protons							
$1g_{7/2}$	1.929	1.928	1.960	1.978	1.984	2.077	2.195
$2d_{5/2}$	1.429	1.426	1.431	1.463	1.515	1.464	1.390
$2d_{3/2}$	0.323	0.336	0.331	0.304	0.283	0.278	0.248
$1h_{11/2}$	0.177	0.144	0.118	0.107	0.052	0.032	0.036
$3s_{1/2}$	0.140	0.165	0.159	0.147	0.164	0.148	0.130
depletion	0.52	0.52	0.51	0.50	0.50	0.48	0.45
neutrons							
$2f_{7/2}$	1.241	1.101	1.140	1.232	1.088	1.124	1.165
$3p_{3/2}$	0.342	0.492	0.418	0.316	0.400	0.351	0.223
$1h_{9/2}$	0.096	0.075	0.077	0.094	0.146	0.128	0.213
$3p_{1/2}$	0.098	0.126	0.149	0.096	0.124	0.136	0.067
$2f_{5/2}$	0.176	0.181	0.198	0.243	0.225	0.247	0.322
$1i_{13/2}$	0.046	0.024	0.016	0.016	0.016	0.014	0.009
depletion	0.38	0.45	0.43	0.38	0.46	0.44	0.42

Acknowledgments
This work is partly supported by the DFNI-T02/19 grant of the Bulgarian Science Fond. This financial support is gratefully acknowledged.

References
[1] C. Bauer *et al.* Phys. Rev. C **88**, 021302(R) (2013) and references therein.
[2] N. Lo Iudice, V. Y. Ponomarev, C. Stoyanov, A. V. Sushkov, and V. V. Voronov, J. Phys. G: Nucl. Part. Phys. **39**, 043101 (2012).
[3] N. Shimizu, T. Otsuka, T. Mizusaki, and M. Honma, J. Phys. Conf. Ser. **49**, 178 (2006).
[4] K. Sieja, G. Martnez-Pinedo, L. Coquard, and N. Pietralla Phys. Rev. C **80**, 054311 (2009).
[5] D. Bianco, F.Andreozzi,N. Lo Iudice, A. Porrino, and F.Knapp, Phys. Rev. C **85**, 034332 (2012).
[6] D. Bianco,N. Lo Iudice, F. Andreozzi,A. Porrino, and F.Knapp, Phys. Rev. C **86**, 044325 (2012).
[7] D. Bianco,N. Lo Iudice, F. Andreozzi,A. Porrino, and F.Knapp, Phys. Rev. C **88**, 024303 (2013).
[8] L. Y. Jia, H. Zhang, and Y. M. Zhao, Phys. Rev. C **75**, 034307 (2007).
[9] F. Andreozzi, A. Porrino and N. Lo Iudice, J. Phys. A **35**, L61 (2002).
[10] F. Andreozzi, N. Lo Iudice and A. Porrino, J. Phys. G **29**, 2319 (2003).
[11] D. Bianco, F. Andreozzi, N. Lo Iudice, A. Porrino and F. Knapp, J. Phys. G: Nucl. Part. Phys. **38**, 025103 (2011).
[12] D. Bianco, F. Andreozzi, N. Lo Iudice, A. Porrino and S. Dimitrova, J. Phys.: Conference Series **205**, 12002 (2010);
[13] M. Hjorth-Jensen, T. T. Kuo, and E. Osnes, Phys. Rep. 261, 125 (1995).
[14] R. Machleidt, Phys. Rev. C **63**, 024001 (2001).
[15] Jun Chen, Nucl. Data Sheets **146**, 1 (2017)
[16] Th. Kröll et al., Eur. Phys. J. Special Topics **150**, 127 (2007).

Radii in the *sd* shell and the $1s_{1/2}$ "halo" orbit: A game changer

J. Bonnard[1,2] **and A. P. Zuker**[3,4]

[1] Institut de Physique Nucléaire, CNRS-IN2P3, Université Paris-Sud, Université Paris-Saclay, 91406 Orsay Cedex, France
[2] Istituto Nazionale di Fisica Nucleare, Sezione di Padova, 35131 Padova, Italy
[3] Dipartimento di Fisica e Astronomia, Università degli Studi di Padova, I-35131 Padova, Italy
[4] Université de Strasbourg, IPHC, CNRS, UMR7178, 23 rue du Loess 67037 Strasbourg, France

Abstract. A new microscopic parametrisation of nuclear radii as a functional of single-particle occupation numbers is presented. Its form is inspired by the Duflo-Zuker phenomenological fit which contains a "correlation" term that recently made it possible to understand the isotope shifts of several species as due to unexpectedly large $1s_{1/2}$ and $1p$ orbits [Bonnard J, Lenzi S M and Zuker A P 2016 *Phys.Rev. Lett.* **116** 212501]. It will be shown that the calculated radii for *sd*-shell nuclei reproduce the experimental data better than the most accurate existing fits. These results reveal a very peculiar behaviour of the $1s_{1/2}$ orbit: It is huge (about 1.6 fm bigger than its d counterparts of about 3.5 fm) up to $N, Z = 14$, then drops abruptly but remains some 0.6 fm larger than the d orbits. An intriguing mechanism bound to challenge our understanding of shell formation.

1. Introduction

Proton radii of nuclei in the *sd* shell depart appreciably from the asymptotic $\rho_0 A^{1/3}$ law. The discrepancy exhibits systematic trends fairly well described by a single phenomenological term in the Duflo-Zuker (DZ) formulation [1], which also happened to explain the sudden increase in slope in the isotope shifts of several chains at neutron number $N = 28$ [2, 3]. It was shown that this term is associated to the filling of abnormally large $1s_{1/2}$ and $1p$ orbits in the *sd* and *pf* shells respectively.

Further to explore the question of "halo" orbits, here we propose a microscopic calculation of *sd*-shell nuclei radii. As the (square) radius is basically a one-body operator, it can be represented as a functional of single-particle occupancies which will be determined within the interacting shell model [4]. The only adjustable parameter models a correction to the mean-field radius of the $1s_{1/2}$ orbit whose evolution will be optimized by demanding agreement of the computed radii with experimental values.

In next section we briefly review the DZ fitting formula giving particular attention to the term that led to the detection of "halo" orbits in Ref. [2]. Section 3 starts by introducing the necessary definitions and clarifying some delicate conceptual points (Subsection 3.1) Then comes subsection 3.2 describing a new shell-model interaction used in subsection 3.3 where the microscopic parametrisation is discussed and the fundamental results obtained. Finally, Section 4 is devoted to explain why these preliminary results amount to a "game changer".

2. The Duflo-Zuker phenomenological formula

As self-bound systems, nuclei have volumes that go as the number of particles A. Therefore their radii go as $A^{1/3}$. As both neutrons and protons are present, an isospin dependence is also expected. Duflo and Zuker (DZ) [1] proposed the followings ($t_z = N - Z$, $\rho_\pi^{\rm n}$ and $\rho_\pi^{\rm c}$ stand for naive and correlated radii respectively):

$$\rho_\pi^{\rm c} = \rho_\pi^{\rm n} + \lambda\left(\frac{n(D_\nu - n)}{D_\nu^2} \times \frac{z(D_\pi - z)}{D_\pi^2}\right) A^{-1/3}, \tag{1}$$

$$\rho_\pi^{\rm n} = A^{1/3}\left(\rho_0 - \frac{\zeta}{2}\frac{t_z}{A^{4/3}} - \frac{\upsilon}{2}\left(\frac{t_z}{A}\right)^2\right) e^{g/A}, \tag{2}$$

where n (z) is the number of active neutrons (protons) between the extruder-intruder (EI) magic numbers [4] $N(Z) = 6, 14, 28, 50\ldots$, while $D_{\nu(\pi)} = 8, 14, 22\ldots$are the corresponding degeneracies. In Eq. (2), the exponential term accounts for the larger size of light nuclei, υ measures the overall dilation or contraction as a function of t_z^2, and ζ is responsible for the difference in radii between the fluids, i.e. the neutron skin. Indeed, assuming isospin conservation implies that the neutron radius of a nucleus equals the proton radius of its mirror: $\rho_\nu^{\rm n,c}(A, t_z) = \rho_\pi^{\rm n,c}(A, -t_z)$, which yield for the neutron skin thickness $\Delta r_{\nu\pi} = \rho_\nu^{\rm n,c} - \rho_\pi^{\rm n,c} = \frac{\zeta}{t} A e^{g/A}$. The quality of the results depends little on ζ in the range $0.0 < \zeta < 1.2$ fm. The value $\zeta = 0.8$ fm represents the DZ estimate for $\Delta r_{\nu\pi}$ that agrees very well with *ab initio* and experimental values [2].

The DZ fits reproduce much better the isotope shifts $\delta\langle r_\pi^2\rangle$—relative to a reference nucleus—than the absolute value of the latter that demands reexamination, carried out only in the original DZ paper. In the present study we shall consider two data sets:

- The compilation used in DZ [1], labelled D. It is based on data used in Duflo's original work [5] and (mostly) on the tables of Najdakov and coworkers [6]. To ensure consistency with the isotope shifts 15 absolute values were corrected.
- The recent Angeli-Marinova (AM) compilation [7].

The numbers speak by themselves: The λ contribution (the Duflo term from now on) makes an enormous difference. Its beneficial effects extend to all regions. In particular in Ref. [2] it explains the—hitherto puzzling—increase of slope in the Ca and K isotope shifts [9, 10]. Row 9 in the table is of particular interest. Indeed, leaving out absolute radii unrelated to isotope shifts allows to reduce very significantly the rmsd in row 8. The necessary reexamination of the AM set is under way [8].

Table 1. Results of fits for D and AM: poor for $\rho_\pi^{\rm n}$ from Eq. (2) (rows 1,5-3,7), improve radically for $\rho_\pi^{\rm c}$ from Eq.(1) (rows 2,6-4,8). There are ν data points for $Z \le Z_M$ and $N \ge Z$. Row 9 keeps only those AM values for which isotope shift measures are available. $\zeta = 0.8$ fm is fixed, ρ_0, λ, υ, and root-mean-square deviations (rmsd) in fm.

I	ρ_0	g	λ	υ	Data	ν	Z_M	rmsd
1	0.946	1.422	-	0.550	D	636	96	0.0299
2	0.942	0.948	6.857	0.526	D	636	96	0.0132
3	0.950	1.232	-	0.312	D	88	30	0.0415
4	0.944	0.985	5.562	0.368	D	88	30	0.0176
5	0.942	1.513	-	0.312	AM	876	96	0.0365
6	0.940	0.879	7.719	0.297	AM	876	96	0.0203
7	0.947	1.370	-	0.295	AM	107	30	0.0419
8	0.940	1.140	5.208	0.334	AM	107	30	0.0246
9	0.930	1.659	4.499	0.379	AMr2	80	30	0.0176

3. Microscopic approach

3.1. Definitions and conceptual points

Let us consider two square radius operators defined through their expectation values in a stationary state as

$$\langle r_{\rm ho}^2\rangle = \frac{41.47}{\hbar\omega}\sum_i m_i(p_i+3/2)/A \qquad (\sim(\rho_\pi^{\rm n})^2) \tag{3a}$$

$$\langle r_\pi^2\rangle = \frac{41.47}{\hbar\omega}\sum_i m_i(p_i+3/2+\delta_i)/A \qquad (\sim(\rho_\pi^{\rm c})^2), \tag{3b}$$

where m_i and p_i represent the total occupancy and principal quantum number of the orbit $i=(p_il_ij_i)$, and $41.47(p_i+3/2)/\hbar\omega$ fm^2 is the harmonic oscillator (ho) value of the square radius for any single-particle wave function belonging to the major shell. Such estimates of radii provide a direct reading of the naive and correlated ($\rho_\pi^{\rm n}$ and $\rho_\pi^{\rm c}$) patterns shown in the left panel of Figure 1 that illustrates the improvement brought about by Duflo's λ term. Its effects are identified with corrections δ_i to the radii of harmonic orbitals. Note that under this guise, i.e. without any additional assumption regarding the δ_i, r_π^2 is a fully general expression for the proton square radius operator.

Eqs. (3) induce two possible expressions for the oscillator frequency

$$\hbar\omega_x = \frac{41.47}{\langle r_x^2\rangle}\sum_i m_i(p_i+3/2)/A \tag{4a}$$

$$\hbar\omega = \frac{41.47}{\langle r_\pi^2\rangle}\sum_i m_i(p_i+3/2+\delta_i)/A, \tag{4b}$$

with x a generic index. $\hbar\omega$ enters shell-model calculations through the two-body matrix elements, taken to scale as $V(\hbar\omega)=V(\hbar\omega_r)\hbar\omega/\hbar\omega_r$ where $\hbar\omega_r$ is an arbitrary reference value. Its evolution with A depends on the choice of the $\langle r_x^2\rangle$ denominator. For $x=$ ho, an asymptotic estimate of the sum in Eq. (4a) with the very schematic form $\langle r_x^2\rangle \propto A^{1/3}$

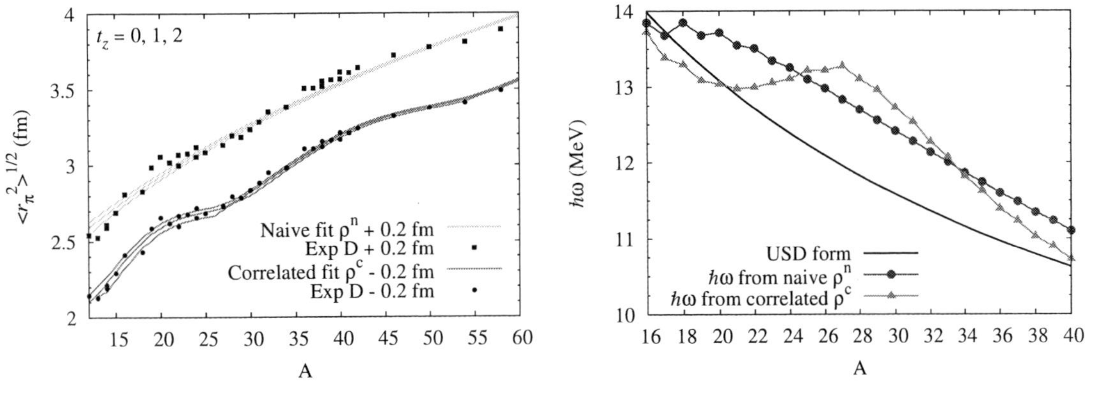

Figure 1. (color online) Left panel: Influence of the Duflo λ term. Fits of the D data, for nuclei with $t_z=0,1,2$ and $Z\leq 30$, using $\rho_\pi^{\rm n}$ and $\rho_\pi^{\rm c}$, labeled naive and correlated respectively. Right panel: Evolution of $\hbar\omega$ for naive, correlated and USD assumptions. Only $t_z=0$ and 1 cases are included, merged in a single curve which leads to the weak staggering observed.

leads to $\hbar\omega = 35.6A^{1/3}/\langle r_x^2\rangle \approx 40A^{-1/3}$, the classic result [11, Eq.(2.157)] on which shell-model calculations traditionally rely to ensure correct saturation properties. A more precise alternative is a nucleus by nucleus evaluation of the sum while resorting to the naive fit Eq. (2), $\langle r_x^2\rangle \equiv (\rho_\pi^{\rm n})^2$. A more audacious one relies on Eq. (1). It amounts to assume that the effect of the δ_i modifications (see Eq. (4b)) is "transferred" to $\hbar\omega_x$ by spreading it among the major oscillator shell, which is achieved by replacing in Eq. (4a) the natural $\langle r_{\rm ho}^2\rangle$ denominator by the (closer to) exact one $\langle r_\pi^2\rangle$ approximated *via* the correlated radii (1). The two variants are plotted in the right panel of Fig. 1, where the USD form $\hbar\omega(A) = \hbar\omega(A = 18)(18/A)^{0.3}$ is also shown [12]. Both the USD-like and correlated propagation will be considered in the following calculations.

The use of the notation $\langle r_\pi^2\rangle$ stresses the obvious: We are dealing exclusively with proton radii. What may not be so obvious is that Eq. (3b) is written in terms of $m_i = n_i + z_i$ and no reference whatsoever is made to z_i or n_i, the number of protons or neutrons, while Eq. (2) does contain a term in $t_z = N - Z$. However (as already shown in [1]) for any value of ζ in Eq. (2) one could find a value of υ associated to t_z^2 that would produce the same rmsd. Furthermore, the all important Duflo term is an isoscalar. Finally the neutron an proton radii are necessarily very close: Otherwise it would be impossible to explain why the structure of the isotope shifts are dictated by the *neutron* filling patterns, while what is measured are radii for *protons* whose number remains fixed [2]. Hence we *must* use an isospin representation. But this means that the expression for $\langle r_\pi^2\rangle$ in Eq. (3b) must be supplemented by an isovector term in $(t_z)_i = n_i - z_i$, which—at this point—cannot be determined, and is therefore ignored: A limitation of this study, unlikely to alter the central result of Section 3.3 as such a term contribute at next order only in t_z/A.

3.2. The monopole corrected interaction

This section is included for the sake of completeness. Indeed, it needs revisiting as our view of the corrective action demanded by realistic interactions will be affected by the findings in next section.

The monopole corrected interaction MCI is derived from the N3LO potential [13] at $\hbar\omega_0 = 14$ MeV, and a $V_{\text{low-}k}$ treatment [14] with a cut-off $\lambda = 2$ fm^{-1} in momentum space, as part of the no-core project outlined in Ref. [2]. Renormalization amounts to an overall 1.1 multiplicative factor and a 30% boost of the quadrupole force [15] plus the monopole corrections defined below. The whole is scaled as mentioned previously with $\hbar\omega$ from the correlated form in the right panel of Fig. 1.

The monopole corrections are given in terms of operators in the "invariant representation" [16], in which the one- and two-body number operators m_t and $m_t m_u$ are separated into a term that contains only the total number operator m, single-particle terms $\Gamma_t^{(1)}$, and two-body terms $\Gamma_{tu}^{(1)}$:

$$m_t \equiv m + \Gamma_t^{(1)}, \quad \frac{m_t(m_u - \delta_{tu})}{1+\delta_{tu}} \equiv \frac{1}{2}m^{(2)} + (m-1)\Gamma_t^{(1)} + \Gamma_{tu}^{(2)}, \tag{5}$$

where $x^{(k)} = x(x-1)\ldots(x-k+1)$. For our purpose it is convenient to associate each orbit t with its complement ct containing all orbits in the space except t. The degeneracy of t is $D_t = 2(2j_t+1)$ and that of its complement $D_{ct} = D - D_t$, D being the total degeneracy of the valence space. In the *sd* shell, orbits $d_{5/2}$, $s_{1/2}$ and $d_{3/2}$ will be called 5, 1, 3 respectively. The complement of 5, say, is $c5 = 1+3$. With this convention:

$$\Gamma_t^{(1)} = \frac{m_t D_{ct} - m_{ct} D_t}{D}, \quad \Gamma_t^{(2)} = \left(\frac{m_t^{(2)}}{D_t^{(2)}} + \frac{m_{ct}^{(2)}}{D_{ct}^{(2)}} - \frac{2m_t m_{ct}}{D_t D_{ct}}\right)\frac{D_t^{(2)} D_{ct}^{(2)}}{(D_t + D_{ct})^{(2)}}. \tag{6}$$

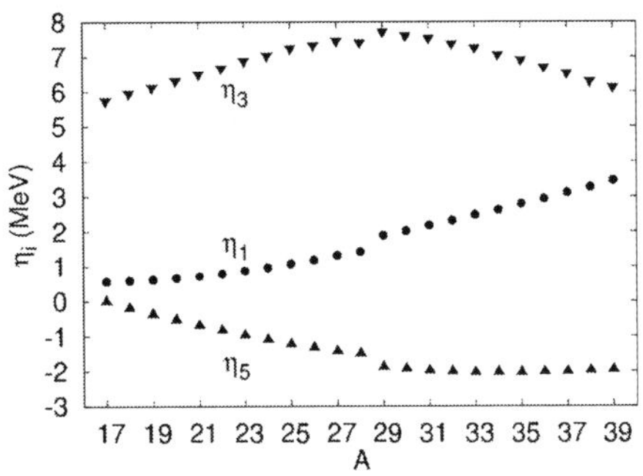

Figure 2. Propagation of the MCI single-particle energies η_i from the first line of Eq. (7).

The monopole corrections finally take the form

$$V_{\rm MC} = e_5\Gamma_5^{(1)}(m-1) + 0.06\Gamma_1^{(1)}(m-1) + 2.0\Gamma_3^{(1)}(m-1)(\overline{m}-1)/(D/2-1)^2 \quad (7)$$

$$+ (1.0\Gamma_5^{(2)} + 0.5\Gamma_3^{(2)})\overline{m}/D, \quad (8)$$

where $e_5 = -0.05$ if $A \leq 28$, -0.11 if $A > 28$. All coefficients in MeV. The sum in the first line defines propagated single-particle operators η_t such that $\sum_t \eta_t D_t = 0$ (see Fig. 2). Note that the terms containing $\overline{m}-1$ and $\overline{m}$ amount to a three body contribution. Under particle-hole transformations $m_t \to \overline{m_t} = D_t - m_t$, $\Gamma_t^{(1)}$ changes sign and $\Gamma_t^{(2)}$ is invariant.

The description of spectra is quite satisfactory for the yrast states, not as good as USDa for the rest, with the exception of ^{24}Mg and ^{28}Si where MCI does better. Figure 2 indicates that the fit detects a—premonitory—abrupt change at $A = 28$.

3.3. Microscopic parametrisation of nuclear radii

We know that the origin of the Duflo term, Eq. (1) must be found in the "halo" orbits. Concentrating on the *sd* shell, the $1s_{1/2}$ orbit is taken to be solely responsible for the fluctuations. Accordingly, setting $\delta_i = 0$ for $i = 0d_{5/2,3/2}$ and $\delta_i = \delta$ for $i = 1s_{1/2}$, and following Eq. (3b), the proposed functional reads

$$(\rho_\pi^{\rm m})^2 = \frac{41.47}{\hbar\omega_0}\sum_i m_i(p_i + 3/2 + \delta_i)/A = \frac{41.47}{\hbar\omega_0}\sum_i m_i(p_i + 3/2)/A + \frac{41.47}{\hbar\omega_0} m_{s_{1/2}}\delta_i/A. \quad (9)$$

To determine $\hbar\omega_0$ we refer to Fig. 3 where the naive part is not obtained by fitting Eq. (2) but Eq. (1) and keeping only the $\lambda = 0$ part, i.e. with $\langle r_\pi^2\rangle = [\rho_\pi^{\rm c}(\lambda = 0)]^2$ in Eq. (4a). Therefore, (9) reduces to

$$(\rho_\pi^{\rm m})^2 = \rho_\pi^{\rm c}(\lambda = 0) + \frac{41.47}{\hbar\omega_0}\frac{m_{s_{1/2}}\delta}{A}. \quad (10)$$

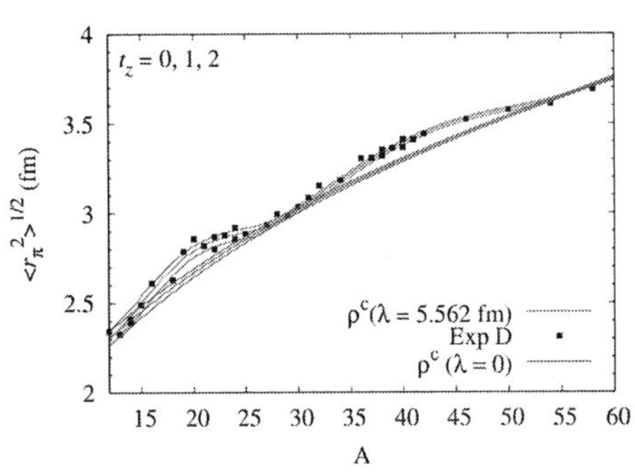

Figure 3. (color online) As Fig. 1 (left panel) with the naive fit ($\lambda = 0$) replaced by the result of using the parameters of the correlated fit setting $\lambda = 0$, which ensures that the corrective term is definite positive.

Exp. set	Duflo		Angeli-Marinova	
Interaction	USDa	MCI	USDa	MCI
$\delta_<$	4.90	4.25	5.50	4.80
$\delta_>$	1.40	1.35	1.45	1.35
rmsd (fm)	0.023	0.020	0.023	0.018

Table 2. Results for the optimal δ's for D and AM sets of data, and for the USDa and MCI interactions. The rmsd are calculated for *sd*-shell nuclei with $t_z = 0, 1$, and 2 (21 in D, 23 in AM).

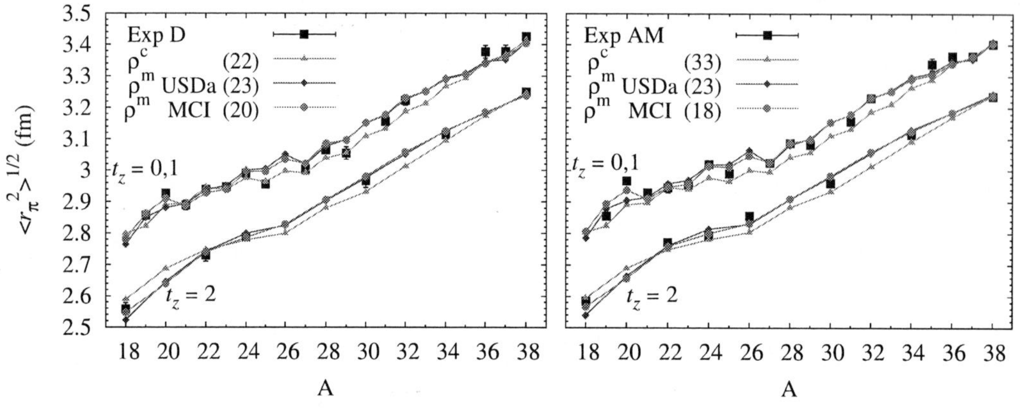

Figure 4. (color online) Solution of Eq. (9): Behaviour of $\rho_\pi^{\rm m}$ for USDa and MCI calculations compared with data sets D (left), AM (right), and associated correlated fits $\rho_\pi^{\rm c}$. The $t_z = 0$ and 1 cases are conflated in a single curve. These and $t = 2$ values are shifted by ± 0.07 fm, respectively, for clarity. In parenthesis the rmsd (10^{-3} fm) for the different cases (see Table 2).

At this stage, the evolution of δ with A and t_z may be extracted from a nucleus-by-nucleus adjustment. Here, instead, we simply adopt a step function discontinuous at the EI closure:

$$\delta = \begin{cases} \delta_< & \text{if} \quad N \text{ and } Z \quad < 14 \\ \delta_> & \text{if} \quad N \text{ or } Z \quad \geq 14. \end{cases} \tag{11}$$

Finally, the occupancies are determined by diagonalization of the USDa and MCI Hamiltonians.

A search for the optimal parameters minimizing the rmsd of radii yields the values in Table 2 leading to Figures 4, in which the microscopically calculated radii are compared with the phenomenological ones obtained with the Duflo term. It appears that the microscopic results reproduce the observed radii at least as well as the phenomenological fit for the D set, and much better in the AM case. The rmsd for MCI is smaller than for USDa, and so are the calculated sizes of the $1s_{1/2}$ orbit deduced from Table 2 and shown in Figure 5. The quantitative differences between the results from the different interactions are not really relevant when compared to the fundamental message.

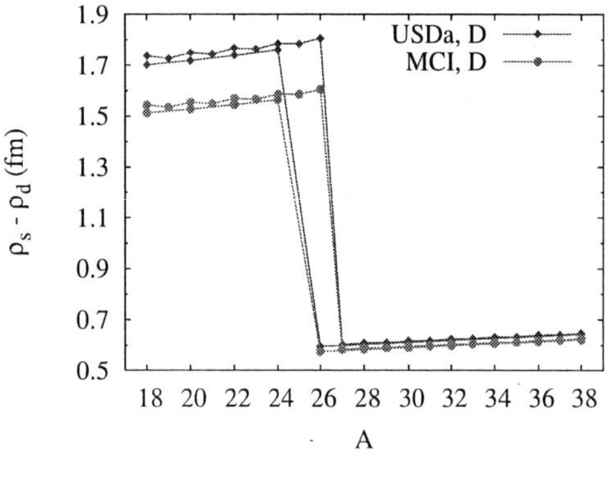

Figure 5. (color online) Differences $\rho_s - \rho_d$, where $\rho_i^2 = \langle im_{j_i}|r^2|im_{j_i}\rangle = \frac{41.47}{\hbar\omega_0}(p_i + 3/2 + \delta_i)$, between the root-mean-square radii of the $1s_{1/2}$ and $0d$ orbits from Eq. (10). The mean values (in fm) associated to the upper and lower δ in Table 2 are (1.56, 0.6) for MCI D, (1.75, 0.62) for USDa D (shown); and (1.76,0.6) for MCI AM, (1.93,0.69) for USD AM (not shown).

4. Conclusion: The game changer

The halo orbits are the essential ingredient in halo nuclei [17]. What we are finding is that they have a pervasive influence throughout. Their description will most likely require non-local potentials [18, 19]. At present only relativistic mean-field calculations seem capable of reproducing the slope increase in isotope shifts (as in Ref. [9] using the DD-ME2 interaction[20]).

The shell-model theory is conducted under the tacit assumption that harmonic oscillator wave functions approximate well enough the single-particle behaviour, which is tenable for conventional Hartree-Fock or Woods-Saxon calculations, but not for halo orbits. This may explain why some USD matrix elements are impossible to reproduce using perturbation theory on realistic ones calculated with oscillator orbits, in particular for the $JT = 20$ state [15] or the coefficient of the $\Gamma_5^{(2)}$ term in Eq. (8).

Which brings us to the game changer: The fundamental problems of saturation and shell formation may not be due to limitations of the potentials but to limitations of the calculations that fail to produce the halo orbits. To fix ideas: Why in our N3LO matrix elements do we use a cut-off $\lambda = 2$ fm^{-1} and not $\lambda = 4$ fm^{-1}? Because then the calculations would become very hard. A problem that is ignored, but not solved with existing, Brueckner or $V_{\text{low-}k}$, approaches.

Therefore, could it not be that the observed magic closures are not associated to self-binding of the orbits of largest j in a major oscillator shell—due to three-body forces—but, instead, to unbinding of the "halo" orbits above them? It sounds improbable, but as Sherlock Holmes put it *Once you eliminate the impossible, whatever remains, no matter how improbable, must be the truth.*

References

[1] Duflo J and Zuker A P 2002 *Phys. Rev. C* **66** 051304
[2] Bonnard J, Lenzi S M and Zuker A P 2016 *Phys. Rev. Lett.* **116** 212501
[3] Heylen H *et al* 2016 *Phys. Rev. C* **94** 054321
[4] Caurier E, Martnez-Pinedo G, Nowacki F, Poves A and Zuker A P 2005 *Rev. Mod. Phys.* **77** 427
[5] Duflo J 1994 *Nucl. Phys. A* **576** 29
[6] Nadjakov E, Marinova K and Gangrsky Y 1994 *Atomic Data and Nuclear Data Tables* **56** 133
[7] Angeli I and Marinova K P 2013 *Atomic Data and Nuclear Data Tables* **99** 69
[8] Zuker A P 2016 *unpublised*
[9] Kreim K *et al* 2014 *Phys. Lett. B* **731** 97
[10] García Ruiz R F *et al* 2016 *Nat. Phys.* **12** 594598
[11] Bohr A and Mottelson B R 1969 *Nuclear Structure, vol. I* (New York: Benjamin)
[12] Brown B A and Richter W A 2006 *Phys. Rev. C* **74** 034315
[13] Entem D R and Machleidt R 2002 *Phys. Lett. B* **524** 93
[14] Bogner S K Kuo T T S and Schwenk A 2003 *Phys. Rep.* **386** 1
[15] Dufour M and Zuker A P 1996 *Phys. Rev. C* **54** 1641
[16] Abzouzi A, Caurier E and Zuker A P 1991 *Phys. Rev. Lett.* **66** 1134
[17] Tanihata I, Savajols H and Kanungo R 2013 *Progress in Particle and Nuclear Physics* **68** 215
[18] Perey F and Buck B 1962 *Nucl. Phys.* **32** 353
[19] Rawitscher G H 2012 *Nucl. Phys. A* **886** 1
[20] Lalazissis G A and Nikšić T, Vretenar D and Ring P 2005 *Phys. Rev. C* **71** 024312

IOP Conf. Series: Journal of Physics: Conf. Series **1023** (2018) 012017 doi:10.1088/1742-6596/1023/1/012017

The Small-Angle Neutron Scattering Data Analysis of the Phospholipid Transport Nanosystem Structure

E.V. Zemlyanaya[1,2,†], M.A. Kiselev[1,2,‡], E.I. Zhabitskaya[1,2], V.L.Aksenov[1,3], O.M. Ipatova[4], O.I.Ivankov[1]

[1]Joint Institute for Nuclear Research, Dubna 141980, Russia
[2]State University "Dubna", Dubna 141980, Russia
[3]NRC "Kurchatov Institute", Gatchina,
[4]V.N. Orekhovich Research Institute for Biomedical Chemistry, Moscow, Russia

E-mail: [†]elena@jinr.ru, [‡]kiselev@jinr.ru

Abstract. The small-angle neutron scattering technique (SANS) is employed for investigation of structure of the phospholipid transport nanosystem (PTNS) elaborated in the V.N.Orekhovich Institute of Biomedical Chemistry (Moscow, Russia). The SANS spectra have been measured at the YuMO small-angle spectrometer of IBR-2 reactor (Joint Institute of Nuclear Research, Dubna, Russia). Basic characteristics of polydispersed population of PTNS unilamellar vesicles (average radius of vesicles, polydispersity, thickness of membrane, etc.) have been determined in three cases of the PTNS concentrations in D_2O: 5%, 10%, and 25%. Numerical analysis is based on the separated form factors method (SFF). The results are discussed in comparison with the results of analysis of the small-angle X-ray scattering spectra collected at the Kurchatov Synchrotron Radiation Source of the National Research Center "Kurchatov Institute" (Moscow, Russia).

1. Introduction

The small-angle scattering of neutrons (SANS) and of X-rays (SAXS) is the well known experimental technique that uses the elastic neutron scattering and the X-ray scattering at small angles to study the structure of various substances at a scale of about 1-100 nm [1, 2]. This technique is widely utilized for investigation of unilamellar phospholipid vesicles (ULVs) with different chemical compositions in large water excess, see [3, 4, 5, 6] and references therein. Investigations of the structure and properties of multilamellar and unilamellar vesicles are of great interest because they can provide an insight to the fundamental properties of biological and artificial lipid membranes. This knowledge can be of practical use in biochemistry and pharmacology [7, 8].

An efficient tool for analysis of the SANS and SAXS spectra is the separated form factors (SFF) method developed in [9, 10]. The SFF method was successfully applied for a study of the structure of dimyristoylphosphatidylcholine (DMPC) ULVs in different phase states of the lipid bilayer [10], at different concentrations of aqueous sucrose solutions [11]. The SFF-based approach was employed to study the multicomponent ULV systems based on ceramide 6 [11, 12, 13, 14]. In [15], the structure of dipalmitoylphosphatidylcholine ULVs in heavy water in the dimethyl sulfoxide presence has been analysed on the basis of the SFF method.

In this work, on the basis of the SFF analysis of SANS experimental data, we study a structure of ULVs of the phospholipid delivery nanosystem (PTNS) obtained in the V.N.Orekhovich Research Institute of Biomedical Chemistry (Moscow, Russia). Due extremely small size of PTNS nanoparticles, the medical drugs incorporation into PTNS ULVs sufficiently improves their therapeutic effectiveness [16, 17]. Estimations of basic parameters of PTNS ULVs within the frame of different approaches are presented in [18, 19, 20]. In our recent work [21], the structure of PTNS polydispersed population of ULVs in water solvent of maltose (concentration 20%, 25%, and 30%) has been analysed on the basis of the SAXS experimental data collected at the Kurchatov Synchrotron Radiation Source of the National Research Center "Kurchatov Institute" (Moscow, Russia).

Here, we investigate the polydispersed population of PTNS ULVs in heavy water solvent of maltose on the basis of the SFF analysis of the SANS spectra collected at the YuMO facility of the IBR-2 reactor (Frank Laboratory of Neutron Physics, Joint Institute for Nuclear Research, Dubna).

2. The Separated Form Factors Method

In case of monodispersed population of ULVs with radius R and the bilayer thickness d, the SANS intensity is determined as follows [22]:

$$I_{\text{mon}}(q) = n\, I_o\, A^2(q)\, S(q), \quad A(q) = 4\pi \int_0^\infty \rho_c(r) \cdot \frac{\sin(qr)}{qr} r^2\, dr. \tag{1}$$

Here I_o is the incident beam intensity, n – number of vesicles in cm^3, A – the small-angle scattering amplitude, q – the scattering vector, $\rho_c(x) = \rho - \rho_0$ – the contrast between the bilayer scattering length density ρ and the solvent scattering length density ρ_0, S – the structural factor depending on n and on the ULV radius R [22].

In case $R >> d$, the factorization of the scattering amplitude formula yields the following form of intensity:

$$I_{\text{SFF}}(q) = n\, I_o\, F_s(q, R)\, F_b(q, d)\, S(q), \tag{2}$$

where $F_s(q, R)$ is the form factor of the spherical surface with radius R

$$F_s(q, R) = \left(4\pi R^2 \frac{\sin(qR)}{qR}\right)^2, \tag{3}$$

and $F_b(q, d)$ – the form factor of the symmetric lipid bilayer of thickness d:

$$F_b(q, \Theta_b) = \left(\int_{-d/2}^{+d/2} \rho_c(x) \cos(qx) dx\right)^2. \tag{4}$$

The SFF method simplifies the numerical analysis and extends a chose of model of the scattering length density distribution across bilayer. Calculations in [21] showed that both "classical" intensity formula (1) and the SFF expression (2) provide the same results of analysis of the PTNS structure, in spite of the small radius of PTNS ULVs. Hence, the SFF approach is shown to be applicable for the further investigations of PTNS ULVs.

Polydispersity σ of the ULV average radius $\widetilde{R}$ is accounted for by means of the Schulz distribution with parameter m:

$$G\left(\widetilde{R}, R\right) = \frac{\widetilde{R}^m}{m!} \left(\frac{m+1}{R}\right)^{m+1} \exp\left[-\frac{(m+1)\widetilde{R}}{R}\right], \quad \sigma = \frac{1}{\sqrt{m+1}} \tag{5}$$

IOP Conf. Series: Journal of Physics: Conf. Series **1023** (2018) 012017 doi:10.1088/1742-6596/1023/1/012017

Then the intensity I_m of the polydispersed ULV system is obtained by means of the standard convolution procedure:

$$I_m(q) = \frac{\int_{R_{\min}}^{R_{\max}} I_{SFF}(q,R) G\left(\widetilde{R}, R\right) dR}{\int_{R_{\min}}^{R_{\max}} G\left(\widetilde{R}, R\right) dR}. \tag{6}$$

The final expression of intensity $I(q)$ has the following form:

$$I(q) = I_m(q) + \frac{1}{2}\Delta^2 \frac{d^2 I_m(q)}{dq^2} + I_B, \tag{7}$$

where I_B is the incoherent background parameter and Δ^2 – the second momentum of the resolution spectrometer function.

The average radius $\widetilde{R}$, the polydispersity parameter m, the number of vesicles per volume unit n, the incoherent background factor I_B, the bilayer thickness d, and the parameters characterizing the internal bilayer structure, are adjusted by minimization of the discrepancy between theoretical and experimental values of intensity:

$$\chi^2 = \frac{1}{N-k} \sum_{j=1}^{N} \left(\frac{I(q_j) - I^{\text{exper}}(q_j)}{\delta^{\text{exper}}(q_j)} \right)^2, \tag{8}$$

where N – number of experimental points, k – number of adjusted parameters, I is determined by Eq. (7), I^{exper} – experimantal values of intensity, δ^{exper} – experimental errors. Depending on value of k, we employ one of two fitting procedures. The first one is based on the generalized least square method [23, 24]. The second approach is the global minimization algorithm of the asynchronous differential evolution, see [25] and references therein for details.

3. Experiment

SANS spectra were measured on a YuMO small-angle neutron spectrometer at the IBR-2 pulsed reactor (Frank Laboratory of Neutron Physics of the Joint Institute for Nuclear Research, Dubna, Russia). The measurements were performed at the room temperature of PTNS ULV sample. The samples of polydispersed populations of PTNS ULVs were prepared by means of the dissolution of the lyophilized PTNS powder in the heavy water at concentrations 5%, 10%, and 25% (w/w). The concentration of maltose after dissolution of the drug was, respectively, 4%, 8%, and 20%.

4. Results and discussion

Results of the SFF-SANS analysis of PTNS ULVs are presented in Fig. 1 and Table 1. In Fig. 1(a-c), the calculated curves $I(q)$ are plotted in comparison with experimental data. Respective values of PTNS ULVs basic parameters obtained within the frame of the SFF approach, are given in Table 1. The calculation has been done with the "step" model of the scattering length density distribution across bilayer (see Fig. 1(d)). In general, the SANS-based estimations are in agreement with the SAXS-based results obtained in [21]. At 20% concentration of maltose where we have both SANS and SAXS data, the discrepancy between SANS-based and SAXS-based values of $\widetilde{R}$ and d is about 10%.

Figure 2 demonstrates the SANS and SAXS estimations of the average radius $\widetilde{R}$ and the bilayer thickness d of PTNS ULVs versus the disaccharide (maltose) concentration. SANS estimations are shown by blue squares, SAXS estimations – by blue circles. For comparison, the same characteristics versus the disaccharide (sucrose) concentration are given for the "classical"

IOP Conf. Series: Journal of Physics: Conf. Series **1023** (2018) 012017 doi:10.1088/1742-6596/1023/1/012017

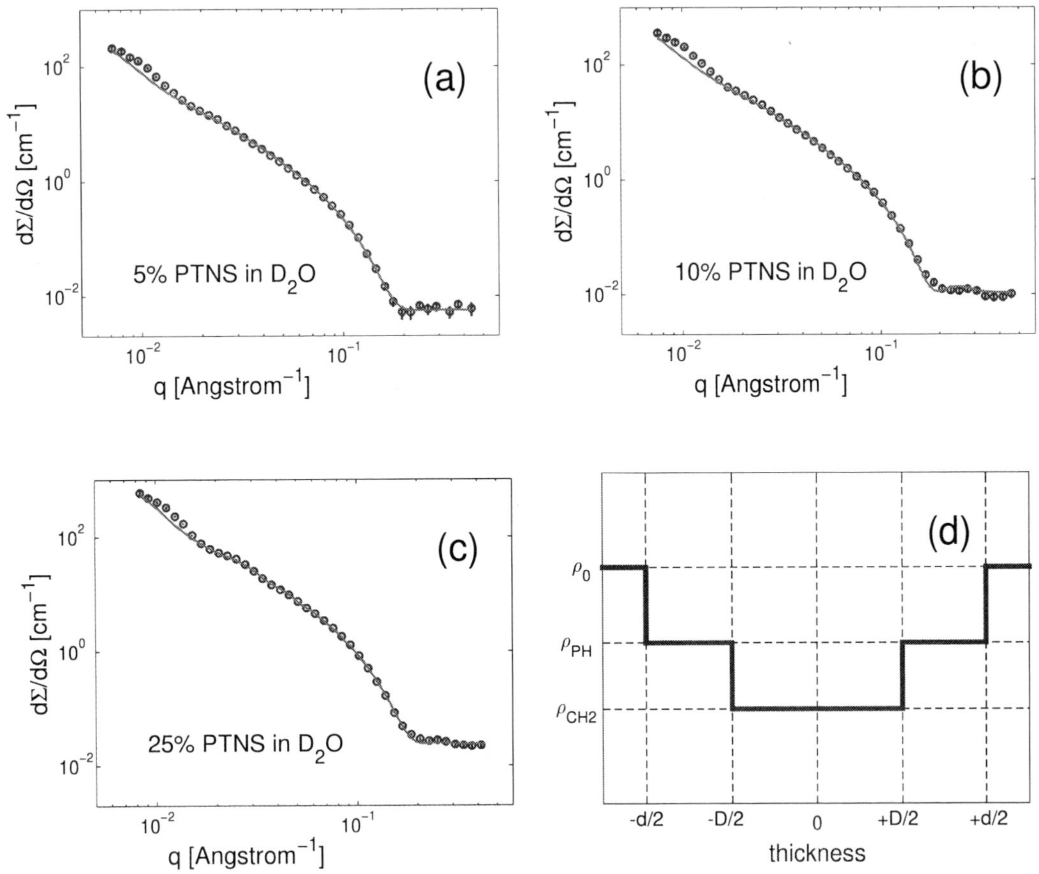

Figure 1. (a–c) The SANS spectra for the PTNS ULVs in D_2O. Circles – experiment, solid lines – theory. (d) The "step" model of the scattering length density distribution across bilayer. Here ρ_{CH2} – density of the hydrocarbon chains region, ρ_{PH} – density of the polar head region, ρ_0 – the solvent density, d – thickness of the ULV bilayer, D – thickness of the hydrocarbon chains region in the bilayer.

Table 1. PTNS ULVs in D_2O: results of the SFF-fitting of SANS data at 5–25% concentration of PTNS.

Concentration of PTNS	Concentration of maltose	R, Å	σ, %	d, Å	D, Å	χ^2
5%	4%	191.5±3.3	41	45.9±1.9	13.0±0.5	6.7
10%	8%	199.6±2.8	41	46.7±1.8	17.0±1.0	7.5
25%	20%	192.2±1.5	34	38.9±3.4	18.0±1.9	8.2

size DMPC ULVs (red trianlges). It is seen that in case of the weak maltose concentration, the PTNS ULV average radius $\widetilde{R}$ is about 200±5 Å that is 1.5 times smaller than the radius of "classical" vesicles of DMPC [11]. Also, one can see that the radius of both DMPC and PTNS ULVs becomes smaller as the disaccharide concentration is growing. As for the PTNS ULV bilayer thickness, its value changes from 45.9±1.9 Å in case of 4% maltose concentration to

37.0±0.2 Å at 30% concentration. Also, it is seen that the values of d_{PTNS} and d_{DMPC} are close to each other.

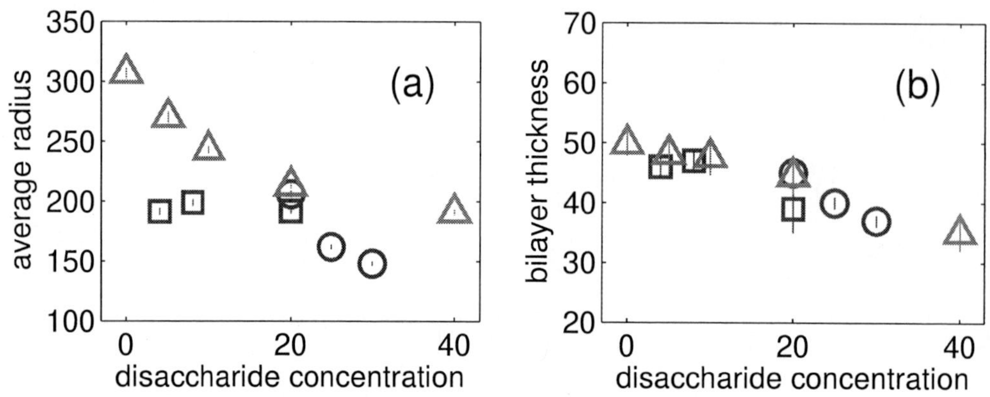

Figure 2. SFF analysis of SAXS and SANS spectra: average radius (a) and the thickness of bilayer (b) of PTNS ULVs (blue squares – SANS estimations, blue circles – SAXS estimations) and of DMPC ULVs (red triangles) versus, respectively, maltose and sucrose concentration (%). Both the average radius and the bilayer thickness are in Å.

5. Summary

- The SFF method has been employed for the SANS analysis of the polydispersed PTNS ULVs structure.
- Basic parameters of PTNS vesicles (average size of vesicles, thickness of bilayer, polydispersity of average radius) have been determined depending on the maltose concentration in the heavy water.
- Estimations on the basis of SANS and of SAXS data are shown to be in the reasonable agreement.
- In case of the weak maltose concentration (5–20%), the average radius of PTNS ULVs is 1.5 times smaller the standard size of the "classical" size vesicles of DMPC.

Acknowledgments

The work has been performed under the grant of Russian Scientific Foundation (project No 14-12-00516).

References

[1] Kiselev M A, Lesieur P, Kisselev A M, Lombardo D, Killany M, Lesieur S, Ollivon M 2001 *Nuclr. Inst. & Method* **A 470** 409
[2] Kiselev M, Lesieur P, Kisselev A, Lombardo D, Killany M, Lesieur S 2001 *J. Alloys and Compounds* **328** 71
[3] Kiselev M A, Wartewig S, Janich M, Lesieur P, Kisselev A M, Ollivon M, Neubert R 2003 *Chemistry and Physics of Lipids***123** (2003) 31
[4] Kiselev M A, Zbytovska J, Matveev D, Wartewig S, Gapienko I V, Perez J, Lesieur P, Hoell A, Neubert R 2005 *Colloids and Surfaces* **A 256** 1
[5] Kiselev M A 2011 *Phys. Part. Nucl.* **42** (2), 302
[6] Nagle J F, Tristram-Nagle S 2000 *Biochim. Biophys. Acta* **1469** 159
[7] Das S, Chaudhury A 2011 *AAPS PharmSci. Tech.* **12** 62

[8] Martins S, Sarmento B, Ferreira D C, Souto E B 2007 *Int. J. Nanomedicine* **2** 595
[9] Kiselev M A, Lesieur P, Kisselev A M, Lombardo D, Aksenov V L 2002 *J. Appl. Phys.* **A 74** S1654
[10] Kiselev M A, Zemlyanaya E V, Aswal V K, Neubert R 2006 *European Biophys. J* **35** 477
[11] Kiselev M A, Zemlyanaya E V, Zhabitskaya E I, Aksenov V L 2015 *Crystallography Reports***60** 143
[12] Zemlyanaya E V, Kiselev M A, Zbytovska J, Almasy L, Aswal V K, Strunz P, Wartevig S, Neubert R H H 2006 *Crystallogr. Rep.* **51** (Suppl 1) S22
[13] Zemlyanaya E V, Kiselev M A, Noibert R, Kohlbrecher J, Aksenov V L 2008 *J. Surface Investigation* **2** (6) 884
[14] Kiselev M A, Zemlyanaya E V, Ryabova N Y, Hauss T, Almasy L, Funari S, Zbytovska J, Lombardo D 2014 *Appl. Phys.* **A 116** 319
[15] Kiselev M A, Zemlyanaya E V 2017 *Crystallogr. Rep.* **62** (5) 795
[16] Archakov A I, Guseva M K, Uchaikin V F, Ipatova O M, Tikhonova E G, Medvedeva N V, Lisitsa A V, Prozorovskij V N, Strekalova O S, Shironin A V. Based on botanical phospholipids nanosystem for activation of biologically active compounds, and method of its manufacture (versions). Patent RU 2391966 1, Russian Federation.
[17] Ipatova O M, Torkhovskaya T I, Medvedeva N V, Prozorovsky V N, Ivanova N D, Shironin A V, Baranova V S, Archakov A I 2010 *Biomedical Chemistry* **4** 82
[18] Kiselev M A, Zemlyanaya E V, Ipatova O M, Gruzinov A Yu, Ermakova E V, Zabelin A V, Zhabitskaya E I, Druzhilovskaya O S, Aksenov V L 2015 *J of Pharmaceutical and Biomedical Analysis* **114** 288
[19] Zemlyanaya E V, Kiselev M A, Zhabitskaya E I, Gruzinov A Yu, Aksenov V L, Ipatova O M, Druzhilovskaya O S 2016 *J. of Phys. Conf. Ser.* **724** 012056
[20] Zhabitskaya E, Zemlyanaya E V, Kiselev M A, Gruzinov A 2016 *EPJ Web of Conf.* **108** 02047
[21] Kiselev M A, Zemlyanaya E V, Gruzinov A Yu, Zhabitskaya E I, Ipatova O M, Aksenov V L 2017 "Analysis of Vesicular Structure of Nanoparticles in the Phospholipid Based Drug Delivery System using SAXS data", JINR Preprint P3-2017-32, submitted to *Crystallogr. Rep.*
[22] Feigin L A, Svergun D I 1987 "Structure analysis by small-angle X-ray and neutron scattering" (Plenum Publishing Corporation, New York)
[23] Sitnik I M 2016 *Comput. Phys. Comm.* **209** 199
[24] Kurbatov V S, Silin I N 1994 *Nucl. Instr. and Meth.* 1994 **A 345** 346
[25] Zhabitskaya E I, Zhabitsky M V 2013 "Asynchronous differential evolution with adaptive correlation matrix" *Proceeding of the 15th Annual Conference on Genetic and Evolutionary Computation* (USA, New York) 455

Role of external neutrons of weakly bound nuclei in reactions with their participation

M A Naumenko[1], Yu E Penionzhkevich[1,2], V V Samarin[1,3], and Yu G Sobolev[1]

[1] Flerov Laboratory of Nuclear Reactions, Joint Institute for Nuclear Research, Dubna, Moscow Region, 141980, Russian Federation

[2] Department of Experimental Methods in Nuclear Physics, National Research Nuclear University MEPhI, Moscow, Moscow Region, 115409, Russian Federation

[3] Department of Nuclear Physics, Dubna State University, Dubna, Moscow Region, 141982, Russian Federation

E-mail: anaumenko@jinr.ru

Abstract. The paper presents the results of measurement of the total cross sections for reactions ^{4,6}He+Si and ^{6,7,9}Li+Si in the beam energy range 5–50 A MeV. The enhancements of the total cross sections for reaction ^{6}He+Si compared with reaction ^{4}He+Si and ^{9}Li+Si compared with reactions ^{6,7}Li+Si have been observed. The performed microscopic analysis of total cross sections for reactions ^{6}He+Si and ^{9}Li+Si based on numerical solution of the time-dependent Schrödinger equation for external neutrons of projectile nuclei ^{6}He and ^{9}Li yielded good agreement with experimental data.

1. Introduction

It is well known that neutron rearrangement may play an important role in nuclear reactions. The aim of this work is the investigation of the reactions with light nuclei having different external neutron shells. The experiments on measurements of total cross sections were performed for reactions ^{4,6}He+Si and ^{6,7,9}Li+Si. The interesting results are the unusual wide enhancement of total cross section for ^{9}Li+Si reaction as compared with ^{6,7}Li+Si reactions. The similar weaker behaviour was found for ^{6}He+Si reaction as compared with ^{4}He+Si reaction. The time-dependent quantum approach combined with the optical model was used for explanation of these effects.

2. Experimental results

The experimental setup for the implementation of the transmission method using a multilayer telescope [1] is shown in figure 1. A system of silicon detectors $\Delta E_i - E$, i = 0-4 (Si-telescope) was surrounded by the CsI(Tl) γ-spectrometer of complete geometry for registration of γ-rays and neutrons. The thin detectors ΔE_0, ΔE_1 were used to identify the beam particles and determine the particle flux incident on the target. The position-sensitive detector ΔE_2 was used as a so-called active collimator [2] which determined the particle flux incident on the central region of the target ΔE_3. The detectors ΔE_4, E were used to analyze the products of reactions occurring in the material of the target ΔE_3.

IOP Conf. Series: Journal of Physics: Conf. Series **1023** (2018) 012018 doi:10.1088/1742-6596/1023/1/012018

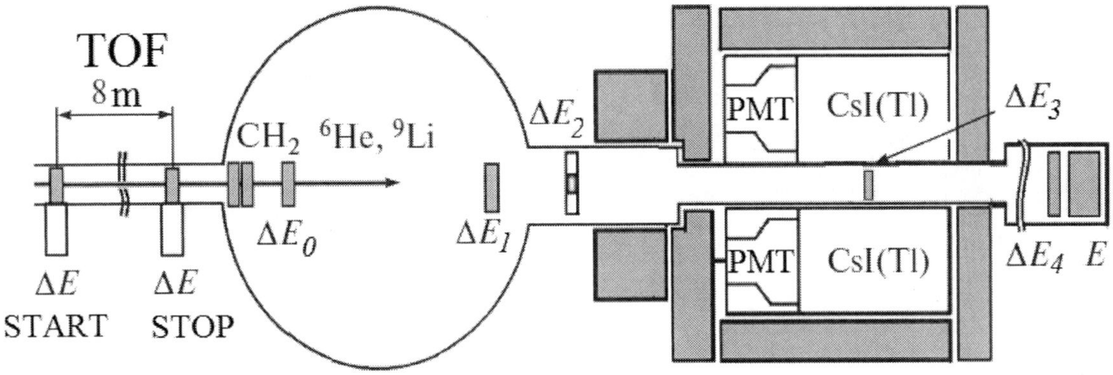

Figure 1. Schematic representation of the experimental setup for measuring the reaction cross sections by the method of the 4π scintillation γ-spectrometer.

The experiment was performed on the accelerator U400M of the Flerov Laboratory of Nuclear Reactions (FLNR), Joint Institute for Nuclear Research (JINR). To obtain the secondary beam the fragmentation reaction of ^{11}B beam with the energy $E_{lab} = 32$ A MeV on the target ^{9}Be was used. The secondary beam consisting of a mixture of particles ^{6}He and ^{9}Li was formed and purified by the magnetic system of the achromatic fragment separator ACCULINNA [3]. The beam energy was varied by a fragment separator magnetic system, the choice of the thickness of the hydrogen-containing plates of CH_2 absorbers in the range $E \approx$15–50 A MeV without significant loss of intensity of the beam of particles. Identification was carried out by energy losses of particles in ΔE_0, ΔE_1 detectors of the telescope and the time of flight. In order to reduce the energy uncertainty, detectors of different thickness (100, 380, or 500 microns) were used in the experiment depending on the beam energy. Detectors of γ-spectrometer recorded γ-quanta and neutrons in coincidence with the start signal from the detector ΔE_1. The number of events of the reaction was determined from the analysis of energy losses in natural Si-target as well as the analysis of gamma and neutron radiation detected by the spectrometer.

The results of measurements of total cross sections for reactions ^{6,7,9}Li+Si and ^{4,6}He+Si are presented in figure 2. Filled symbols are the results of measurements by various methods performed in FLNR, JINR; open symbols are experimental data of other authors. Two peculiarities of the energy dependence of the total cross sections for the reactions with neutron-rich nuclei can be observed in the experimental data. The first peculiarity is the increased cross section for the reaction with the neutron-rich halo nucleus ^{6}He with respect to its core ^{4}He in the whole studied energy range. The second peculiarity is the local increase of the values of the cross section in the form of a bump in a limited energy range 10–30 A MeV for the case of ^{9}Li+Si. The first peculiarity may be explained by the larger size of the ^{6}He nucleus, while the second peculiarity may be a manifestation of the dynamic effects associated with rearrangement of external weakly bound nucleons or their clusters in the ^{9}Li nucleus.

The analysis of these effects using the microscopic complex folding potential in Ref. [10] as well as within the optical model in Ref. [11] did not provide satisfactory explanation of the observed features in the behaviour of the energy dependence of the total cross section. In this study, the potentials of the optical model were modified to take into account the dynamic rearrangement of two external neutrons of projectile nuclei ^{6}He and ^{9}Li.

IOP Conf. Series: Journal of Physics: Conf. Series **1023** (2018) 012018 doi:10.1088/1742-6596/1023/1/012018

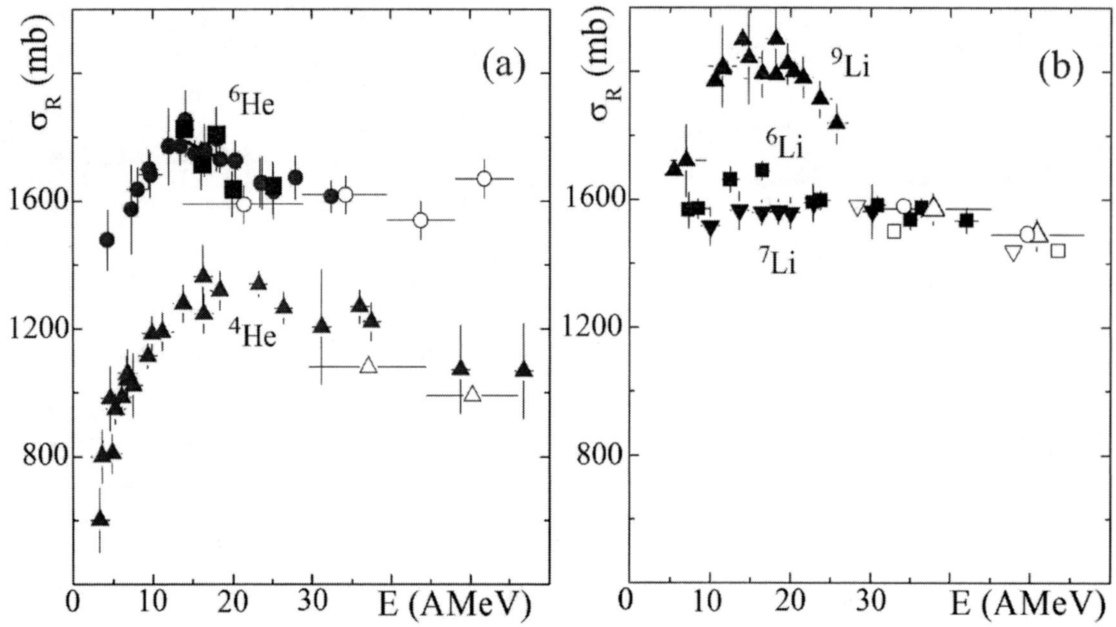

Figure 2. The total cross sections for the reactions ^{4,6}He+Si (a) and ^{6,7,9}Li+Si (b) as a function of the energy of the incident particles. Symbols are the experimental data from Refs. [1, 4–9].

3. Theoretical analysis

The presence of two external neutrons forming a weakly bound neutral cluster (dineutron) is common for the most probable configurations of nuclei ^{6}He ($\alpha + n + n$) and ^{9}Li ($\alpha + t + n + n$). In a simplified approximation, neutrons can be regarded as moving independently in the mean field of the shell model. The rearrangement of external neutrons between the projectile nucleus and the target nucleus, for example, their presence with a high probability between the surfaces of the colliding nuclei, can significantly enhance the mutual attraction of the nuclei and lead to an increase in the total cross section of the reaction.

For theoretical description of neutron rearrangement during collisions of heavy atomic nuclei we used the time-dependent Schrödinger equation (TDSE) approach [12–14] for the external neutrons combined with the classical equations of motion of atomic nuclei. The evolution of the components ψ_1, ψ_2 of the spinor wave function $\Psi(\vec{r}, t)$ for the neutron with the mass m during the collision of nuclei is determined by equation (1) with the operator of the spin-orbit interaction $\hat{V}_{LS}(\vec{r}, t)$

$$i\hbar\frac{\partial}{\partial t}\Psi(\vec{r},t) = \left\{-\frac{\hbar^2}{2m}\Delta + W(\vec{r},t) + \hat{V}_{LS}(\vec{r},t)\right\}\Psi(\vec{r},t). \quad (1)$$

The centres of nuclei $\vec{r}_1(t), \vec{r}_2(t)$ with the masses m_1, m_2 move along classical trajectories. We may assume that before contact of the surfaces of spherical nuclei with the radii R_1, R_2 the potential energy of a neutron is equal to the sum of its interaction energies with both nuclei. The initial conditions for the wave functions were obtained based on the the shell model calculations with the parameters providing neutron separation energies close to the experimental values.

Examples of the evolution of the probability density of the external neutrons of ^{9}Li nucleus when colliding with the nucleus ^{28}Si at different energies are shown in figure 3. During a slow (adiabatic) relative motion of the colliding nuclei the external neutrons (dineutron cluster) of

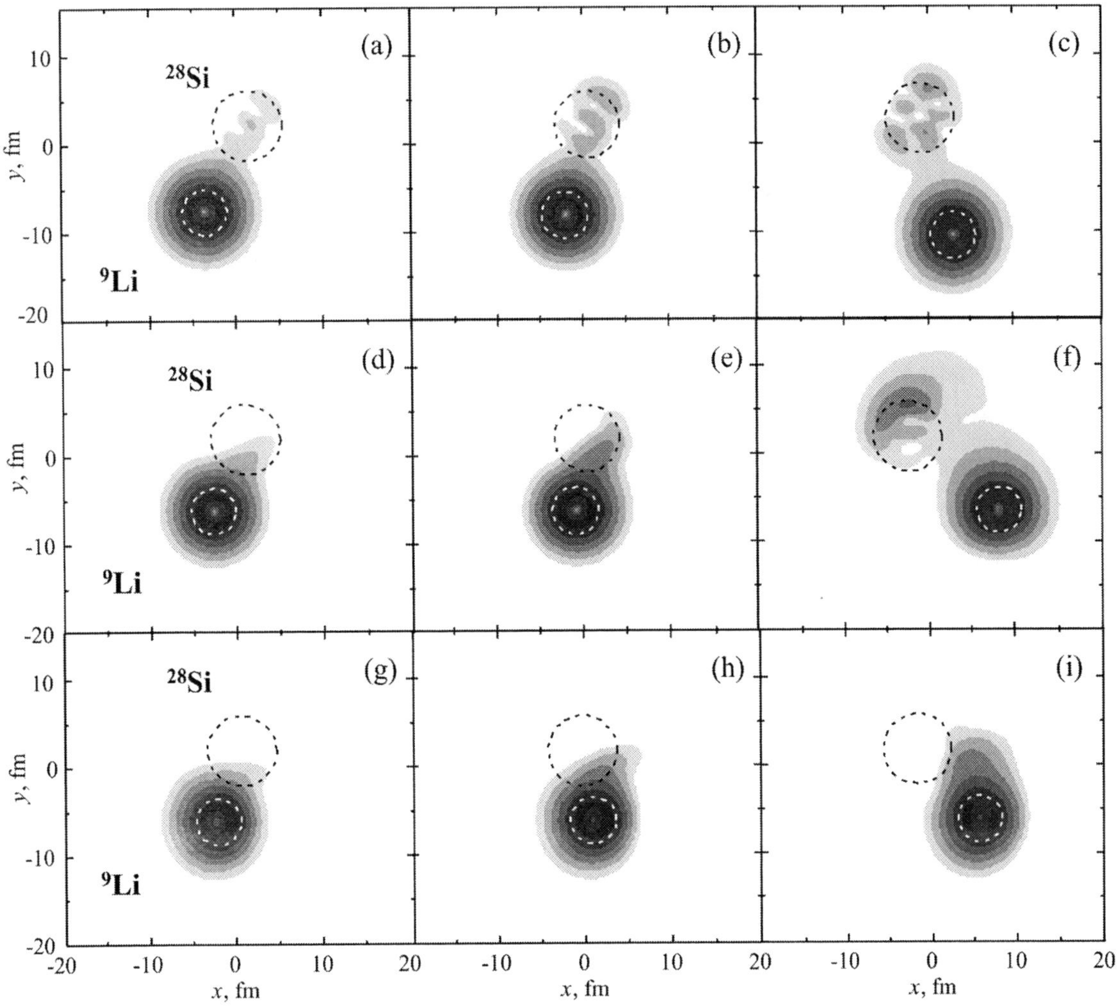

Figure 3. Examples of the evolution of the probability density of external neutrons of the ^{9}Li nucleus during the collision with the ^{28}Si nucleus at energies $E_{\rm lab} = 2$ A MeV ((a), (b), (c), $\gamma = 0.24$, adiabatic regime), 15 A MeV ((d), (e), (f), $\gamma = 0.7$, intermediate regime), and 60 A MeV ((g), (h), (i), $\gamma = 1.4$, diabatic regime). The course of time corresponds to the panel locations from left to right.

^{9}Li nucleus are penetrating the ^{28}Si nucleus and populating the slowly changing two-centre ("molecular") states, the probability density for which fills a large part of the volume of the target nucleus. During the rapid (diabatic) relative motion the probability density of neutrons does not have time to fill all the target nucleus and its change is more local. After the separation of the nuclei the wave packet in the surface region of the target nucleus remains spreading and rotating with large angular momentum. At intermediate velocities there is a transition from the adiabatic regime to the diabatic one.

The qualitative character of the rearrangement of external neutrons during the approach of nuclei depends on the ratio of the average velocity $\langle v \rangle$ of the external neutron and the relative velocity $v_{\rm rel}$ of the nuclei in the process of collision. The average kinetic energy $\langle \varepsilon \rangle$ of external weakly bound neutrons in the nuclei ^{6}He and ^{9}Li may be approximately calculated within the shell model. Using estimation $v_{\mathrm{rel}} \approx v_1 = \sqrt{2E_{\mathrm{lab}}/m_1}$, where E_{lab} is the energy of the projectile nucleus with the mass $m_1 = Am_0$, m_0 is the atomic mass unit, we obtain the ratio of velocities

$$\frac{v_1}{\langle v \rangle} \approx \gamma \equiv \left(\frac{E_{\mathrm{lab}}}{\langle \varepsilon \rangle A} \right)^{1/2}. \tag{2}$$

At low energies, when $\langle v \rangle >> v_1$, $\gamma << 1$, during the flight of the projectile nucleus close to the target nucleus the weakly bound neutrons may, relatively speaking, make many turns around the cores of both nuclei. In the extremely diabatic case (at intermediate energies), when $\langle v \rangle << v_1$, $\gamma >> 1$, the neutron may not be able to move to the target nucleus during the time of flight. The value of the parameter γ may be used to estimate the degree of adiabaticity of the collision.

The real part of the potential $\bar{V}(R)$ for nuclei with "frozen" neutrons was supplemented with the diabatic correction arising from an increase in neutron probability density between the surfaces of the nuclei as they approach

$$V_{\mathrm{d}}(R, E_{\mathrm{lab}}) = \bar{V}(R) + \eta(E_{\mathrm{lab}})\delta V_{\mathrm{d}}(R, E_{\mathrm{lab}}) \tag{3}$$

with the function $\delta V_{\mathrm{d}}\left(R(t), E_{\mathrm{lab}}\right)$

$$\delta V_{\mathrm{d}}\left(R(t), E_{\mathrm{lab}}\right) = \int\limits_{\Omega} d^3 r_3 \delta\rho_1(r_3, t) U_{\mathrm{T}}\left(|\vec{r}_3 - \vec{r}_2(t)|\right), \tag{4}$$

where $U_{\mathrm{T}}(r)$ is the mean field for neutrons in the target nucleus, $\delta\rho_1(r_1, t) = \rho_1(r_1, t) - \rho_1^{(0)}(r_1)$, $\rho_1(r_1, t)$ is the probability density of the external neutrons of the projectile nucleus, $\rho_1^{(0)}(r_1)$ is the same density calculated in the absence of interaction of these neutrons with the target nucleus, Ω is the region between the surfaces of the nuclei,

$$\eta(E_{\mathrm{lab}}) = \frac{1}{1 + \exp\left[\frac{1}{\alpha}\left(\langle\varepsilon\rangle - \left(\frac{E_{\mathrm{lab}}}{A}\right)\right)\right]} \tag{5}$$

with the variable parameters $\langle \varepsilon \rangle \approx 10$ MeV determining the position $\bar{E}_{\mathrm{lab}} = \langle\varepsilon\rangle A$ of the transition region and $\alpha \approx 2$ MeV determining its width. The diabatic correction $\delta V_{\mathrm{d}}\left(R, E_{\mathrm{lab}}\right)$ reduces the height $B\left(E_{\mathrm{lab}}\right)$ and shifts to the right the position $R_B\left(E_{\mathrm{lab}}\right)$ of the Coulomb barrier

$$R_B(E) = R_{B,0} + \delta R_B(E). \tag{6}$$

For the imaginary part of the potential, we used the approximation with the exponential dependence

$$W(r) = \begin{cases} -W_1, r < R_b \\ W_1 \exp\left(-\frac{r - R_b}{b}\right), r \geq R_b \end{cases} \tag{7}$$

and the radius R_b increasing together with the shift of the barrier position

$$R_b(E) = R_a + k\delta R_B(E), \tag{8}$$

where $b = 1$ fm, $k = 2$, $R_a = 5.8$ fm for the reaction ^{9}Li+Si. In the case of reactions with nuclei ^{4}He, ^{7}Li, for the real and the imaginary parts of the nuclear potential the Woods-Saxon form was used

$$\mathrm{Re}\left\{V_{\mathrm{N}}(R)\right\} \equiv V(R) = -V_0[1 + \exp\left((R - R_V)/a_V\right)]^{-1}, \tag{9}$$

$$\mathrm{Im}\left\{V_{\mathrm{N}}(R)\right\} \equiv W(R) = -W_0[1 + \exp\left((R - R_W)/a_W\right)]^{-1}. \tag{10}$$

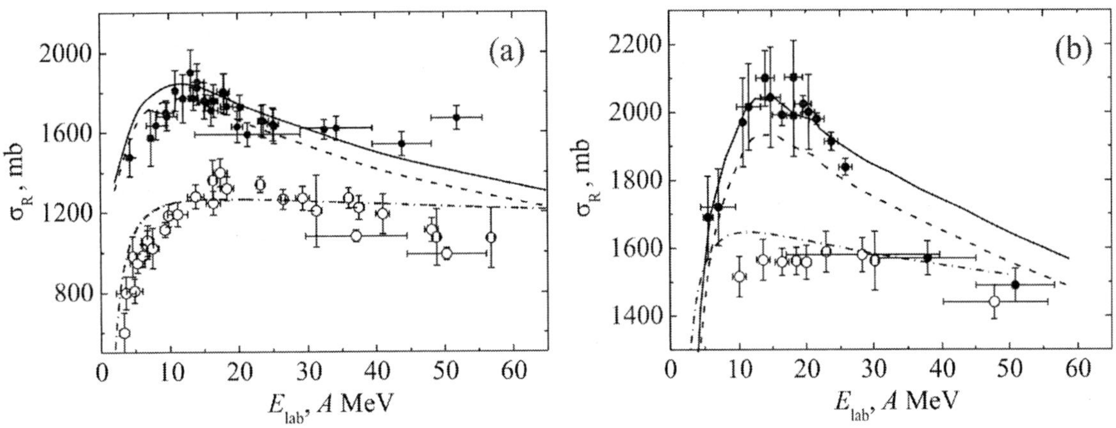

Figure 4. The total cross sections for reactions ^{4,6}He+^{28}Si (a) and ^{7,9}Li+^{28}Si (b), symbols are the experimental data from Refs. [1, 4–9]: ^{6}He+^{28}Si and ^{9}Li+^{28}Si (filled circles), ^{4}He+^{28}Si and ^{7}Li+^{28}Si (empty circles), curves are the results of calculation within the optical model with the potentials (3), (7): (a) for $R_a = 5.0$ fm (solid line) and $R_a = 4.8$ fm (dashed line), (b) for $R_a = 5.8$ fm (solid line) and $R_a = 5.6$ fm (dashed line); dash-dotted lines are the results of calculations with the potentials (9), (10) for the reactions ^{4}He+^{28}Si (a) and ^{7}Li+^{28}Si (b).

For collisions ^{6,7}Li+^{28}Si, the parameters of the potential were obtained by fitting the angular distributions for elastic scattering of ^{6,7}Li on ^{28}Si at several collision energies with the initial approximation in the Akyüz-Winther parametrization [15]. It should be mentioned that for a wide energy range the same values of the parameters of the optical potential were obtained. Because of the absence of experimental data on elastic scattering of ^{9}Li on ^{28}Si, the parameters for the reaction ^{9}Li+^{28}Si were obtained by extrapolation of the parameters for ^{6,7}Li+^{28}Si. Similarly, because of the absence of experimental data on elastic scattering of ^{4,6}He on ^{28}Si, the values of the parameters for reactions ^{4,6}He+^{28}Si were found by extrapolation of the values of the parameters obtained by fitting experimental differential cross sections for the elastic scattering of ^{4,6}He on ^{12}C.

4. Results and discussion

The results of calculation of the total cross section for reactions ^{4,6}He+^{28}Si and ^{6,7,9}Li+^{28}Si are shown in figure 4. As can be seen, they are in good agreement with the experimental data in the whole energy range. Thus, the calculations indicated a possible reason for the manifestation of the enhancement of the total cross sections for reactions ^{6}He+^{28}Si and ^{9}Li+^{28}Si. It consists in the formation of a region of increased neutron density between the surfaces of the nuclei during the entire time of the collision. This increases the attraction of the nuclei, shifts the position of the barrier toward higher values, and increases the size of the region where processes leading to an exit from the elastic channel take place with considerable probability. Within the framework of the optical model, the total contribution of such processes is taken into account by the imaginary part of the potential and the dependence on the energy of its parameters: radius and/or diffuseness. The smaller manifestation of the local maximum in the energy dependence of the cross section for the ^{6}He nucleus in comparison with ^{9}Li can be explained by the greater extent and sparseness of the neutron halo of the ^{6}He nucleus compared to the more compact neutron layer (skin) of the ^{9}Li nucleus.

5. Conclusions

In this paper we presented experimental results of a direct measurement of the total cross sections for the reactions ^{4,6}He+Si and ^{6,7,9}Li+Si in the beam energy range 5–50 A MeV. The enhancements of the total cross sections for reactions ^{6}He+Si and ^{9}Li+Si have been observed. The theoretical analysis of the possible causes of these effects in the collisions of nuclei ^{6}He and ^{9}Li with Si nuclei was performed including the influence of external neutrons of weakly bound projectile nuclei. The time-dependent model proposed in the paper shows that the rearrangement of external weakly bound neutrons of nuclei ^{6}He and ^{9}Li during the collision changes the real and the imaginary parts of the interaction potential, which may cause a local enhancement in the total reaction cross sections. This enhancement is most noticeable in the range of energies where the relative velocity of the nuclei is close in magnitude to the average velocity of external neutrons of the studied light weakly bound nuclei.

Acknowledgments

We express our deep gratitude to the team of researchers of the ACCULINNA experimental facility (FLNR, JINR) as well as to the team of the U400M accelerator (FLNR, JINR) for maintenance of experiments. We also thank A S Denikin and A V Karpov for fruitful discussions. The work was supported by Russian Science Foundation (RSF), grant No. 17-12-01170.

References

[1] Sobolev Yu G *et al* 2005 Energy dependence of total reaction cross section for ^{4,6}He, ^{7}Li+^{28}Si interaction at E = 5–50 MeV/nucleon *Bull. Russ. Acad. Sci.: Phys.* **69** 1790–5
[2] Sobolev Yu G, Ivanov M P, Kondratiev N A and Penionzhkevich Yu E 2011 Active collimators in experiments with exotic nuclear beams *Instr. Exp. Techn.* **54** 449
[3] Rodin A M *et al* 2003 Status of ACCULINNA beam line *Nucl. Instr. Meth. Phys. Res.* B **204** 114–8
[4] Ugryumov V Yu *et al* 2005 Energy dependence of the total cross section for the reaction of ^{4}He ions with silicon nuclei *Phys. Atom. Nucl.* **68** 16–20
[5] Sobolev Yu G *et al* 2017 Experimental study of the energy dependence of the total cross section for the ^{6}He+^{nat}Si and ^{9}Li+^{nat}Si reactions *Phys. Part. Nucl.* **48** 922–6
[6] Warner R E *et al* 1996 Total reaction and 2n-removal cross sections of 20-60 A MeV ^{4,6,8}He, $^{6-9,11}$Li, and ^{10}Be on Si *Phys. Rev.* C **54** 1700–9
[7] Warner R E *et al* 2006 Reaction and proton-removal cross sections of ^{6}Li, ^{7}Be, ^{10}B, ^{9,10,11}C, ^{12}N, ^{13,15}O, and ^{17}Ne on Si at 15 to 53 MeV/nucleon *Phys. Rev.* C **74** 014605
[8] Ingemarsson A *et al* 2000 New results for reaction cross sections of intermediate energy α-particles on targets from ^{9}Be to ^{208}Pb *Nucl. Phys.* A **676** 3–31
[9] Penionzhkevich Yu E, Sobolev Yu G, Samarin V V and Naumenko M A 2017 Peculiarities in total cross sections of reactions with weakly bound nuclei ^{6}He, ^{9}Li *Phys. Atom. Nucl.* **80** 928–41
[10] Lukyanov K V, Zemlyanaya E V, Lukyanov V K, Kukhtina I N, Penionzhkevich Yu E and Sobolev Yu G 2008 Microscopic analysis of the energy dependence of the ^{6}He, ^{6}Li+^{28}Si total reaction cross sections in the energy range E = 5–50 A MeV *Bull. Russ. Acad. Sci.: Phys.* **72** 356–60
[11] Kabdrakhimova G D, Sobolev Yu G, Kukhtina I N, Kuterbekov K A, Mendibaev K O and Penionzhkevich Yu E 2017 Investigation of total cross sections for reactions induced by ^{6}He interaction with silicon nuclei at energies between 5 and 50 MeV/A *Phys. Atom. Nucl.* **80** 32–7
[12] Samarin V V and Samarin K V 2012 Dynamic tunnel effect at low-energy nuclear reactions with neutron-rich nuclei *Bull. Russ. Acad. Sci.: Phys.* **76** 450–3
[13] Naumenko M A, Samarin V V, Penionzhkevich Yu E and Skobelev N K 2016 Near-barrier neutron transfer in reactions with ^{3}He nucleus *Bull. Russ. Acad. Sci.: Phys.* **80** 264–72
[14] Naumenko M A, Samarin V V, Penionzhkevich Yu E and Skobelev N K 2017 Near-barrier neutron transfer in reactions ^{6}He+^{45}Sc, ^{64}Zn, and ^{197}Au *Bull. Russ. Acad. Sci.: Phys.* **81** 710–6
[15] Winther A 1994 Grazing reactions in collisions between heavy nuclei *Nucl. Phys.* A **572** 191–235

IOP Conf. Series: Journal of Physics: Conf. Series **1023** (2018) 012019 doi:10.1088/1742-6596/1023/1/012019

Low collectivity of the first 2^+ states of ^{212,210}Po

D. Kocheva[1], G. Rainovski[1], J. Jolie[2], N. Pietralla[3], A. Blazhev[2], A. Astier[4], R. Altenkirch[2], M. Bast[2], M. Beckers[2], S. Ansari[2], Th. Braunroth[2], M. Cappellazzo[2], M.L. Cortés[3], A. Dewald[2], F. Diel[2], M. Djongolov[1], C. Fransen[2], K. Gladnishki[1], A. Goldkuhle[2], A. Hennig[2], V. Karayonchev[2], J.M. Keatings[5], E. Kluge[2], Th. Kröll[3],J. Litzinger[2], K. Moschner[2], C. Müller-Gatermann[2], P. Petkov[6], M. Rudigier[2], M. Scheck[5], P. Spagnoletti[5], Ph. Scholz[2], T. Schmidt[2], M. Spieker[2], C. Stahl[3], R. Stegmann[3], A. Stolz[2], A. Vogt[2], M. Stoyanova[1], P. Thöle[2], N. Warr[2], V. Werner[3], W. Witt[3], D. Wölk[2], J.C. Zamora[3], K.O. Zell[2], P. Van Isacker[7], V.Yu. Ponomarev[3]

[1] Faculty of Physics, St. Kliment Ohridski University of Sofia, 1164 Sofia, Bulgaria
[2] Institut für Kernphysik, Universität zu Köln, 50937 Cologne, Germany
[3] Institut für Kernphysik, Technische Universität Darmstadt, 64289 Darmstadt, Germany
[4] CSNSM, IN2P3/CNRS and Universite Paris-Sud, F-91405 Orsay Campus, France
[5] University of the West of Scotland, PA1 2BE Paisley, UK and SUPA, Glasgow G12 8QQ, UK
[6] National Institute for Physics and Nuclear Engineering,77125 Bucharest-Magurele, Romania
[7] Grand Accélérateur National d'Ions Lourds, CEA/DRF-CNRS/IN2P3, Bd. Henri Becquerel BP 55027, F-14076 Caen, France

E-mail: dkocheva@phys.uni-sofia.bg

Abstract. The lifetimes of the first 2^+ excited states of ^{212,210}Po were measured in two transfer reactions ^{208}Pb(^{12}C,^{8}Be)^{212}Po and ^{208}Pb(^{12}C,^{10}Be)^{210}Po by the Recoil Distance Doppler Shift (RDDS) method and by the Doppler Shift Attenuation method (DSAM), respectively. The derived absolute $B(E2)$ values of 2.6(3) W.u. for ^{212}Po and 1.83(28) W.u. for ^{210}Po indicate low collectivity. It is shown that the properties of the yrast 2_1^+, 4_1^+, 6_1^+ and 8_1^+ states in both nuclei cannot be described consistently in the framework of nuclear shell models. It is also demonstrated in the case of ^{210}Po that Quasi-particle Phonon Model (QPM) calculations cannot overcome this problem thus indicating the existence of a peculiarity which is neglected in both theoretical approaches.

1. Introduction

Understanding nuclear structure in terms of single-particle motion and collective behaviour is one of the challenges of nuclear physics. The nuclear shell model provides the basic framework for understanding of the single-particle motion in nuclei [1].The collective behaviour is understood as a result of coherent movement of valence nucleons caused by the residual interaction [2]. In this regard, properties of open-shell nuclei in the immediate vicinity of double-magic cores are of particular importance. Such nuclei can often be understood well within the framework of the shell models and at the same time their valence particles can induce the onset of collective behaviour. With two and four valence nucleons in respect to the double-magic core ^{208}Pb the nuclei ^{210}Po and ^{212}Po provide a good testing ground for studying the beginning of the evolution from single-particle to collective motion in the mass $A \approx 208$ region.

In a recent study, it has been shown that a phenomenological single-j shell model accounts very well for the energies of the low-lying states of ^{212}Po (cf. Fig. 3 in Ref. [3]). In addition, Auerbach and Talmi [4] have suggested that the yrast sequence $2_1^+ - 4_1^+ - 6_1^+ - 8_1^+$ of ^{212}Po follows a seniority-like energy pattern resulting in an isomeric 8_1^+ state. All of this raises the question whether shell models can describe consistently the electromagnetic properties of the $2_1^+ - 4_1^+ - 6_1^+ - 8_1^+$ sequence as well. To address these questions experimental information on the absolute $B(E2)$ strengths for the $4_1^+ \rightarrow 2_1^+$ and $2_1^+ \rightarrow 0_1^+$ transitions is needed.

In the case of ^{210}Po, data on its absolute strengths for $E2$ transitions between its yrast states are available [5, 6]. Large-scale shell-model studies of ^{210}Po using realistic interactions [7, 8] well reproduce the energies of the yrast 2^+, 4^+, 6^+, and 8^+ levels. The $E2$ strengths for the transitions between the 4^+, 6^+ and 8^+ states are also almost perfectly reproduced (cf. Table VII in Ref. [7] and Table III in Ref. [8]). However, in both studies [7, 8] the $B(E2; 2_1^+ \rightarrow 0_1^+)$ value is overestimated by a factor of six. Such a significant discrepancy between the shell-model calculations and the data is an indication for either an inaccurate experimental value [7, 8] or for deficiencies in the model, as suggested in Ref. [8].

All of this has motivated us to perform two experiments especially designed to measure the lifetimes of the 2_1^+ states of ^{212,210}Po. Both experiments were performed at the FN Tandem facility at the University of Cologne, Germany.

2. Lifetime measurement of the first 2^+ state of ^{212}Po

2.1. Experimental set-up

The lifetime of the 2_1^+ state of ^{212}Po was measured by utilizing the RDDS method [9, 10]. The excited states of ^{212}Po were populated using the α-transfer reaction ^{208}Pb(^{12}C,^{8}Be)^{212}Po. The target consisted of a 0.6 mg/cm^2 thin layer of Pb evaporated on a 2 mg/cm^2 thick Au backing foil and was placed with the Au facing the beam. The beam energy of 64 MeV was chosen in such a way that the energy at which the reaction takes place after the Au backing to be about $\sim$ 62 MeV. The reaction was induced in the reaction chamber of the Cologne coincidence plunger device [11]. The stopper was a self-supporting 2 mg/cm^2 thick Au foil. Data were taken at six plunger distances: 25μm, 35μm, 43μm, 55μm, 70μm and 100μm.

For detecting the light reaction fragments six solar cells (10 mm $\times$ 10 mm) were used. The array of solar cells was mounted in the plunger chamber at backward angles with respect to the beam axis, covering an angular range between 116.8° and 167.2°. The solar cells were placed at a distance of about 15 mm between their centres and the target. The γ rays from the decay of the excited states of ^{212}Po were registered by 11 HPGe detectors mounted outside the plunger chamber in two rings at distance of, on average, 12 cm from the target. Five detectors were positioned at backward angles (142° with respect to the beam axis) and the other six detectors were placed at forward angles (45° with respect to the beam axis). Data were taken in coincidence mode of at least one solar cell and one HPGe detector (particle-γ) or when at least two HPGe detectors (γ-γ) were in coincidence.

2.2. Data analysis and results

The particle-γ coincidence data were sorted in twelve matrices depending on the positions of the HPGe detectors and the plunger distances. A particle projection of the particle-γ matrix is shown in Fig. 1(a) as an example. The γ rays in coincidence with the group of particles indicated as "^{212}Po & ^{200}Tl" in Fig. 1(a) are shown in Fig. 1(b). This spectrum is dominated by transitions from excited states of ^{200}Tl which is produced by the ^{197}Au(^{12}C,2αn) transfer reaction in the backing or in the stopper. However, the 727-, 405-, and 223-keV lines which are the γ-ray transitions depopulating the first three yrast states of ^{212}Po [12, 13, 3] are also clearly

IOP Conf. Series: Journal of Physics: Conf. Series **1023** (2018) 012019 doi:10.1088/1742-6596/1023/1/012019

visible. Moreover, it is also visible from Fig. 2 (a) that the 727-keV transition between the 2_1^+ state of ^{212}Po and its ground state, has a well pronounced shifted component which evolves as a function of plunger distance.

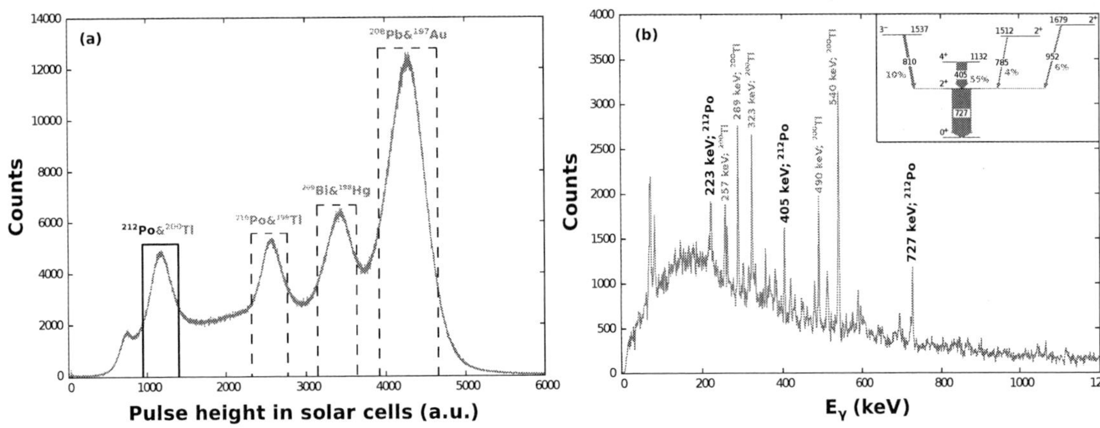

Figure 1. (Color online) (a) The projection of the particle-γ matrix obtained at plunger distance (D=43 μm) by coincident detection of charged particles in the solar-cell array and γ rays at a polar angle $\Theta_\gamma = 142°$. The marked ranges represent parts of the particle spectrum found to be in coincidence with the γ rays from the indicated nuclei. (b) The γ-ray spectrum in coincidence with the group of particles indicated as "^{212}Po & ^{200}Tl" in panel (a). (inset) Partial level scheme of ^{212}Po showing the excited states which feed the first excited 2^+ state directly. The thicknesses of the arrows (and the shown percentages) are proportional to the observed γ -ray intensities [14].

The RDDS data for this transition was analysed by utilizing the Differential Decay Curve Method (DDCM) [15, 16]. The standard application of DDCM requires the I_γ^{sh} and I_γ^{un} components (for each distance) to be measured from spectra in coincidence with Doppler-shifted components of transitions that feed directly the excited state of interest. Then the lifetime τ_i of the level of interest for the i-th target-to-stopper distance depends on I_{sh} and I_{un} in the simple way [15, 16]:

$$\tau_i(x) = \frac{I_{un}(x)}{\langle v \rangle \frac{d}{dx} I_{sh}(x)} \qquad (1)$$

as here the derivative of the Doppler shifted intensities as a function of the target-to-stopper distance, $\frac{d}{dx}I_{sh}$, is determined by a piecewise polynomial fit to the measured intensities I_{sh}.

The presented particle-gated γ-ray spectra in Fig. 2(a) are, in fact, γ-ray singles spectra which, in principal, only contain information for the effective lifetime of the 2_1^+ state of ^{212}Po. Therefore, the intensities of the I_{sh} and I_{un} derived from these spectra cannot be used directly for extracting the lifetime of the 2_1^+ state by means of the DDCM. Special care has to be taken into account for analyzing the impact of the transitions directly feeding the 2_1^+ state on the I_{sh} and I_{un} components of the 727-keV transition (see the inset in Fig. 1(b)).

The lifetimes of the $2_{2,3}^+$ states of ^{212}Po were measured in our previous study [3] to be below 1 ps which means that they contribute only to the fast feeding of the 2_1^+ state. The lifetime of the 3^- state at 1537 keV is not known and in order to simplify the discussion at this moment we assume that its lifetime is sufficiently short so that it decays only in flight. Under this assumption the only essential feeder to the 2_1^+ state remains the 405-keV transition which depopulates the 4_1^+ state of ^{212}Po. It is expected that the 4_1^+ state has a long lifetime of about 140 ps, or longer [3]. Indeed, as can be seen from the insets in Fig. 2(a) the 405-keV transition has only a stopped component for all plunger distances, *i.e.* the decay of the 4_1^+ state contributes

IOP Conf. Series: Journal of Physics: Conf. Series **1023** (2018) 012019 doi:10.1088/1742-6596/1023/1/012019

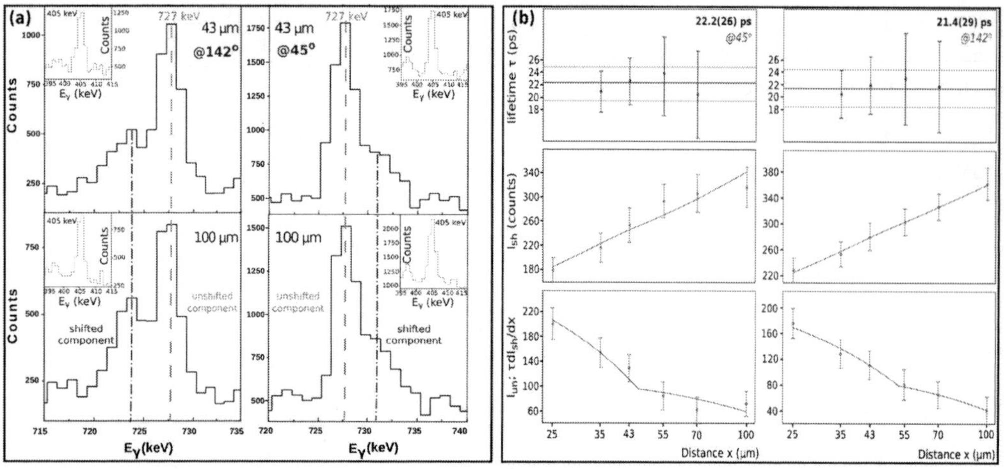

Figure 2. (Color online) (a) Examples of the evolution of the intensities of the Doppler shifted peaks of the 727-keV ($2^+_1 \rightarrow 0^+_1$) transition observed at backward angles and at forward angles for two different target-to-stopper distances. The dot-dashed lines (blue) represent the positions of the Doppler-shifted peak; the dashed lines (red) represent the unshifted peak positions. At the upper corners are shown the peaks of the 405-keV ($4^+_1 \rightarrow 2^+_1$) transition at the same detector angle and distance as those for the $2^+_1 \rightarrow 0^+_1$ transition. (b) The lifetime of the first excited 2^+ state of ^{212}Po determined at forward and backward angles. The middle panels show the shifted intensities at different distances. Continuous curves are fitted through the points to calculate the derivative. In the bottom panels, curves that represent the product between the time derivatives of the shifted intensities and the lifetime of the level are compared with the experimental unshifted intensities. Out of this comparison, the lifetimes corresponding to each distance in the region of sensitivity are extracted, as seen in the upper panel. The horizontal lines represent the weighted mean values.

only to the stopped component of the 727-keV transition. Hence, that extra contribution to the stopped component of the 727-keV transition has to be eliminated. In our analysis this was achieved by subtracting the efficiency-corrected number of counts in the 405-keV line out of the efficiency-corrected number of counts in the stopped component of the 727-keV transition (cf. Fig. 2(a)). Under the considerations above, it is obvious that the intensities of the shifted components of the 727-keV transition being directly determined from the particle-gated spectra are also related only to the lifetime of the 2^+_1 state of ^{212}Po. In this respect, both I^{un}_γ and I^{sh}_γ can be considered as effectively derived from γ-ray spectra in coincidence with the shifted components of all transitions directly feeding the state of interest. Therefore, they can be used directly in Eq. (1).

To proceed with the DDCM analysis the mean velocity of the recoiling nuclei $\langle v \rangle$ has to be known. This value was experimentally determined from the centroids of the shifted and the unshifted components of 727-keV transition to be $\langle v \rangle$ =0.72(5)%c. The DDCM fits using $\langle v \rangle$ =0.72(5)%c and intensities (I^{un}_γ and I^{sh}_γ) extracted with the procedure described above are shown in Fig. 2(b). These fits result in a weighted mean value for the lifetime of the 2^+_1 state of 21.8(19) ps.

It has to be noted that the only assumption in the derivation of the above result which is not directly supported by experimental observations, is that the feeding from the 3^-_1 state is fast. To investigate the influence of this feeding to the lifetime of the 2^+_1 state further, we have also considered the alternative limit, *i.e.* we have assumed that the feeding from the 3^-_1 state is very slow and contributes only to the unshifted component of the 727-keV transition. In this case, besides the intensity of the 405-keV transition, the intensity of the unshifted component of the 727-keV transition has to be reduced by additional 10% which accounts for the intensity of the 810-keV transition ($3_1^- \to 2_1^+$, cf. the inset of Fig. 1(b)). This alternative approach reduces the deduced lifetime of the 2_1^+ to 19.2(18) ps. For the final value for the lifetime of the 2_1^+ state we conservatively adopt the average value between the two limits which is: $\tau(2_1^+, E_x = 727$ keV$) = 20.5(26)$ ps. Taking into account the known electron conversion coefficient for the $2_1^+ \to 0_1^+$ transition of ^{212}Po [12] and α-branching ratio [13], the newly derived lifetime of the 2_1^+ state translates to absolute transition strength $B(E2; 2_1^+ \to 0_1^+) = 193(24)$ e^2fm$^4 = 2.6(3)$ Wu.

3. Lifetime measurement of the first 2^+ state of ^{210}Po

3.1. Experimental set-up

For measuring the lifetime of the first 2^+ state of ^{210}Po by means of DSAM (cf. Ref. [9] and references therein) the ^{208}Pb(^{12}C,^{10}Be)^{210}Po transfer reaction was used. The target was a self-supporting 10 mg/cm^2 thick Pb foil enriched to 99.14 % with the isotope ^{208}Pb. Besides the placement of the ring at forward angles at 35°, the experiment was performed with the same detectors' set-up as the previous one.

3.2. Data analysis and results

The particle-γ coincidence data were sorted in two matrices depending on the position of the HPGe detectors. A projection of the particle-γ matrix obtained with γ-ray detection at 142.3° is shown in Fig. 3(a). The γ rays in coincidence with the group of particles indicated as "^{210}Po" in Fig. 3(a) are shown in Fig. 3(b). This spectrum is dominated by the 1181-keV and the 245-keV lines which are the γ-ray transitions depopulating the first two yrast states of ^{210}Po [17]. Besides some contaminants from ^{211}Po (which are shown in purple), all other γ rays originate from the decay of excited states of ^{210}Po. The 1181-keV γ-ray line shows Doppler shape which allow us to extract the lifetime of the 2_1^+ state of ^{210}Po (cf. Fig 4).

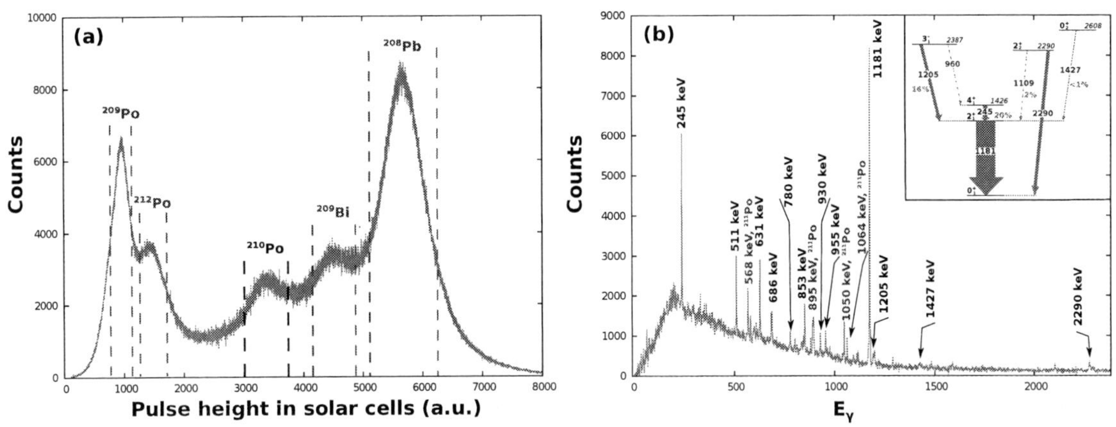

Figure 3. (Color online) (a) The projection of the particle-γ matrix obtained by coincidence detection of charged particles in the solar-cell array and γ rays at $\Theta_\gamma = 142°$ polar angle. The vertical dashed lines represent parts of the particle spectrum found to be in coincidence with γ rays from the indicated nuclei. (b) The γ-ray spectrum in coincidence with the group of particles indicated as "^{210}Po" in panel (a). (inset) Partial level scheme of ^{210}Po showing the excited states which feed the first excited 2^+ state directly. The thicknesses of the arrows (and the shown percentages) are proportional to the observed γ -ray intensities [18]. The transitions not observed in our data are presented with dashed arrows.

IOP Conf. Series: Journal of Physics: Conf. Series **1023** (2018) 012019 doi:10.1088/1742-6596/1023/1/012019

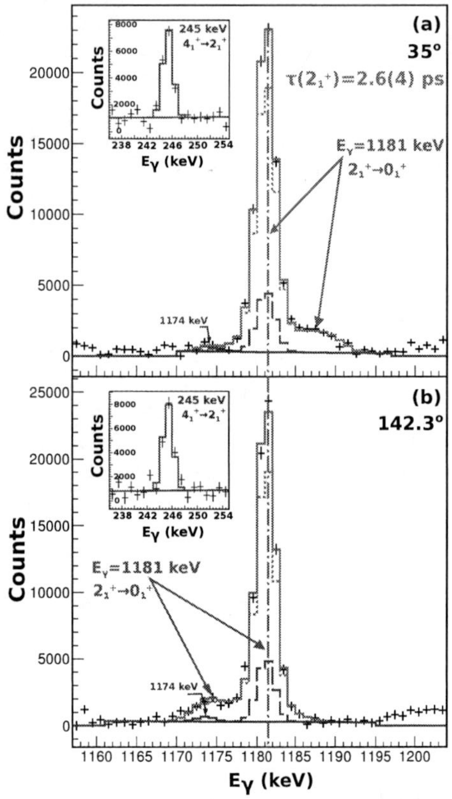

Figure 4. (Color online) Simultaneous line-shape fits of the 1181-keV ($2_1^+ \to 0_1^+$) transition observed at forward (a) and at backward (b) angles. The solid (red) line represents the total fit. The 1181-keV line is fitted simultaneously with the 245-keV ($4_1^+ \to 2_1^+$) line which is always emitted from a stopped nucleus (the insets). The dotted and dashed lines represent the individual contributions of 1181-keV (red) and 245-keV (purple) lines, respectively, to the total fit. An unidentified stopped contaminant with $E_\gamma = 1174$ keV is taken into account (brown).

The line-shape analysis was performed with the integrated software package APCAD (Analysis Program for Continuous Angle DSAM) [19]. Details about the used approach can be found in Ref. [3] where the software was verified.

The lifetime of the first excited 2^+ state of ^{210}Po was obtained from the line shape of the 1181-keV ($2_1^+ \to 0_1^+$) transition. The feeding history of the state (see the inset in Fig. 3(b)) was accounted. A special care was taken to account for the impact of the 245-keV ($4_1^+ \to 2_1^+$) transition. The 4_1^+ state of ^{210}Po is a long-lived state with lifetime $\tau = 2.21(10)$ ns [5]. Consequently, it always decays at rest in the present experiment. Indeed, as can be seen from the insets in Fig. 4 the 245-keV γ-ray line show no indication of Doppler-shifted components in its shape. Hence, when the 2_1^+ state is fed from the $4_1^+ \to 2_1^+$ transition it also always decays at rest which gives extra counts into the fully stopped component of the 1181-keV transition. In order to extract correctly the lifetime of the first 2^+ state of ^{210}Po by means of the Doppler-shift attenuation method, the contribution of the γ rays coming from the 245-keV transition to the fully stopped component of the 1181-keV transition has to be eliminated. That procedure could be automatically carried out with APCAD by simultaneously fitting the 1181- and the 245-keV lines.

Under this assumption, the final value of the lifetime of the 2_1^+ is extracted to be 2.6(4) ps. Taking into account the known electron conversion coefficient for the $2_1^+ \to 0_1^+$ transition of ^{210}Po [20], the revised lifetime of the 2_1^+ state translates to absolute transition strength $B(E2; 2_1^+ \to 0_1^+) = 136(21)$ e^{2}fm^4 = 1.83(28) Wu.

4. Discussion

The derived absolute $B(E2)$ values between the first 2^+ and ground states in both (^{212,210}Po) nuclei indicate very low collectivity. In fact, the revised value for ^{210}Po is a factor of three times larger than the adopted one but still two times smaller than the calculated one in the framework of the single-j shell model [14]. In the case of ^{212}Po, the obtained value is more than a factor of two times smaller than the calculated one in the framework of the single-j shell model [3]. The results from the single-j shell-model calculations for ^{210,212}Po and ^{210}Pb (for completeness) are presented in Table 1. The labelling of the columns reflects the approach in choosing the effective charges. In the case of SM1-gh, the effective proton and neutron charges in the $E2$ transition operator were determined from the measured $B(E2; 8_1^+ \to 6_1^+)$ values for ^{210}Pb and ^{210}Po [20] This approach yields effective charges of $e_\nu = 1.04e$ and $e_\pi = 1.52e$. Another approach is to determine the effective charges from the measured $B(E2; 2_1^+ \to 0_1^+)$ values for ^{210}Pb and ^{210}Po which leads to effective charges of e_ν=0.83e and e_π=1.09e. The results from these calculations are presented in the column labelled as SM2-

Table 1. Comparison between the experimental and calculated properties of the yrast states in ^{210}Po, ^{210}Pb and ^{212}Po. Except the values from this study, all other experimental $B(E2)$ values are from Ref. [20]. $B(E2)$ values are given in e^2fm^4.

Transition	^{210}Po			^{210}Pb			^{212}Po		
$J_i^\pi \to J_f^\pi$	Expt	SM1-*gh*	SM2-*gh*	Expt	SM1-*gh*	SM2-*gh*	Expt	SM1-*gh*	SM2-*gh*
$2_1^+ \to 0_1^+$	136(21)	263	137	105(30)	166	106	193(24)	464	271
$4_1^+ \to 2_1^+$	331(13)	302	157	360(68)	191	121	–	535	313
$6_1^+ \to 4_1^+$	229(7)	209	109	158(60)	132	84	293(83)	301	178
$8_1^+ \to 6_1^+$	84(3)	84	44	53(23)	53	34	173(68)	103	62

Table 2. Results from the realistic shell-model calculations (SM) for the low-lying yrast states of ^{210}Po and ^{210}Pb, and the QPM calculations for the nucleus ^{210}Po in comparison with the experimental data taken from Ref. [20]. $B(E2)$ values are given in e^2fm^4.

	^{210}Pb		^{210}Po				^{210}Pb		^{210}Po		
J_i^π	E_x, (MeV)		E_x, (MeV)			J_f^π	$B(E2; J_f^\pi \to J_i^\pi)$		$B(E2; J_f^\pi \to J_i^\pi)$		
	Expt	SM	Expt	SM	QPM		Expt	SM	Expt	SM	QPM
2_1^+	0.80	0.84	1.18	1.20	1.10	0_1^+	105(30)	109	136(21)	133	135
4_1^+	1.10	1.10	1.43	1.47	1.16	2_1^+	360(68)	144	331(13)	169	41
6_1^+	1.20	1.19	1.47	1.48	1.20	4_1^+	158(60)	101	227(5)	116	28
8_1^+	1.28	1.23	1.56	1.53	1.21	6_1^+	53(23)	43	83(3)	46	11

gh. The problem is evident – if the effective charges are fixed to the $B(E2; 8_1^+ \to 6_1^+)$ values (SM1-*gh*), the $B(E2; 2_1^+ \to 0_1^+)$ values are overestimated, otherwise, if the effective charges are fixed to the $B(E2; 2_1^+ \to 0_1^+)$ values (SM2-*gh*), the $B(E2; 8_1^+ \to 6_1^+)$ and the $B(E2; 6_1^+ \to 4_1^+)$ values are underestimated. This analysis suggests that agreement between experimental and simple single-j shell model cannot be achieved for the $B(E2)$ rates for ^{210,212}Po and ^{210}Pb by adjusting the effective charges.

At this point, it can be speculated that the failure of the single-j shell model originates from the extremely limited model space. To check this hypothesis we have performed realistic shell-model calculations in the cases of ^{210}Po and ^{210}Pb. The results are presented and compared to the experimental data in Table 2 and the details about these calculations are presented in [14]. The realistic shell model (SM) reproduces almost perfectly the energies of the yrast states in ^{210}Pb and ^{210}Po. However, in both cases the description of the $B(E2)$ values is only marginally improved with respect to the ones obtained in the single-j shell-model calculation (SM2-*gh*).

It is also interesting to check whether the problem is specific for shell models only. For this purpose we have performed Quasi-particle Phonon Model (QPM) calculations [21] for ^{210}Po. The results are presented in Table 2 and the details about these calculations can be found in [18]. The energies of the states of interest are reasonably well reproduced. It has to be noted that in the chosen approach to fix the strength parameters to the electric strengths of the 2_1^+ state, the result for the energies of the states should be considered as a prediction of the model. The major discrepancy between the QPM calculations and the experimental data appears in the $E2$ transition strengths for the cascade $8_1^+ \to 6_1^+ \to 4_1^+$. Overall, the model underestimates these values by a factor of 8. The problem existing in the shell model description also appears in a different form in the present QPM calculations.

5. Summary

In the present study we have measured the lifetimes of the 2_1^+ states of ^{212,210}Po. The derived absolute $B(E2)$ values indicate low collectivity in the structure of these states. No consistent description of the properties of the yrast 2_1^+, 4_1^+, 6_1^+ and 8_1^+ states in both nuclei is observed in the framework of nuclear shell models. The additional QPM calculations which have been done for the nucleus ^{210}Po show that the problem existing in the shell model description also appears in a different form in the QPM calculations. More thorough theoretical investigations of this problem are needed.

Acknowledgements

D.K. acknowledges the support from Deutscher Akademischer Austauschdienst. This work was supported by the partnership agreement between the University of Cologne and University of Sofia, by the BgNSF under grant DN08/23/2016, by STFC, by the German-Bulgarian exchange program under grants PPP57082997 & DNTS/01/05/2014, by the DFG under grant SFB 1245, and by the BMBF under grants 05P12RDCIB, 05P15RDFN1,9 and 05P15RDCIA.

References

[1] M.G. Mayer and J.H.D. Jensen,*Elementary Theory of Nuclear Shell Structure* (John Wiley & Sons, Inc., 1955).
[2] R.F. Casten, Phys. Lett. B **152**, 145 (1985).
[3] D. Kocheva *et al.*, Phys. Rev. C **93**, 011303(R) (2016).
[4] N. Auerbach, I. Talmi, Phys. Lett. **10**, 297 (1964).
[5] O. Häusser *et al.*, Nucl. Phys. A **273**, (1976) 253.
[6] C. Ellegaard *et al.*, Nucl. Phys. A **206**, (1973) 83.
[7] L. Coraggio *et al.*, Phys. Rev. C **60**, (1999) 064306.
[8] E. Caurier, M. Rejmund, H. Grawe, Phys. Rev. C **67**, (2003) 054310.
[9] T.K. Alexander and J.S. Forster, Adv. Nucl. Phys. **10**, 197 (1978).
[10] A.Z. Schwarzschild, E.K. Warburton, Ann. Rev. Nucl. Sci. **18**, 265 (1968).
[11] A. Dewald, O. Möller, and P. Petkov, Prog. Part. Nucl. Phys. **67**, 786 (2012).
[12] A.R. Poletti *et al.*, Nucl. Phys. A **473**, 595 (1987).
[13] A. Astier, P. Petkov, M.-G. Porquet, D.S. Delion, and P. Schuck, Phys. Rev. Lett. **104**, 042701 (2010); A. Astier, P. Petkov, M.-G. Porquet, D.S. Delion, and P. Schuck, Eur. Phys. J. A **46**, 165 (2010).
[14] D. Kocheva *et al.*, Phys. Rev. C **96**, 044305 (2017)
[15] A. Dewald, S. Harissopulos and P. von Brentano, Z. Phys. A**334**, 163 (1989).
[16] G. Böhm, A. Dewald, P. Petkov and P. von Brentano, Nucl. Inst. Meth. A**329**, 248 (1993).
[17] L. G. Mann *et al.*, Phys. Rev. C **38**, (1988) 74.
[18] D. Kocheva *et al.*, Eur. Phys. J. A **53**, 175 (2017)
[19] C. Stahl, J. Leske, M. Lettmann, N. Pietralla, Comp. Phys. Com. **214**, (2017) 174.
[20] M. Shamsuzzoha Basunia, Nucl. Data Sheets **121**, (2014) 561.
[21] V. G. Soloviev, Theory of Atomic Nuclei, Quasiparticles and Phonons (IOP, London, 1992).

IOP Conf. Series: Journal of Physics: Conf. Series **1023** (2018) 012020 doi:10.1088/1742-6596/1023/1/012020

Dynamic chirality in mass regions A=105 and A=130

M.S. Yavahchova[1], D. Tonev[1], G. de Angelis[2], P. Petkov[1,3], N. Goutev[1], R.K. Bhowmik[4], R. P. Singh[4], S. Muralithar[4], N. Madhavan[5], R. Kumar[6], M. Kumar Raju[8], J. Kaur[9], G. Mohanto[10], A. Singh[11], N. Kaur[12], R. Garg[5,7], A. Shukla[13], R. Mateva[1], E. Geleva[1], B. Slavchev[1], and S. Brant[14,†]

[1] *Institute for Nuclear Research and Nuclear Energy, Bulgarian Academy of Sciences, 1784 Sofia, Bulgaria*
[2] *INFN, Laboratori Nazionali di Legnaro, I-35020 Legnaro, Italy*
[3] *Horia Hulubei NIPNE, R-76900, Bucharest-Magurele, Romania*
[4] *Inter University Accelerator Centre, New Delhi-110067, India*
[5] *Inter University Accelerator Centre, Aruna Asaf Ali Marg, New Delhi 110067, India*
[6] *Department of Physics, Indian Institute of Technology Bombay, Powai, Mumbai 400076, India*
[7] *Department of Physics and Astrophysics, University of Delhi, Delhi 110007, India*
[8] *Nuclear Physics Department, Andhra University, Visakhapatnam 530003, India*
[9] *Department of Physics, Guru Nanak Dev University, Amritsar, Punjab 143005, India*
[10] *Nuclear Physics Division, Bhabha Atomic Research Centre, Mumbai 400085, India*
[11] *Indian Institute of Technology, Kharagpur 721302, India*
[12] *Department of Physics, Panjab University, Chandigarh, India*
[13] *Department of Physics, Banaras Hindu University, Varanasi, India*
[14] *Department of Physics, Faculty of Science, Zagreb University, Zagreb, Croatia*

Corresponding author's e-mail address: m.yavahchova@gmail.com

Abstract. Chiral symmetry is dichotomic and its spontaneous breaking by the axial angular momentum vector leads to a pair of degenerate $\Delta I = 1$ rotational bands, called chiral doublet bands. In many cases the energy degeneracy of the chiral candidate bands has almost been observed but the transition probabilities of the analogue states have been found to be different. In the angular momentum region where chirality sets in the *B(E2)* values of the electromagnetic transitions deexciting analogue states of the chiral twin bands should be almost equal. Correspondingly the *B(M1)* values should exhibit staggering. Experimental data for two nuclei from different mass regions are compared with theoretical predictions. Exited states in ^{134}Pr were populated in the fusion-evaporation reaction ^{119}Sn(^{19}F, 4*n*)^{134}Pr, while excited states in ^{102}Rh were populated in the fusion-evaporation reaction ^{94}Zr(^{11}B, 3n)^{102}Rh. Our lifetime measurements and level-scheme investigations show that there are similarities in both chiral candidates nuclei related to dynamic chirality.

1. Introduction

Chirality" is recognized as an interesting phenomenon in different branches of science. Some of the most popular cases, which are easy to understand, are identified in chemistry. The chiral molecule is non-superimposable on its mirror image. The mirror images of a chiral molecule are called enantiomers and they could be right- and left-handed. The terminology emphasizes the analogy to the human hands, for which one hand can be superimposed onto the mirror image of the other. The term "chirality" originates from the Greek word for hand and is a synonym to "handedness".

In the last decades chirality in nuclei is among the most studied phenomena. A spontaneous breaking of chiral symmetry can take place for configurations where the angular momenta of the valence proton, the valence neutron, and the core are mutually perpendicular [1]. Under such conditions, the angular momenta of the valence particles are aligned along the short and long axes of the triaxial core, while the angular momentum of the rotational core is aligned along the intermediate axis. The resulting aplanar

IOP Conf. Series: Journal of Physics: Conf. Series **1023** (2018) 012020 doi:10.1088/1742-6596/1023/1/012020

total angular momentum can be arranged into a right- or a left-handed systems, which differs by intrinsic chirality [2]. Because the chiral symmetry is dichotomic, its spontaneous breaking by the axial angular momentum vector leads to doublets of closely lying rotational bands of the same parity [1,2,3].

Due to the underlying symmetry, the pairs of chiral twin bands should exhibit systematic properties [4, 5, 6, 7]. The first property is the existence of a couple of yrast and side bands, which are nearly degenerate. In the angular momentum region where chirality sets in, the *B(E2)* values of the electromagnetic transitions deexciting analogue states of the chiral twin bands should be almost equal. Correspondingly the *B(M1)* values should exhibit odd-even staggering, being much bigger for transitions deexciting states with odd spins than for transitions deexciting states with even spin [6]. The *B(M1)* values for $\Delta I=1$ transitions connecting the side to the yrast band should have odd-even staggering which is out of phase with respect to the *B(M1)* staggering for transitions deexciting states in the yrast band and the side band. This is a fingerprint of the ideal static chirality as discussed in [6,8,9]. To investigate the presence of chirality in a certain nucleus, it is crucial to determine the *B(E2)* and *B(M1)* values.

In many cases, the energy degeneracy of the chiral candidate bands is nearly observed but the corresponding transition probabilities are different, as in the cases of ^{134}Pr [7,10] and ^{102}Rh [11]. We will study in the present paper the similarities in these two nuclei and investigate the existence of dynamic chirality in real nuclei.

2. Experiments

We have performed three experiments to determine reduced transition probabilities in ^{134}Pr and ^{102}Rh. The goal of the present work is to investigate similarities in the level-schemes and transition probabilities in the chiral twin bands in these two nuclei from the mass region A=130 and A=105.

3. *The case of ^{134}Pr*

The two chiral candidate bands are displayed in Fig.1. Excited states in ^{134}Pr were populated using the fusion-evaporation reaction ^{119}Sn(^{19}F, 4*n*) at a beam energy of 87 MeV for the recoil distance Doppler-shift (RDDS) measurement. The beam was delivered by the Vivitron accelerator at IReS in Strasbourg. The target consisted of 0.5 mg/cm^2 ^{119}Sn foil enriched to 89.9%. It was evaporated on a 1.8 mg/cm^2 ^{181}Ta foil facing the beam. A 6.0 mg/cm^2 gold foil was used to stop the recoils that were leaving the target with a mean velocity of 0.98(2)% of the velocity of light, c . The γ rays de-exciting the recoiling ^{134}Pr nuclei were detected using the EUROBALL IV [12] detector array composed of 26 clover and 15 cluster Ge detectors and an inner BGO (bismuth germanate) ball. The cluster and clover detectors of EUROBALL were grouped into 10 rings corresponding to approximately the same polar angle with respect to the beam axis.

In order to measure the lifetimes in the femtosecond region the Doppler-shift attenuation method (DSAM) was utilized. For the DSAM experiment, a beam energy of 83 MeV was used. The target consisted of 0.7 mg/cm^2 ^{119}Sn evaporated on a 9.5 mg/cm^2 ^{181}Ta backing used to stop the recoils.

Detailed description of the experiments performed to investigate the structure of ^{134}Pr is presented in work [7].

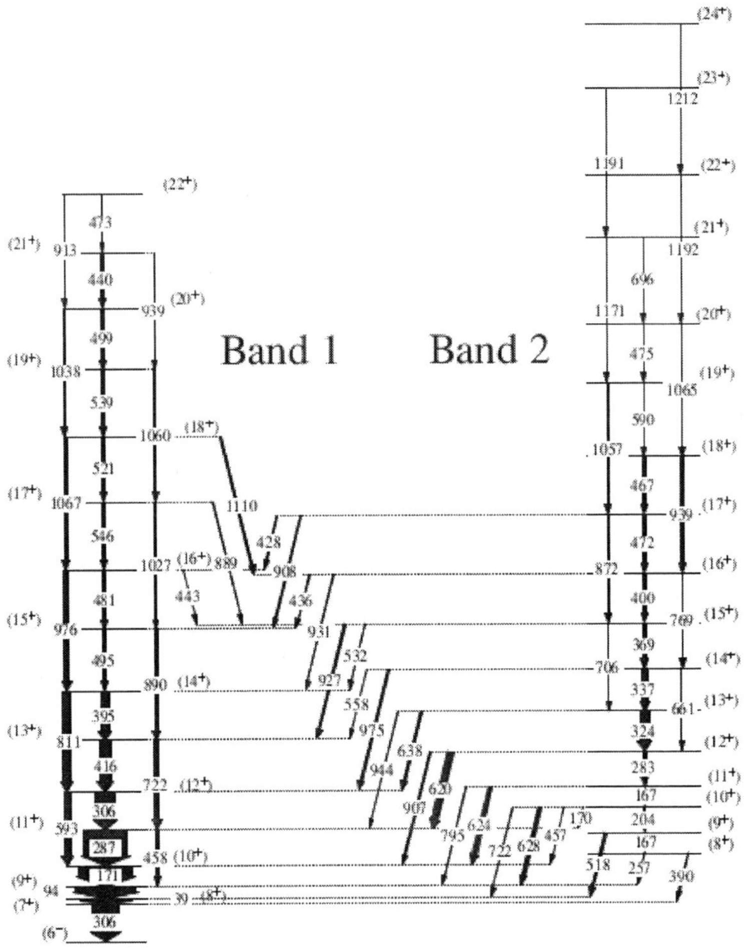

Figure 1. Partial level scheme of ^{134}Pr from Ref. [7].

4. *The case of ^{102}Rh*

The partial level scheme of ^{102}Rh is displayed in Fig. 2. Excited states in ^{102}Rh were populated using the reaction ^{94}Zr(^{11}B, 3*n*)^{102}Rh at a beam energy of 36 MeV. The beam was delivered by the 15-UD Pelletron accelerator at the Inter University Accelerator Center (IUAC) in New Delhi. The target consisted of 0.9 mg/cm^2 ^{94}Zr, enriched to 96.5%, evaporated onto a 8 mg/cm^2 gold backing. The recoils were leaving the target with the mean velocity v of about 0.9% of the velocity of light, c. The deexciting γ rays were registered by the Indian National Gamma Array (INGA), whose 15 clover detectors are accommodated in a 4π geometry [13]. The detectors of INGA were grouped into rings with approximately the same position with respect to the beam axis. For more information see [11].

New accelerator centers like the National cyclotron center in Sofia [14] will open new opportunities for nuclear structure investigations.

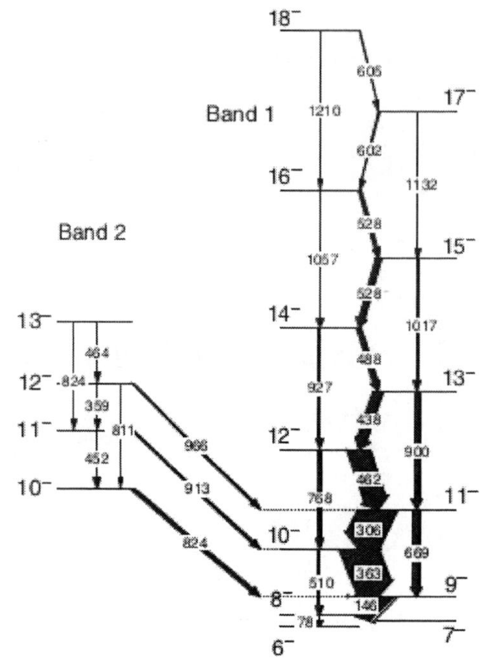

Figure 2. Partial level scheme of ^{102}Rh from Ref. [11].

5. Data analysis

6. *^{134}Pr*

The two chiral candidate bands of ^{134}Pr are displayed in Figure 1 and in the angular momentum region where chirality sets in, they are almost degenerate. In addition, we have performed lifetime measurements to compare the transition probabilities. For the lifetime determination we have used the differential decay curve method (DDCM), proposed in [15,16], using the approach described in [17]. According to this method, at each target-to-stopper distance x, the lifetime $\tau(x)$ of the level of interest is determined from quantities obtained directly from the measured data. For the DSAM measurement new advanced methods have been used [18,19]. Fourteen lifetimes of excited nuclear states were measured in both bands, 13 of them for the first time.

In addition to the lifetimes, branching ratios and the electric or magnetic character of some transitions was also investigated. Using the clover detectors of the EUROBALL spectrometer linear polarization measurements were performed [7].

7. *^{102}Rh*

The Doppler-shift attenuation method was utilized to determine lifetimes of excited states in ^{102}Rh. The analysis was carried out within the framework of the differential decay curve method, according to the procedure outlined in [18,19]. In order to investigate the level scheme and electromagnetic properties of the transitions of interest in ^{102}Rh we performed four types of data analysis. The ordering of the transitions in the level scheme was determined according to γ – ray relative intensities, γ-γ coincidence relationships, and γ-ray energy sums. The electric or magnetic character and multipolarity of the transitions were deduced by linear polarization and angular correlations measurements [11]. For the first time this four types of analysis were used in one experiment with the INGA spectrometer. Thus the level-scheme was extended by a new ΔI=1 band with negative parity and new eight lifetimes were determined for the first time, six in the yrast and two in the side band [11].

8. Discussion

Dynamic chirality in nuclear physics is introduced in the work [7]. The term *dynamical chirality*, refers to the possibility that the angular momenta of the proton, neutron, and core in the odd-odd nucleus, find themselves in the favorable geometry, as if they would in the equivalent triaxially deformed rotor. The IBFFM calculations in [7] show that such a possibility is present, but it is far from being dominant. The condition for the appearance of twin bands with wave functions realistic enough to reproduce the electromagnetic decay of the bands is that the core is a γ -unstable rotor whose effective γ is in the range of triaxial values.

In the article [20], for shortness, the structure in which a pair of twin bands is close in excitation energy, but the electromagnetic decay properties do not show the chiral pattern, is denoted as *case A*. The structure where the pair of twin bands is close in excitation energy and the electromagnetic decay properties display the chiral pattern, is denoted as *case B*. We will use the same types of chiral description in the present work. Odd-odd nuclei in the A ~ 130 mass region can be classified as *case A* or *case B* nuclei. In all these nuclei the cores are γ-soft, their odd-proton odd-mass neighbors have also a similar structure and their odd-neutron odd-mass neighbors have a similar structure, too. Therefore, there is *a priori* no evident reason why should they be different in structure, some of them being chiral (*case B*) and some being not chiral (*case A*). Either the structure of all these nuclei is not chiral, or there is a mechanism that dynamically induces chirality, in such a way that in *case B* it is far more pronounced than in *case A*. For the first time we will investigate the possibility to have a nucleus from mass region A~105 with the same characteristics like ^{134}Pr. The level-schemes of both nuclei as well as their electromagnetic properties will be compared with the theoretical expectations.

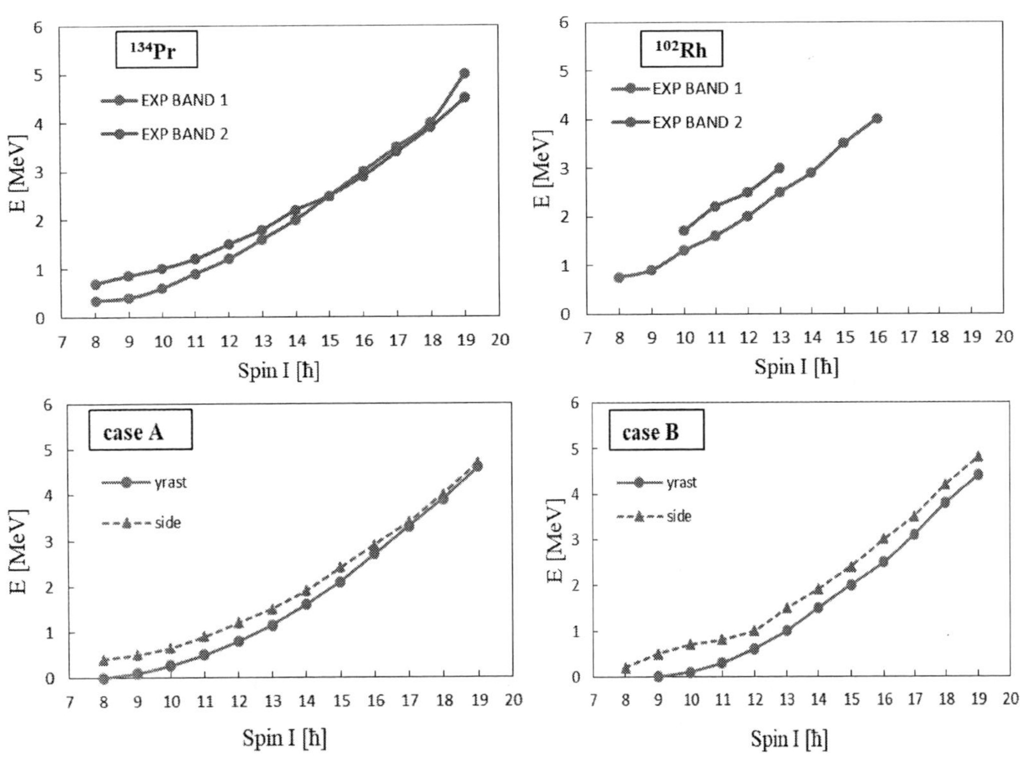

Figure 3. In the upper panels the excitation energies of levels in the yrast and side band of ^{134}Pr (left) and ^{102}Rh (right) are presented. In the second raw of panels sister bands calculated in the IBFFM for *case A* (left) and *case B* (right) are shown.

In Figure 3 are presented the excitation energies of levels of the twin chiral bands in ^{134}Pr and ^{102}Rh as well as the predictions obtained from IBFFM for *case* A and *case* B [20]. The level energies for the yrast and side bands are becoming closer with increasing spin and cross at high spins, while in *case* B they are equidistant from the spin region where chirality sets up. In all cases, experiment and theory, the level energies of the side band are higher than these of the yrast band for low and middle spin regions. For the nucleus of ^{134}Pr the behavior of the level energies is following *case* A. For ^{102}Rh excitation energies are similar to both *cases A* and *B*. Due to the limited statistics we have identified only states up to 13^- for the side band of ^{102}Rh. Obviously an extension to higher spins for the second band will answer the question whether the twin bands cross in excitation energies of the states. We need to note that two bands could be called degenerate at least when the energy difference for analogue states is no more than 500 keV.

The predictions for case *A* are close to the two real nuclei in mass regions A~105 and A~130. From the point of view of the comparison of the level-schemes we could note that the two bands are almost degenerate and similarities for the nucleus of ^{102}Rh with *cases A* and *B* are good for the low spins. For the nucleus of ^{134}Pr we observe the behavior like in *case A*. Obviously we need more experimental data for band 2 in ^{102}Rh.

In order to make conclusions about the existence of chirality in nuclei it is crucial to perform lifetime measurements.

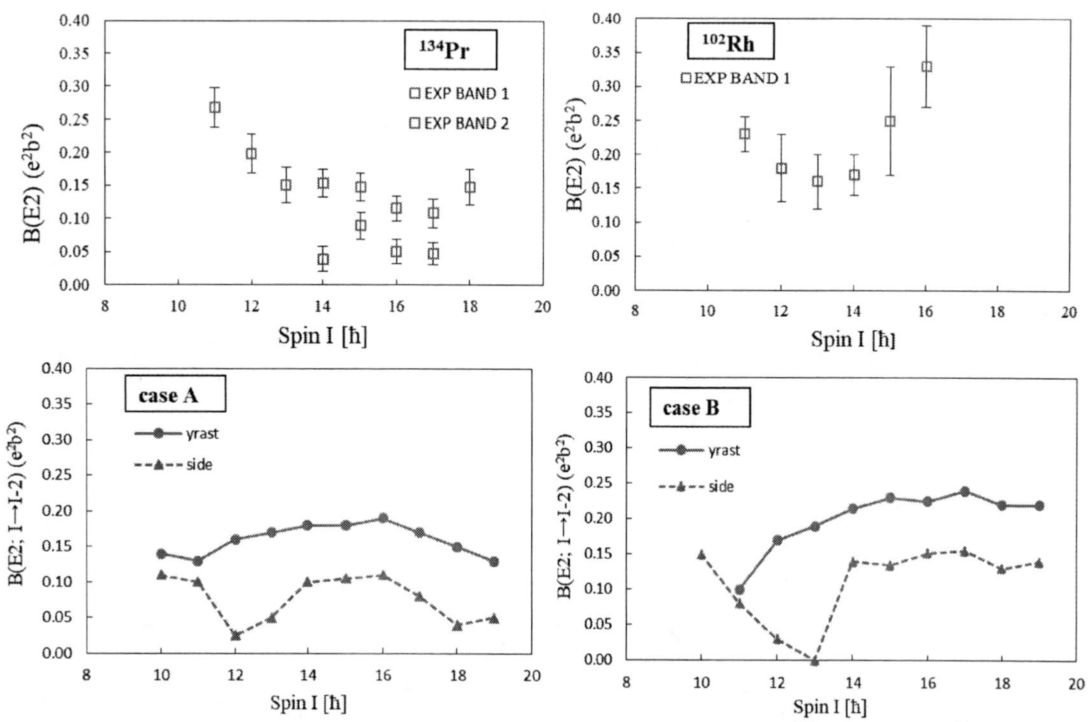

Figure 4. In the upper panels experimentally determined B(E2) values for ^{134}Pr (left) and ^{102}Rh (right) are shown. In the second raw of panels B(E2) values for twin bands calculated in the IBFFM for *case A* (left) and *case B* (right) are shown.

In Figure 4 are compared B(E2) values obtained for ^{102}Rh and ^{134}Pr with the predictions from the work [20]. For the nucleus of ^{134}Pr determined B(E2) values in the chiral candidate bands are different and in addition they have different behavior. This fact clearly shows that static chirality is not present in this nucleus.

For the nucleus of ^{102}Rh we have determined only B(E2) values in Band 1. These B(E2) values show a similar behavior like in the yrast band of ^{134}Pr. Concerning Band 2 of ^{134}Pr the B(E2) values show similarities like in *case* A.

In order to conclude about the presence of chirality we need to check for existence of staggering of B(M1) values determined in both nuclei. This test is may be the strongest one for existence of static chirality. Figure 5 shows experimentally determined B(M1) values for ^{134}Pr and ^{102}Rh. Chiral interpretation in IBFFM also like in the case of two quasiparticle plus triaxial rotor model (TQPTR) predicts the existence of staggering for B(M1). There is a pronounce similarity for B(M1) in Band 1 and Band 2 of ^{134}Pr, but the staggering is not seen. The same is the case with the nucleus of ^{102}Rh, where B(M1) staggering - the fingerprint for static chirality, is not present. The B(M1) staggering predicted for static chirality is not present in both nuclei, which means that we observe dynamic chirality only.

Even-even nuclei in the A~130 mass region are well described by the O(6) symmetry in IBM, that corresponds to γ-unstable rotor [21], and the accepted interpretation is that they are γ-soft. The potential energy surface is flat in the γ direction and rather broad in β direction for γ-unstable nuclei in the O(6) limit or close to it. The wave functions are not (β,γ) dependent, but an *effective* γ_{eff} can be calculated [22] and for O(6) is 30°. γ_{eff} is not the static geometrical deformation and the fluctuations of γ are very large [23]. In this context the term dynamical chirality refers to the possibility that the angular momenta of the proton, neutron, and core in the odd-odd nucleus, find themselves in the favorable geometry [7]. This means that the core is a γ-unstable rotor whose effective γ is in range of triaxial values.

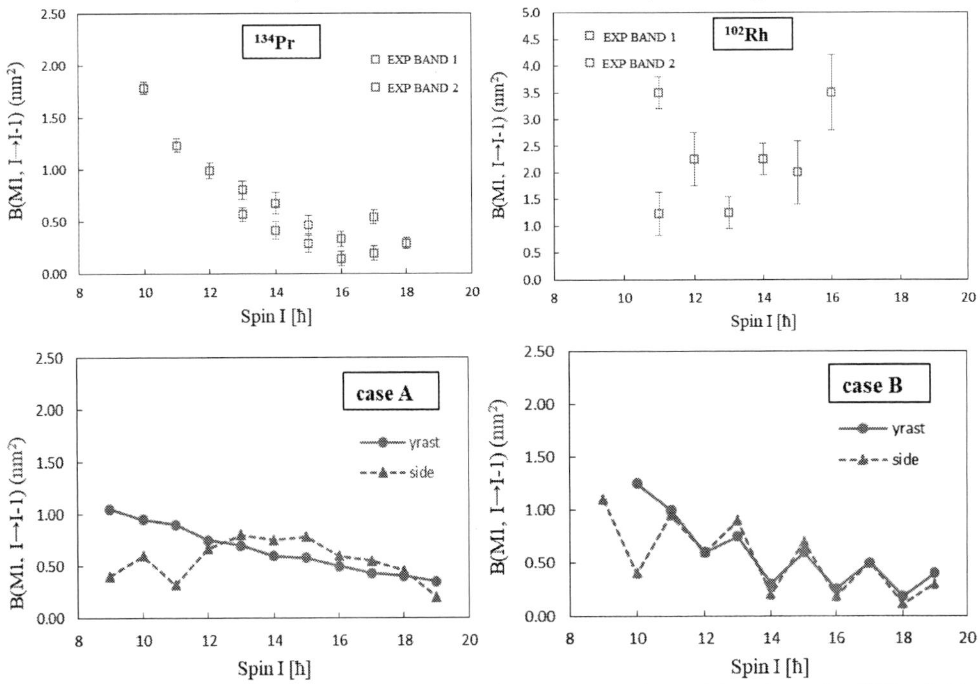

Figure 5. In the upper panels experimentally determined B(M1) values for ^{134}Pr (left) and ^{102}Rh (right) are shown. In the second raw of panels B(M1) values for twin bands calculated in the IBFFM for *case A* (left) and *case B* (right) are shown.

For the nucleus of ^{134}Pr the absence of the staggering in the IBFFM is due to fluctuations of the shape with contributions from near axial shapes that wash out the staggering at high spin. Therefore, the experimental difference between the $B(E2)$ values in Bands 1 and 2 and the absence of the pronounced staggering of the $B(M1)$ values, indicate that the coupling due to shape fluctuations plays a central role in the structure of the twin bands in ^{134}Pr [7].

The similar absence of an appreciable staggering of the data in Band 1 of ^{102}Rh indicates that the expectations for the observation of a static chirality in ^{102}Rh are not realized [11]. Also, the TQPTR calculations reveal that the optimum value of the triaxiality parameter γ=20° differs from the value of 30° characterizing the static chiral case. The differences in the core contributions to the yrast and yrare negative-parity bands mentioned above point that dynamic effects as coupling of the quasiparticles to fluctuations of the shape of the core may lead to differences in the properties of these two bands.

The same mechanisms are supposed to determine the structures of the twin bands in ^{134}Pr and ^{102}Rh, revealing dynamic character of chirality.

9. Conclusions

All models applied to describe chirality in nuclei, like the tilted axis cranking model [2], two quasiparticle+triaxial rotor model [10], core-particle-hole coupling model [1,3] suppose a rigid triaxial core. On the contrary all odd-odd nuclei in which twin bands have been observed are in regions of masses where even-even nuclei are γ-soft, effectively triaxial but not rigid. Their potential energy surface is rather flat in the γ-direction and the coupling with other core structures, not only the ground state band, are significant. It is evident that odd-odd nuclei in these mass regions could not reach the full requirements needed for the existence of chirality, but they can approach them. Our theoretical and experimental investigations clearly show that nuclei of ^{134}Pr and ^{102}Rh support the presence of dynamic chirality. In the cases of nuclei ^{128}Cs [24], ^{135}Nd [25] and ^{194}Tl [26] next to the existence of degenerate bands, B(M1) and B(E2) are similar and could be that conditions for static chirality are fulfilled. More theoretical and experimental efforts are needed to investigate chirality in nuclear physics [27].

10. Acknowledgments

This research has been supported by Bulgarian Science Fund under Contract No. DN 08/6, 13.12.2016.

References

[1] Frauendorf S, Meng Jie 1997 *Nucl. Phys. A* **617** 131
[2] Dimitrov V I, Frauendorf S, and Dönau F, 2000 *Phys. Rev. Lett.* **84** 5732
[3] Starosta K, Koike T, Chiara C J, Fossan D B, and LaFosse D R, 2001 *Nucl. Phys. A* **682** 375c
[4] Starosta K, et al., 2002 *Phys. Rev. C* **65** 044328
[5] Koike T, et al., 2003 *Phys. Rev. C* **67** 044319
[6] Koike T, Starosta K, and Hamamoto I 2004 *Phys. Rev. Lett.* **93** 172502
[7] Tonev D, et al., 2007 *Phys. Rev. C* **76** 044313
[8] Wang S Y, Zhang S Q, Qi B, and Meng J 2007 *Chin. Phys. Lett.* **24** 664
[9] Meng J and Zhang S Q 2010 *J. Phys. G* **37** 064025
[10] Tonev D, et al. 2006 *Phys. Rev. Lett.* **96** 052501
[11] Tonev D et al. 2014 *Phys. Rev. Lett.* **112** 052501
[12] Simpson J 1997 *Z. Phys. A* **358** 139
[13] Muralithar S, et al. 2010 *Nucl. Instrum. Methods Phys. Res., Sect A* **622** 281
[14] Tonev D, Goutev N, Georgiev L S 2016 *J. Phys. Conf. Proc.* **724** 012049
[15] Dewald A, Harissopoulos S, and von Brentano P 1989 *Z. Phys. A* **334** 163
[16] Böhm G, Dewald A, Petkov P, and von Brentano P 1993 *Nucl. Instrum. Methods Phys. Res. A* **329** 248
[17] Petkov P, Tonev D, Gableske J, Dewald A, Klemme T, von Brentano P 1999 *Nucl. Instrum. Methods Phys. Res. A* **431** 208

IOP Conf. Series: Journal of Physics: Conf. Series **1023** (2018) 012020 doi:10.1088/1742-6596/1023/1/012020

[18] Petkov P, Tonev D, Gableske J, Dewald A, von Brentano P 1999 *Nucl. Instrum. Methods Phys. Res. A* **437** 274
[19] Petkov P, Tonev D, Dewald A, von Brentano P 2002 *Nucl. Instrum. Methods Phys. Res. A* **488** 555
[20] Brant S, Tonev D, de Angelis G, Ventura A 2008 *Phys. Rev. C* **78** 034301
[21] Wilets L and Jean M 1956 *Phys. Rev.* **102** 788
[22] Vogel O, et al. 1996 *Phys. Rev. C* **53** 1660
[23] Jolos R V, et al. 1997 *Nucl. Phys. A* **618** 126
[24] Grodner E, et al. 2006 *Phys. Rev. Lett.* **97** 172501
[25] Mukhopadhyay S, et al. 2007 *Phys. Rev. Lett.* **99** 172501
[26] Masiteng P L, et al. 2014 *Eur. Phys. J A* **50** 119
[27] Starosta K and Koike T 2017 *Phys. Scr.* **92** 093002

IOP Conf. Series: Journal of Physics: Conf. Series **1023** (2018) 012021 doi:10.1088/1742-6596/1023/1/012021

Extraction of the ANC from the ^{10}Be$(d,p)^{11}$Be transfer reaction using the ADWA method

J.Yang[1,2], P.Capel[1,3]

[1]Physique Nucléaire et Physique Quantique, Université Libre de Bruxelles, C.P.229, B-1050 Bruxelles, Belgium

[2]Afdeling Kern-en Stralingsfysica, Celestijnenlaan 200d-bus 2418, 3001 Leuven, Belgium

[3]Institut für Kernphysik, Technische Universität Darmstadt, 64289 Darmstadt, Germany

E-mail: `jiecyang@ulb.ac.be`

Abstract. A halo nucleus is built from a core and at least one weakly bound neutron or proton. To understand this unique cluster structure, lots of efforts have been undertaken. During the past decades, the (d,p) reaction has been widely used in experiments and has become an important tool for extracting single-particle properties of nuclei. In this work, our goal is to obtain the Asymptotic Normalization Coefficient (ANC) of the halo nuclei ^{11}Be using the ADWA method. We perform the analysis for the ^{10}Be$(d,p)^{11}$Be stripping reaction at E_d=21.4, 18, 15, and 12MeV for the ground state and first excited state of the composite nucleus ^{11}Be. The experimental measurement was carried out at Oak Ridge National Laboratory by Schmitt *et al.*[1] The sensitivity of the calculations to the optical potential choice is also checked. Overall, the transfer process becomes more peripheral at lower energies and forward angles. Investigation in this area is the best way to extract a reliable ANC from the experimental data. For the ground state of ^{11}Be, the ANC obtained using our method (C=$0.785^{+0.029}_{-0.030}$ fm$^{-1/2}$) shows perfect agreement with the one obtained by Ab initio calculations ($C_{\rm Ab}$=0.786 fm$^{-1/2}$) [2].

1. Introduction

Halo nuclei [3] constitute a unique class of exotic systems, which are mostly found in the neutron-rich region of the nuclear chart and exhibit a much larger matter radius than the stable nuclei do. The halo is a threshold effect arising for some weakly bound nuclei. Since the discovery of the first halo nucleus ^{6}He in 1986 [4], great interest has been put in the study of the halo phenomenon in nuclei both experimentally and theoretically. Up to now, the extraction of accurate structure information of halo nuclei remains as a big challenge due to their very short lifetime, which requires unusual experimental methods and accurate theoretical analyses.

Experimentally, the upgrade of rare isotope beam facilities worldwide provides us with new ways to explore these halo systems. As one of the most classical transfer reaction [5], (d,p) reaction plays an important role in extracting single-particle properties of nuclei with one neutron populating the unoccupied state of the target [6].

On the theoretical side, many reaction theories have been developed to study deuteron-induced reactions. The Distorted Wave Born Approximation (DWBA) [7] is one of the representative methods and received considerable success in describing transfer reactions and extracting spectroscopic information on nuclear structure. However, since this method does not present an adequate treatment of the exact three-body interaction, such as the effects caused

IOP Conf. Series: Journal of Physics: Conf. Series **1023** (2018) 012021 doi:10.1088/1742-6596/1023/1/012021

by the breakup of the deuteron, it is desirable to develop an alternative formulation, which does contain the dominant contributions from all the channels without losing the relative simplicity of the DWBA method. Johnson and Soper initially introduced an zero-range adiabatic prescription (ZR-ADWA) [8] to efficiently take into account the contributions from the breakup channels in transfer processes. Following this work, a finite-range version of the ADWA method (FR-ADWA) was proposed by Johnson and Tandy [9]. The accuracy and effective range of ADWA has been studied by direct comparison with the continuum-discretized coupled channels (CDCC) method and exact Faddeev-AGS calculations (FAGS) [10, 11]. For transfer cross sections at low energy ($E_d \approx 20$MeV), there is a remarkable agreement among the three models. Moreover, when the angular momentum transfer is small and the neutron-nucleus system is loosely bound, ADWA is a good candidate to predict the transfer cross sections.

In the present work, we are interested in studying the nucleus ^{11}Be. It is well known for its distinct n+^{10}Be halo structure [12]. With the aim to understand the halo nature of ^{11}Be, many different experimental methods (e.g. β decay [13], breakup [14], neutron knockout [15] and transfer reactions [1]) have already been carried out. Meanwhile, there has also been considerable effort put forward on the theoretical side to describe the structure and help explain the reactions involving ^{11}Be. Belyaeva *et al.* extracted several representative parameters [16] about the formation of neutron halo in the ground state as well as the excited state of this nuclei within the Coupled Reaction Channels (CRC) calculations. Recently, Ab initio calculations [2] got improved by including the three-nucleon forces and the ^{10}Be-n continuum effects when studying this nucleus. Thanks to these pioneering investigations, now we are able to make some attempts to check the structure information of ^{11}Be. The main purpose of this work is to obtain the ANC by comparing the experimental data with theoretical results, and to find out if our method could increase the reliability of such obtained values. We are going to look at the experimental data from Schmitt *et al.*'s work [1]. In their experiment, the ^{10}Be(d,p) reaction has been used in inverse kinematics at four beam energies to study the structure of ^{11}Be.

The structure of this paper is as follows: In Sec.2, we briefly present the ADWA theoretical framework. In Sec.3, we introduce the numerical inputs and different descriptions of the n-^{10}Be bound state. Finally we show the calculation results and have some discussions in Sec.4. The conclusion is given in Sec.5.

2. Theoretical Framework

Considering a stripping reaction of the form $A(d,p)B$, we adopt the three-body model to help build the theoretical framework. In a simple physical picture, this transfer reaction $A+d \to p+B$ can be viewed as a process in which the valence neutron n from the incident deuteron d populates an unoccupied state in the target nucleus A, producing a composite nucleus $B = n + A$. The proton, neutron and the core A form a three-body system. Within the three-body model, the interactions between them can be described by two-body potentials V_{np}, V_{nA} and U_{pA} during the collision process. U means the optical potential is complex in general. The transfer matrix elements of the (d,p) stripping reaction can be expressed in the post form as [5]

$$T_{post}(pB, dA) = \langle \chi_{pB}^{(-)} \varphi_{nA} | V_{pn} + U_{pA} - U_{pB} | \psi_{dA}^{(+)} \rangle, \tag{1}$$

where $\chi_{pB}^{(-)}$ is the distorted wave function generated by the potential U_{pB} which is responsible for reproducing the elastic scattering on the p-B channel, φ_{nA} represents the wave function of the n-A bound state, and $\psi_{dA}^{(+)}$ corresponds to the exact wave function of this deuteron-target system. For the interaction terms, V_{pn}, the binding potential of deuteron, is the dominating term, and the rest part $(U_{pA} - U_{pB})$ is considered as the remnant term.

The primary concern here is the exact solution of the function $\psi_{dA}^{(+)}$ with respect to the three-body system. Since the DWBA method does not include the breakup component of the wave

IOP Conf. Series: Journal of Physics: Conf. Series **1023** (2018) 012021 doi:10.1088/1742-6596/1023/1/012021

function, we chose to use the adiabatic prescription that was initially introduced by Johnson and Soper [8] to take into account the contributions from the breakup channels in transfer processes. In the ADWA model, the three-body wave function is approximated by

$$\psi_{dA}^{(+)}(\boldsymbol{r'}, \boldsymbol{R}) \simeq \chi_{dA}^{(+)}(\boldsymbol{r'}, \boldsymbol{R})\varphi_{pn}(\boldsymbol{r'}), \tag{2}$$

where $\chi_{dA}^{(+)}$ is the deuteron-target scattering wave function. It depends on the relative coordinates $\vec{r'}$ between the neutron and the proton and $\vec{R}$ between the n-p-centre of mass and the target A. φ_{pn} is the deuteron wave function. Once the transfer amplitude $T_{post}(pB, dA)$ is obtained, we can calculate the differential cross section

$$\left(\frac{d\sigma}{d\Omega}\right)^{th} = \frac{\mu_{dA}\mu_{pB}}{(2\pi^{-}\hbar)^2}\frac{k_p}{k_d}|T_{post}(pB, dA)|^2, \tag{3}$$

in which μ_{iX} and k_i are the reduced mass and wave number for the relative motion between i and X in each of the two channels dA (elastic entrance channel) and pB (transfer channel).

Referring to the wave function for the relative motion between n and A that forms B in Eq. 1, let us consider a specific nucleus B that has a strong single-particle character. In this case, the valence neutron orbit has a well-defined asymptotic behavior at large distances where nuclear forces are vanishingly small [17],

$$\varphi_{nA}(r) \xrightarrow{r\to\infty} b_{n_r lj}\frac{W_{-\eta,\ l+\frac{1}{2}}(2\kappa r)}{r}, \tag{4}$$

in which $b_{n_r lj}$ is the single-particle ANC (SPANC) defining the strength of the exponential tail of the n-A bound-state wave function, W is the Whittaker function, η is the Sommerfeld parameter for the bound state, and κ is the wave number. Note in order to well simulate what has been observed in experiments from the microscopic calculations, the bound-state wave function φ_{nA} in Eq. 1 should be replaced by the overlap function I_{nA}^{B}, which will take into account nuclear structure effects realistically. The overlap integrals are not longer normalized to unity, but to the spectroscopic factors (SF) $S_{n_r lj}$ while experiencing a similar asymptotic behavior [6]. In the single-particle approximation, the ANC and SFANC are related by $C_{lj}^2 = S_{n_r lj}b_{n_r lj}^2$. When the reaction is completely peripheral, the transfer differential cross sections should only be sensitive to the value of ANC without retaining information on the SF.

3. Transfer calculation

3.1. Numerical inputs

For the ^{10}Be$(d,p)^{11}$Be transfer reaction, the equivalent deuteron incident energy E_d is taken to be 21.4, 18, 15, and 12MeV refer to the experiment of Schmitt *et al.*[1] In our calculations, a finite-range version of the adiabatic potential developed by Johnson and Tandy is used for transfers from deuteron [18]. The nucleon-nucleus optical potentials are obtained from the global parametrization CH89 [19] without including the spin-orbit terms. Using ADWA approach, $U_{p-^{10}\mathrm{Be}}$ and $U_{n-^{10}\mathrm{Be}}$ are calculated at half the incident energy $E_d/2$, while the auxiliary potential $U_{p-^{11}\mathrm{Be}}$ is obtained at the proton energy corresponding to the exit channel. For the p-n interaction, the Reid soft-core interaction [20] is chosen to get the appropriate wave function of the deuteron. All transfer calculations in this study were performed with FRESCO [21], and the adiabatic potentials are calculated using the front-end code of TWOFNR [22].

3.2. Description of ^{11}Be

As mentioned, ^{11}Be can be modeled as a neutron loosely bound to a ^{10}Be core. With the assumption that the ^{10}Be core is in its ground state (0^+), the $1/2^+$ ground state (g.s) of ^{11}Be

IOP Conf. Series: Journal of Physics: Conf. Series **1023** (2018) 012021 doi:10.1088/1742-6596/1023/1/012021

can be described by a $2s_{1/2} \otimes {}^{10}\mathrm{Be}(0^+)$ configuration, and the $1/2^-$ excited state (ex.s) by a $1p_{1/2} \otimes {}^{10}\mathrm{Be}(0^+)$ configuration. In this study, we adopt Gaussian potentials to describe the interaction between ^{10}Be and the valence neutron. The potential is defined as $V(r) = V_0 e^{-r^2/2r_0^2}$. We test nine sets of Gaussian potentials with different widths r_0 starting from 0.4fm to 2.0fm. The depth V_0 is adjusted based on the neutron binding energy (0.502MeV for the g.s; 0.182MeV for the ex.s). They are all listed in Tables 1 with the corresponding SPANCs ($b_{2s_{1/2}}$, $b_{1p_{1/2}}$) provided. Using such potentials, we can calculate the wave function of the n-^{10}Be bound state.

Table 1. Gaussian parameters of the n-^{10}Be potentials. The potential depths and SPANCs of the $2s1/2$ ground state $b_{2s1/2}$ and the $1p1/2$ first excited state $b_{1p1/2}$ are shown as well.

Potential	r_0 (fm)	$V_0(g.s)$ (MeV)	$b_{2s1/2}$ ($\mathrm{fm}^{-1/2}$)	$V_0(ex.s)$ (MeV)	$b_{1p1/2}$ ($\mathrm{fm}^{-1/2}$)
V_1	0.4	1314.6	0.601	869.4	0.068
V_2	0.6	592.3	0.632	387.3	0.085
V_3	0.8	337.8	0.664	218.4	0.100
V_4	1.0	219.2	0.697	140.2	0.114
V_5	1.2	154.4	0.732	97.7	0.127
V_6	1.4	115.1	0.769	72.1	0.140
V_7	1.6	89.3	0.807	55.4	0.152
V_8	1.8	71.6	0.846	44.0	0.165
V_9	2.0	58.8	0.888	35.8	0.177

The results of the g.s and the ex.s wave functions are shown in Fig. 1. According to the definition of SPANC in Eq.(4), all the wave functions shown here will exhibit a same tail after scaled by their SPANCs. This phenomenon can help us define the confidence interval to extract ANC.

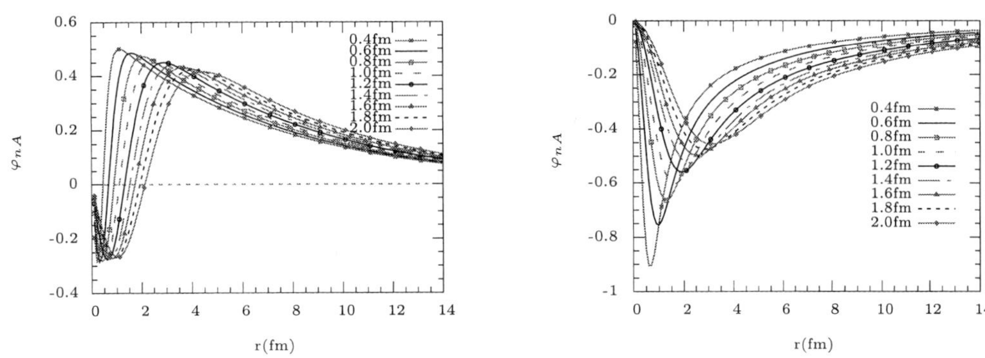

Figure 1. (Color online). Wave functions φ_{nA} obtained with nine different sets of Gaussian potentials listed in Tables 1 for the $1/2^+$ g.s of ^{11}Be (left) and for the $1/2^-$ excited state (right).

4. Results and Discussion

In this section, a series of calculation results is presented. Here we take two cases at different energies to clarify our analysis work and findings. On the left side of Fig.2, the top plot (a) shows the angular distribution of the cross sections calculated at 21.4MeV. With narrower

Gaussian potentials, the probability of occurrence of the reaction gradually decreases. At forward angles, there exist big discrepancies among the calculated cross sections with different potentials. Dividing the cross sections by the square of the SPANC, we obtain plot (b). From (a) to (b), these discrepancies are much reduced. Most curves begin to get on top of each other in the forward region except those of r_0=0.4fm, and 0.6fm. In order to gain quantitative insight into the variation between different distributions after scaling, we compare them taking the calculation at r_0=1.4fm as a reference:

$$\mathcal{R}_{r_0/1.4fm} = \frac{(\frac{d\sigma}{d\Omega})^{th}_{r_0}/(b_{n_rlj})^2_{r_0}}{(\frac{d\sigma}{d\Omega})^{th}_{1.4fm}/(b_{n_rlj})^2_{1.4fm}} - 1 \tag{5}$$

in which the index r_0 corresponds to a set of Gaussian potential used for calculation. The results are presented in plot (c). Then we choose the empirical value ±5% set as the tolerance band to define the peripheral area of the reaction. One sees from this plot that only a small angular region meets our peripheral definition except the variations with respect to 0.4fm, and 0.6fm. For the ^{10}Be$(d,p)^{11}$Be(g.s) reaction at 21.4MeV, we have $\theta_{c.m.}$=0°–7° extracted as the angular range where the $b^2_{n_rlj}$ scaling works.

Coming to the right side of Fig.2, the reaction is carried out at E_d=12MeV. All the plots are organized following the same logical structure shown for the case of 21.4MeV. Through comparison between the results at two different energies, we can find that $b^2_{n_rlj}$ scaling works better at 12MeV, leading to a wider and more consistent angular region. The peripheral region in this case is $\theta_{c.m.}$=0°–17°. Similar analyses are also made at 18MeV and 15MeV, which are not presented here, to check the variation of the angular agreement range with the deuteron energy. The peripheral part is 0°–10° for 18MeV, and 0°–20° for 15MeV without considering the results of 0.4fm, and 0.6fm. After investigating all the results, we have two basic conclusions: a)the peripheral area of this transfer reaction is always found at forward angles; b)when the incident energy decreases, the reaction exhibits a more pronounced peripheral property.

4.1. ANC Extraction of the n-^{10}Be bound state

With the peripheral information obtained in the last section, we can extract the ANCs by performing the χ^2 analysis for the scaling of the theoretical results to the experimental ones. The related results are shown in Fig. 3. It can be seen that the ANC extraction is more reliable at lower energy for both cases since the reaction becomes more peripheral there. For 21.4MeV, we don't observe a plateau since at this energy the reaction is less peripheral and the interior part of the bound-state wave function plays a non-negligible role. Therefore this is not an ideal sample for ANC extraction. There is a problem with the results at 18MeV which are always smaller than the others. It has been seen in the analysis of Schmitt *et al.* [1] as well. However, the most reliable sets of data at 15MeV, and 12MeV lead to consistent results. The value of the ANC we extract for ^{11}Be is $0.785^{+0.029}_{-0.030}$ fm$^{-1/2}$ ($C_{\rm Ab}$=0.786 fm$^{-1/2}$) for the g.s and $0.136^{+0.005}_{-0.005}$ fm$^{-1/2}$ ($C_{\rm Ab}$=0.129 fm$^{-1/2}$) for the ex.s. In general, our results are in agreement with the ANC predicted from an Ab initio calculation [2] and also close to that found by Belyaeva *et al.* [16] with the CRC model. Moreover, the extracted ANC reproduces a good agreement at forward angles between theoretical and experimental results despite the 'always' overestimation of the calculations at 18MeV. Due to the fact that this analysis relies a lot on the accuracy of the experimental data, the conclusion of this analysis is that experimenters should focus on low energies and forward angles to extract precise ANCs.

4.2. The sensitivity to the optical potential choice

On the other hand, the angular distributions of cross sections calculated with ADWA model can depend strongly on the optical potential parameters. In this section, we would like to check the

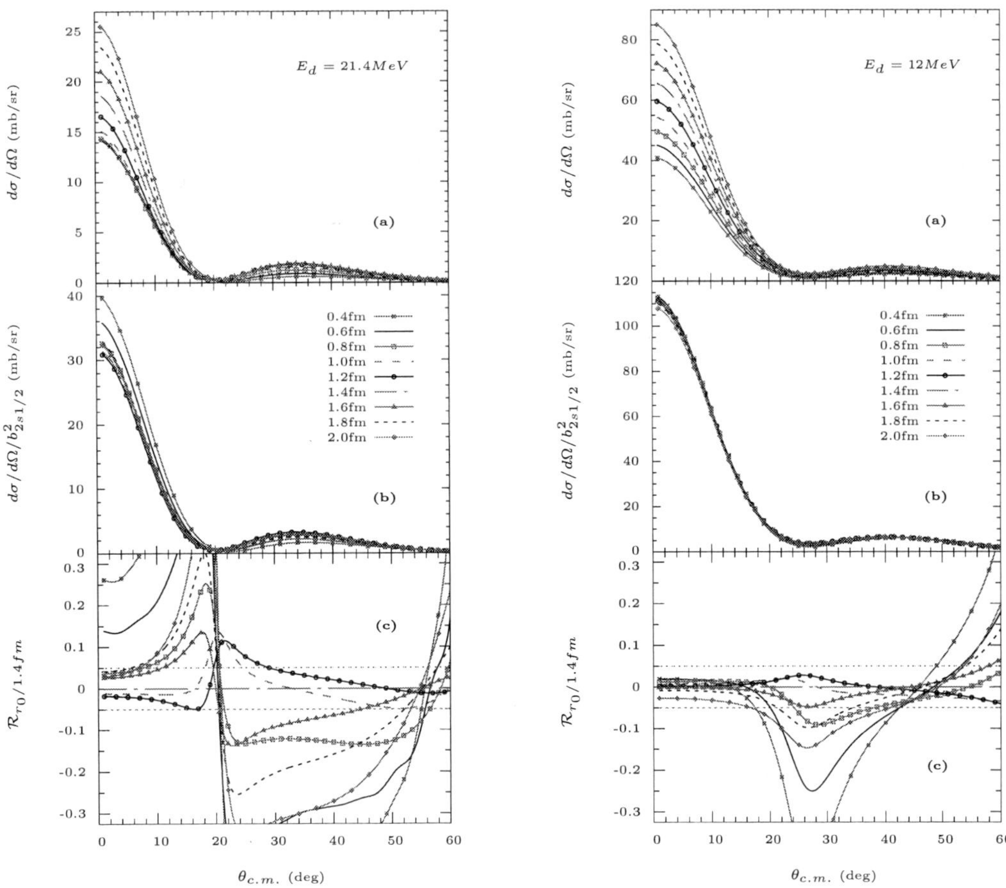

Figure 2. (Color online). Analysis of the cross sections of ^{10}Be$(d,p)^{11}$Be to the ground state of ^{11}Be with incoming deuteron energy of 21.4 MeV (left side) and 12MeV (right side).

influence of the nucleon-nucleus optical potentials on the calculations. In particular, we choose Koning & Delaroche potential (KD) [23] instead of CH89 to perform this series of analyses. Since the cross sections calculated with KD potentials lead systematically to larger cross sections than those with CH89, we can expect a smaller ANC obtained using the same method. The value of the extracted ANC is $0.755^{+0.028}_{-0.029}$ fm$^{-1/2}$ with a 3.9% deviation from the Ab initio one.

5. Conclusion

An analysis of the theoretical results of Schmitt *et al.*'s experiment has been performed with the aim to extract the ANC of the halo nucleus ^{11}Be using the ADWA method in this paper. Using Gaussian potentials of different widths to generate the wave functions of the $n+^{10}$Be bound states, we are able to define a peripheral area where the SPANC^2 scaling works well. By comparing between theoretical results at four energies, it is noticed that transfer reaction becomes more peripheral at lower energies and forward angles. Based on this knowledge, we proposed a systematic analysis to extract ANCs from experimental data in this region. The ANC extracted for the ground state of ^{11}Be (C_{lj}=$0.785^{+0.029}_{-0.030}$ fm$^{-1/2}$) is in excellent agreement with the value predicted by Ab initio calculations (C'_{lj}=0.786 fm$^{-1/2}$). The sensitivity of the calculations to the optical potential choice is also checked.

IOP Conf. Series: Journal of Physics: Conf. Series **1023** (2018) 012021 doi:10.1088/1742-6596/1023/1/012021

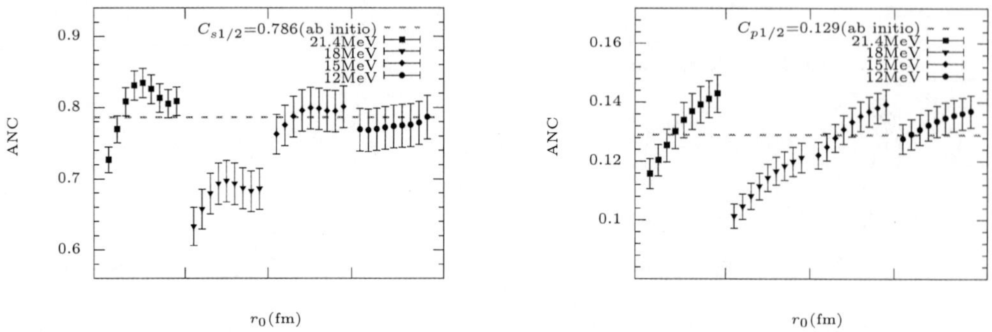

Figure 3. ANC extracted for the g.s (left side) and ex.s (right side) of ^{11}Be. The Ab initio result (C_{Ab}=0.786 fm$^{-1/2}$ for g.s or 0.129 fm$^{-1/2}$ for ex.s) [2] for comparison is presented by a red dashed line. From left to right in each grid of the horizontal axis, the data points start with the ANC extracted for r_0=0.4fm and end with that for r_0=2.0fm.

Overall, our study indicates that investigation at lower energies and forward angles for transfer reaction can ensure us the peripherality of the reaction and is the best way to obtain a reliable ANC from experimental data. Our method also suggests a good practice for this purpose.

Acknowledgments

We thank Antonio M. Moro for the support in doing the calculation. This work was supported by the China Scholarship Council (CSC), the Belgian Research Initiative on eXotic nuclei (BriX) network and in part by the European Union's Horizon 2020 research and innovation programme under Grant Agreement No. 654002.

References

[1] Schmitt K T, Jones K L, Bey A *et al.* 2012 *Phys. Rev. Lett.* **108**(19) 192701
[2] Calci A, Navrátil P, Roth R *et al.* 2016 *Phys. Rev. Lett.* **117**(24) 242501
[3] Al-Khalili J 2004 *An Introduction to Halo Nuclei* (Berlin, Heidelberg: Springer Berlin Heidelberg) pp 77–112
[4] Tanihata I, Hamagaki H, Hashimoto O *et al.* 1985 *Physics Letters B* **160** 380 – 384
[5] Camacho J G and Moro A M 2014 *The Euroschool on Exotic Beams* **Vol. IV** 39–66
[6] Thompson I J and Nunes F M 2009 *Nuclear Reactions for Astrophysics, by Ian J. Thompson, Filomena M. Nunes, Cambridge, UK: Cambridge University Press, 2009*
[7] Satchler G 1964 *Nuclear Physics* **55** 1 – 33
[8] Johnson R C and Soper P J R 1970 *Phys. Rev. C* **1**(3) 976–990
[9] Johnson R and Tandy P 1974 *Nuclear Physics A* **235** 56 – 74
[10] Nunes F M and Deltuva A 2011 *Phys. Rev. C* **84**(3) 034607
[11] Upadhyay N J, Deltuva A and Nunes F M 2012 *Phys. Rev. C* **85**(5) 054621
[12] Tanihata I, Kobayashi T, Yamakawa O *et al.* 1988 *Physics Letters B* **206** 592 – 596
[13] Madurga M, Borge M, Alcorta M *et al.* 2009 *Physics Letters B* **677** 255 – 259
[14] Fukuda N, Nakamura T, Aoi N *et al.* 2004 *Phys. Rev. C* **70**(5) 054606
[15] Aumann T, Navin A, Balamuth D P *et al.* 2000 *Phys. Rev. Lett.* **84**(1) 35–38
[16] Belyaeva T L, Perez-Torres R, Ogloblin A A *et al.* 2014 *Phys. Rev. C* **90**(6) 064610
[17] Pang D Y, Nunes F M and Mukhamedzhanov A M 2007 *Phys. Rev. C* **75**(2) 024601
[18] Wales G and Johnson R 1976 *Nuclear Physics A* **274** 168 – 176
[19] Varner R, Thompson W, McAbee T, Ludwig E and Clegg T 1991 *Physics Reports* **201** 57 – 119
[20] Reid R V 1968 *Annals of Physics* **50** 411 – 448
[21] Thompson I J 1988 *Computer Physics Reports* **7** 167 – 212
[22] Igarashi M and Toyama M 2008 *computer program TWOFNR, University of Surrey version*
[23] Koning A and Delaroche J 2003 *Nuclear Physics A* **713** 231 – 310

IOP Conf. Series: Journal of Physics: Conf. Series **1023** (2018) 012022 doi:10.1088/1742-6596/1023/1/012022

The boundary of the N=90 shape phase transition: ^{148}Ce

P. Koseoglou [1,2], V. Werner [1,3], N. Pietralla [1], S.Ilieva [1], M. Thürauf [1], C. Bernards [3], A. Blanc [4], A. M. Bruce [5], R. B. Cakirli [6,7], N. Cooper [3], L. M. Fraile [8], G. de France [9], M. Jentschel [4], J. Jolie [10], U. Koester [4], W. Korten [11], T. Kröll [1], S. Lalkovski [12], H. Mach [13,†], N. Mărginean [14], P. Mutti [4], Z. Patel [15], V. Paziy [9], Z. Podolyák [15], P. H. Regan [15,16], J.-M. Régis [10], O. J. Roberts [5], N. Saed-Samii [10], G. S. Simpson [17], T. Soldner [4], C. A. Ur [18], W. Urban [4], D. Wilmsen [9], E. Wilson [14]

[1] IKP TU-Darmstadt, Germany / [2] GSI, Germany / [3] Yale University, USA
[4] ILL Grenoble, France / [5] University of Brighton, UK / [6] MPIK Heidelberg, Germany
[7] Department of Physics, University of Istanbul, Turkey
[8] Universidad Complutense, Spain / [9] GANIL Caen, France
[10] IKP University of Cologne, Germany
[11] CEA, IRFU, Université Paris-Saclay, France
[12] University of Sofia "St. Kliment Ohridski", Bulgaria
[13] National Centre for Nuclear Research, Poland / [14] Horia Hulubei NIPNE, Romania
[15] University of Surrey, UK / [16] National Physical Laboratory, UK
[17] LPSC Grenoble, France / [18] INFN Legnaro, Italy

pkoseoglou@ikp.tu-darmstadt.de

Abstract. The even-even N=90 isotones with Z=60-66 are known to undergo a first order phase transition. Such a phase transition in atomic nuclei is characterized by a sudden change of the shape of the nucleus due to changes in the location of the potential minimum. In these proceedings we report a measurement of the $B_{4/2}$ ratio of ^{148}Ce, which will probe the location of the low-Z boundary of the N=90 phase transitional region. The measured $B_{4/2}$ value is compared to the prediction from the X(5) symmetry within the interacting boson model at the critical point between the geometrical limits of vibrators and rigid/axial rotors. The EXILL&FATIMA campaign took place at the high-flux reactor of the Institut Laue Langevin, Grenoble, were ^{235}U and ^{241}Pu fission fragments were measured by a hybrid spectrometer consisting of high-resolution HPGe and fast $LaBr_3$(Ce)-scintillator detectors. The fast $LaBr_3$(Ce) detectors in combination with the generalized centroid difference method allowed lifetime measurements in the picosecond region. Furthermore, this kind of analysis can serve as preparation for the FATIMA experiments at FAIR.

1. Introduction-Motivation

The even-even N=90 isotones with Z=60-66, present an interesting phenomenon, known as shape-phase transition. This phase transition in atomic nuclei is characterized by a sudden change of the shape of the nucleus due to changes in the location of the potential minimum [1]. Both the characteristic level schemes and transition strengths, or their ratios, are fingerprints of the location of an isotope within a phase transition region. As such, the $R_{4/2}=E(4^+_1)/E(2^+_1)$ and $B_{4/2}=B(E2; 4^+_1\rightarrow 2^+_1)/B(E2; 2^+_1\rightarrow 0^+_1)$ values are sensitive benchmarks for this type of structure of a given nucleus.

In F. Iachello's Physical Review Letter [4] a schematic representation of the lowest portion of the spectrum of X(5) symmetry is provided, which can be compared to experimental data. For ^{152}Sm and ^{150}Nd this is presented in R.F. Casten's Physical Review Letter [1] and reveals the X(5) flavor of the nuclei. This X(5) symmetry describes nuclei that are located on the critical point between spherical, U(5) and axially deformed shapes, SU(3). Using the adopted experimental data, the discussed phase transition can be observed in a $R_{4/2}$ plot of these isotopes over the neutron number (figure 1). The sharp transition in the gadolinium and samarium isotopic chains [1, 2, 3] from spherical nuclei ($R_{4/2}$=2-2.4) to deformed ones ($R_{4/2}$=3-3.33) around N=90 is less pronounced in the neodymium and cerium chains. The latter chains undergo a transitional $R_{4/2}$ ratio from spherical to deformed nuclei ($R_{4/2}$=2.93 and 2.86 for neodymium and cerium respectively) at N=90 and the transition from N=88 to N=90 is smooth. The $B_{4/2}$ ratio can give additional information on the shape of the nucleus ($B_{4/2}$=2 for spherical symmetry, $B_{4/2}$=1.4 for γ-rigid and γ-soft deformed). Figure 2 presents the $B_{4/2}$ ratio for gadolinium, samarium and neodymium isotopes as a function of the neutron number. The adopted data were used. While the transition from N=88 to N=90 from near spherical symmetry to quadrupole deformed shapes is sharp for gadolinium and samarium it is less so for neodymium.

As indicated in figure 3, the N=90 isotones lie near the X(5) prediction [4]. In the case of ^{148}Ce, located at the low-Z boundary of this phase transition, the $R_{4/2}$ ratio (2.86) matches the X(5) prediction of 2.91 [4]. The P-factor, introduced in [5], defined in terms of number of valence protons and neutrons (N_p,N_n) by: $P=(N_p{\cdot}N_n)/(N_p+N_n)$, reflects the strength of the valance p-n interaction (numerator) and the strength of the pairing interaction (denominator). As noted in [2], a typical p-n interaction in heavy nuclei is ~ 200 keV and the pairing interaction is of the order of 1 MeV. Empirically it is observed that the p-n interaction begins to dominate for P ~ 4 - 5, which provides a pathfinder to possible X(5) candidates. For ^{148}Ce, with 8 valence protons and 8 valence neutrons, P=4, making this isotope a candidate for an X(5)-like structure. One signature for such an X(5) symmetry, as mentioned above, would be the $B_{4/2}$ ratio, for ^{148}Ce that has not been measured to date.

IOP Conf. Series: Journal of Physics: Conf. Series **1023** (2018) 012022 doi:10.1088/1742-6596/1023/1/012022

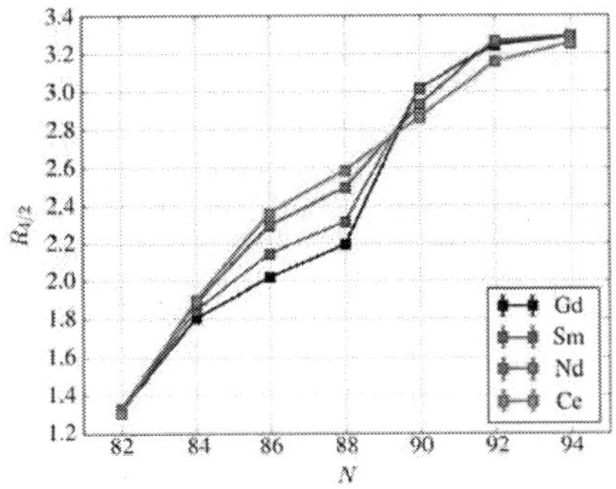

Figure 1. $R_{4/2}$ ratio over neutron number. The sharp transition of Gd and Sm from spherical nuclei ($R_{4/2}$=2) to deformed ones ($R_{4/2}$=3.33) is not present in Nd and Ce chains. Data taken from [6].

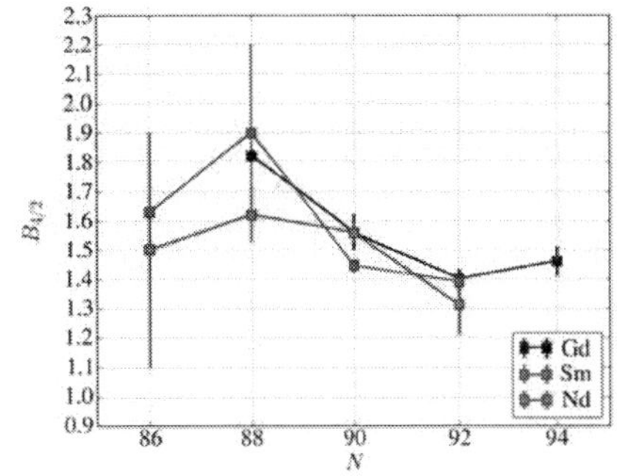

Figure 2. $B_{4/2}$ ratios for Gd, Sm and Nd isotopes as a function of neutron number. The transition from N=88 to N=90 from near spherical symmetry to γ-rigid and γ-soft symmetry is sharp for Gd and Sm but not for Nd. Data taken from [6].

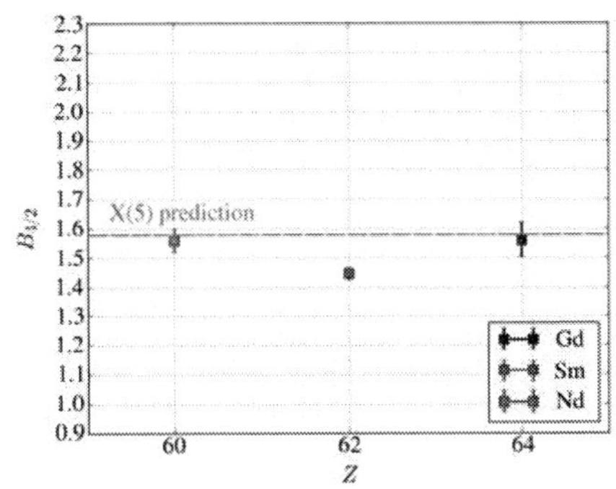

Figure 3. The $B_{4/2}$ ratio for N=90 isotones. All isotopes lie near the X(5) prediction. Data taken from [6].

2. Experiment

The EXILL&FATIMA campaign provided data for neutron-rich species in the vicinity of N=90. Using cold-neutrons fission of ^{235}U and ^{241}Pu was induced at the Institut Laue-Langevin (ILL) of Grenoble, France, and the prompt γ-rays coming from the nuclei in interest were detected using a hybrid array of HPGe and Ce-doped $LaBr_3$ detectors, the EXILL&FATIMA spectrometer. The EXILL array was composed of 8 BGO-shielded EXOGAM Clover detectors, each one consisting of 4 HPGe crystals, with a target-to-detector distance of 14.5 cm, placed at 90° relative to the beam direction, in ring arrangement. FATIMA consisted of 16 (5% Ce-doped) $LaBr_3$ detectors in two rings, at 40° and 140° relative to the beam direction. The efficiency of the $LaBr_3$ detectors was crucial for the experiment, so they were placed as close as possible to the target, resulting in a target-to-detector distance of 8.5 cm, almost touching each other. The neutron flux at the target position was approximately $9 \cdot 10^7$ n/cm²·s. More details about the experimental setup can be found in [7].

IOP Conf. Series: Journal of Physics: Conf. Series **1023** (2018) 012022 doi:10.1088/1742-6596/1023/1/012022

All the fission fragments were stopped by the thick backing of the targets within ~1ps. More than 100 different isotopes were produced from the fission, thus multiple coincidences were utilized to separate an isotope of interest. Ge-$LaBr_3$-$LaBr_3$ coincidences were used to gate on the cascade of interest. The HPGe high-resolution detectors of the EXILL array allowed a precise energy gate to be set on a γ-ray cascade of interest and, hence, the nuclei of interest. The excellent timing performance of the $LaBr_3$ detectors in combination with the generalized centroid difference method allowed to measure lifetimes down to the ps range [8].

In this work the investigation at ^{148}Ce is presented.

3. Fast-timing measurements

As mentioned in the previous section the hybrid nature of the detector array allows both clean energy gates (using the HPGe detectors of EXILL) and fast-timing measurements (using the the fast $LaBr_3$ detectors of FATIMA). The Ge selection gates were placed in the same cascade as the level of interest but not on its feeder or decay gammas. The FATIMA detectors were used to measure the lifetimes of these levels. The $LaBr_3$ detectors were connected to the TACs in such a way that it allowed to know which of the two prompt gammas (decay or feeder) was the one providing the start or stop signal. In total 15 TACs were used (N-1, with N the number of $LaBr_3$ detectors) in such a way that TAC_i was able to be started by detector i and to be stopped by detector j, with $j>i$. A detailed description of the electronics setup used in the experiment can be found in Ref. [7].

Two time-difference spectra can be produced depending on whether the decay gamma ray is providing the start (start spectrum) or stop (stop spectrum) signal (whether the feeder gamma ray is providing the stop or start signal respectively). In the start spectrum the centroid (C_{start}) is shifted, by one lifetime, on the left direction, in the stop spectrum the centroid (C_{stop}) is shifted on the right direction. In figure 4 (adapted from [8]), a simpler case of this experimental setup together with the two corresponding time-distribution spectra are presented. The time spectra presented corresponds to a short lifetime in the ps range, this is the reason for the two prompt-shaped time distributions. The lifetime derives from the centroid difference (ΔC), see section 3.2. In a case of a long-lived state, a slope appears on one side of the spectrum, see section 3.1, and the slope corresponds to the lifetime.

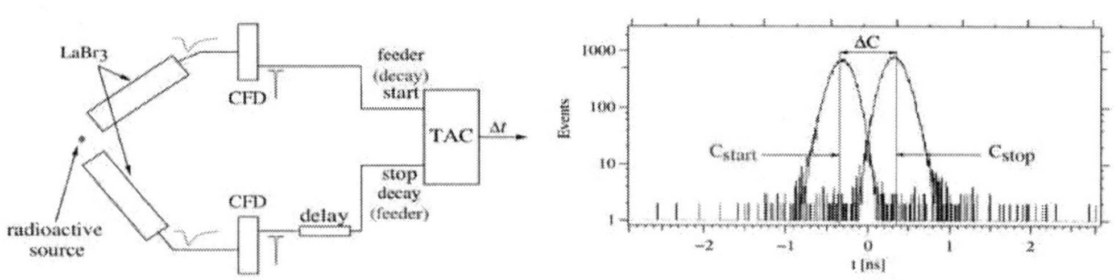

Figure 4. Simpler case of this experimental setup of just two detectors and the two time-distribution spectra are presented. Two time spectra can be produced depending on whether the decay gamma is providing the start (with the corresponding centroid, C_{start}) or stop (with the corresponding centroid, C_{stop}) signal (whether the feeder-gamma is providing the stop or start signal respectively).

IOP Conf. Series: Journal of Physics: Conf. Series **1023** (2018) 012022 doi:10.1088/1742-6596/1023/1/012022

3.1 2^+_1 lifetime using the slope method

For the 2^+_1 state, the slope method [9,10] was used due to the long lifetime in the order of ns. A Ge energy gate, on the decay gamma ray of the $6^+_1 \rightarrow 4^+_1$ transition, has been used for the selection of the cascade of interest. The start and the stop signals to the TACs were given by the feeder and the decay gamma of the 2^+_1 state and detected with FATIMA. After the Ge gate was performed, the feeder gamma ray ($4^+_1 \rightarrow 2^+_1$ transition) was set as the stop signal and the decay gamma ray ($2^+_1 \rightarrow 0^+_1$ transition) as start signal. Two background cuts on the right and left of the feeder gamma ray performed for Compton and random background subtraction from the time-difference spectrum. In figure 5 the resulting time-difference spectrum is shown. A clear slope was observed, corresponding to the lifetime of the state. The lifetime measured [11] was in agreement with the literature value [12], giving confidence in our method of handling the Compton and random background.

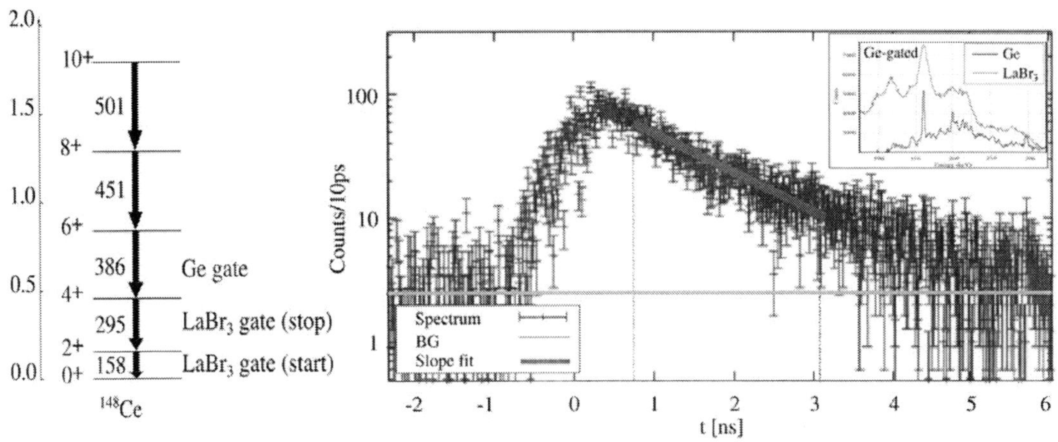

Figure 5. Start time spectrum, after performing the two background cuts. A clear slope can b seen. The slope corresponded to the lifetime of the state. The lifetime measured tagreed with the literature value [12]. The Ge-gated $LaBr_3$ (green) and Ge (blue) spectra can be seen on the top right. On the left side the level scheme is presented and the gates performed are indicated.

3.2 4^+_1 lifetime using the Generalized Centroid Difference method

For the 4^+_1 state the generalized centroid shift method [13] has been used for the determination of the lifetime (in the ps range). A Ge energy gate was set on the decay gamma of the 2^+_1 state to select the cascade of interest. The start and the stop spectra were produced by gating the decay gamma ray on the start and stop signal respectively (gatting the feeder gamma ray on the stop and start signal respectively). From the centroid difference (ΔC), of the centroids of the two time-difference spectra produced (see figure 7), the lifetime (τ) of the state could be measured. From using

$$\tau = (\Delta C - PRD) / 2 \ ,$$

where the prompt response difference ($PRD=PRD(E_{feeder}-E_{decay})=PRF(E_{feeder})-PRF(E_{decay})$) is reflecting the different time that the setup requires to record gammas with different energies (the so-called time walk). The prompt response function (PRF) was calibrated by measuring known cascades from a ^{152}Eu source and the $^{48}Ti(n, \gamma)^{49}Ti$ reaction. In figure 6 the produced PRF can been seen, adjusted for the reference energy of 342 keV. A precise description of the PRD calibration procedure is given in [14]. The Compton contribution to the time spectrum has been eliminated by performing two background cuts on the right and the left of the feeding gamma peak and subtract the mean time spectrum from the time spectrum produced by cutting on the peak. The two time-spectra (start and stop) are shown in figure 7, together with the centroids.

From the measured lifetimes the B(E2; $4^+_1 \rightarrow 2^+_1$) and B(E2; $2^+_1 \rightarrow 0^+_1$) were calculated to be 140 (7) W.u. and 88 (3) W.u. [11] respectively, the $B_{4/2}$ ratio was calculated to be 1.6 (1), very near to the X(5) prediction (figure 8). In Ref. [15], an early state of the analysis was presented; quadruple coincidences have been used (Ge-Ge-$LaBr_3$-$LaBr_3$) for the measurement of the 4^+_1 lifetime. Beside the low statistics on the quadruple-gated case, which resulted in a large uncertainty, the $B_{4/2}$ ratios detected through the two methods (triple and quadruple gates) are in agreement, within the errors.

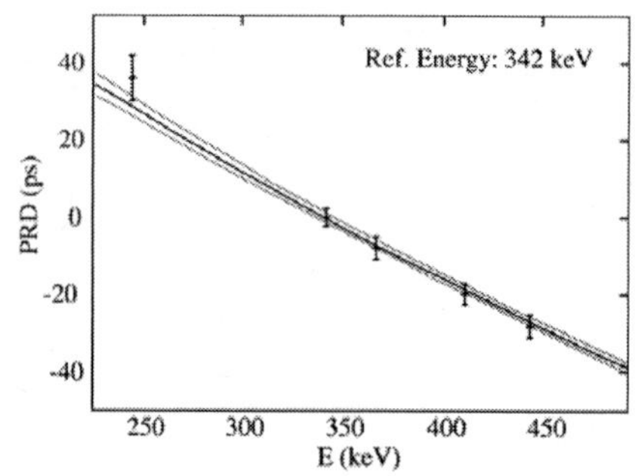

Figure 6. The prompt response function (PRF), adjusted for the reference energy o 342 keV, has been calibrated by measuring known cascades from a ^{152}Eu source and the ^{48}Ti(n, γ)^{49}Ti reaction.

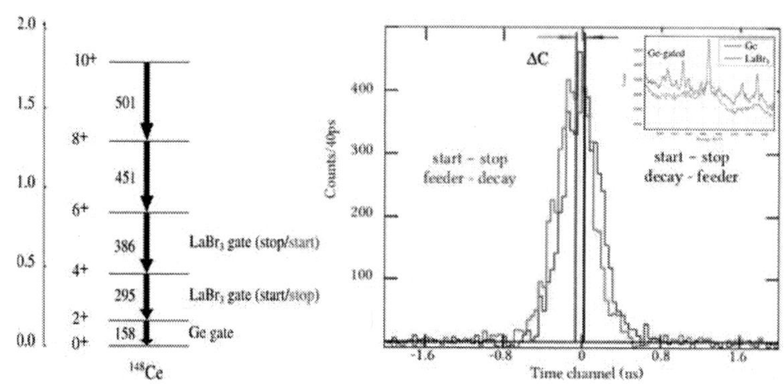

Figure 7. The two time-spectrum (start and stop), the centroids are also pictured. B measuring the lifetime the $B_{4/2}$ ratio was calculated. Also the Ge-gated $LaBr_3$(green) and Ge (blue) spectra can be seen on the top right. On the left side the level scheme and and the gates selected are indicated.

4. $B_{4/2}$ ratio – X(5) character of ^{148}Ce

By measuring the 4^+_1 lifetime for first time, the the $B_{4/2}$ ratio was obtained, after correction for internal conversion. The calculated $B_{4/2}$ ratio is very near to the X(5) symmetry prediction for nuclei on the transition phase between spherical, U(5) and axially deformed shapes, SU(3). Plotting the energy levels emanating from X(5) and the experimentally measured in ^{148}Ce side by side (figure 8, left), one can perceive the good agreement between them.

^{148}Ce seems to follow the N=90 isotones' picture of the X(5) symmetry. In figure 8 one can witness the trend over the isotones. Together, all the above can designate ^{148}Ce as a good X(5) candidate.

Further indicators for the X(5) symmetry would be additional transition strengths' ratios, that were not obtained from the experimental data due to low statistics on higher excited levels.

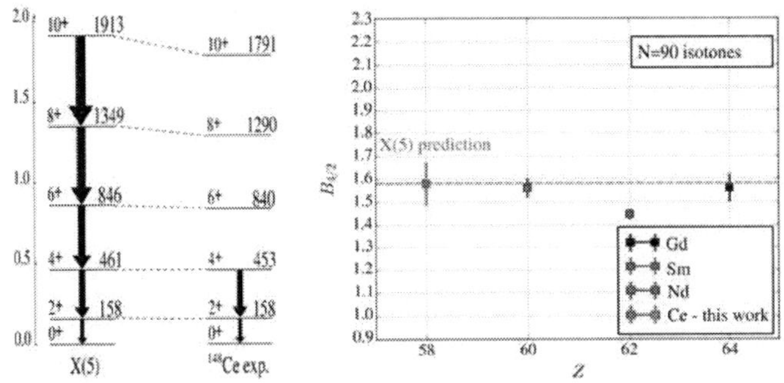

Figure 8. $B_{4/2}$ ratio of N=90 isotones. ^{148}Ce seems to follow the X(5) trend of the rest isotopes. Data, except for ^{148}Ce, taken from [8]. On the left the comparison between the X(5) prediction for the energy levels (normalized to the 2^+_1 state) and the experimental values is presented.

Acknowledgments

The EXILL&FATIMA campaign would not have been possible without the support of several services at the ILL and the LPSC. We are grateful to the EXOGAM collaboration for the loan of the detectors,to GANIL for assistance during installation and dismantling, and to the FATIMA collaboration for the provision of LaBr3 (Ce) detectors and analogue electronics.

This work was supported by the cooperation between TU Darmstadt and the GSI Helmholtz Center for Heavy Ion Research and by HGS-HIRe, by US DOE under No. DE-FG02-91ER-40609, by STFC (UK) under No. ST/L005743/1 and ST/G000751/1, by Istanbul University Scientific Research Project No. 54135, by DFG under No. KR 1796/2-1 and KR 1796/2-2 and by BMBF under No. 05P12RDNUP 05P15RDFN1, 05P15PKFNA and 05P12PKNU.

References

[1] R. F. Casten and N. V. Zamfir, Phys. Rev. Lett. **87**, 052503 (2001).
[2] R. F. Casten, Nature Physics **2**, 811 (2006).
[3] N. Pietralla and O.M.Gorbachenko, Phys. Rev. C **70**, 011304 (2004).
[4] F. Iachello, Phys. Rev. Lett. **87**, 052052 (2001).
[5] R. F. Casten, D. S. Brenner and P. E. Haustein, Phys. Rev. Lett. **58** 658 (1987).
[6] Nuclear Data Sheets, http://www.nndc.bnl.gov/ensdf/
[7] J.-M. Régis, et al., Nucl. Instr. and Meth., A **763** 210 (2014).
[8] J.-M. Regis, et al., Nucl. Instr. and Meth., A **622** 83 (2010).
[9] P. C. Simms, N. Benczer-Koller and C. S. Wu, Phys.Rev. **121** 1169 (1961).
[10] V. Werner, et al., Phys. Rev. C **93**, 034323 (2016).
[11] P. Koseoglou, et al., in preparation.
[12] N. Nica, Nuclear Data Sheets **117**, 1 (2014).
[13] J.-M. Regis, et al., Nucl. Instr. and Meth. A **726** 191 (2013).
[14] J.-M. Regis, et al., Nucl. Instr. and Meth., A **684** 36 (2012).
[15] P. Koseoglou, et al., Lifetime of the 4^+_1 state of ^{148}Ce and its $B_{4/2}$ ratio. In: 3rd one-day workshop on New Aspects and Perspectives in Nuclear Physics. [online] Athens: Hellenic Institute of Nuclear Physics, pp.43 (2016).

IOP Conf. Series: Journal of Physics: Conf. Series **1023** (2018) 012023 doi:10.1088/1742-6596/1023/1/012023

Signatures of triaxiality in low-spin spectra of ^{86}Ge

M. Lettmann[1], V. Werner[1], N. Pietralla[1], P. Doornenbal[2], A. Obertelli[3,2,1], T.R. Rodríguez[4], K. Sieja[5], G. Authelet[3], H. Baba[2], D. Calvet[3], F. Château[3], S. Chen[2,6], A. Corsi[3], A. Delbart[3], J.-M. Gheller[3], A. Giganon[3], A. Gillibert[3], V. Lapoux[3], T. Motobayashi[2], M. Niikura[7], N. Paul[2,3], J.-Y. Roussé[3], H. Sakurai[2,7], C. Santamaria[3], D. Steppenbeck[2], R. Taniuchi[2,7], T. Uesaka[2], T. Ando[2,7], T. Arici[8], A. Blazhev[9], F. Browne[10], A. Bruce[10], R.J. Caroll[11], L.X. Chung[12], M.L. Cortés[1,2,8], M. Dewald[9], B. Ding[13], F. Flavigny[14], S. Franchoo[14], M. Górska[8], A. Gottardo[14], A. Jungclaus[15], J. Lee[16], B.D. Linh[12], J. Liu[16], Z. Liu[13], C. Lizarazo[1,8], S. Momiyama[2,7], K. Moschner[9], S. Nagamine[7], N. Nakatsuka[17], C. Nita[18], C.R. Nobs[10], L. Olivier[14], Z. Patel[11], Zs. Podolyák[11], M. Rudigier[11], T. Saito[7], C. Shand[11], P.-A. Söderström[1,2,8], I. Stefan[14], V. Vaquero[15], K. Wimmer[7], Z. Xu[16,19]

[1] Institut für Kernphysik, Technische Universität Darmstadt, 64289 Darmstadt, Germany
[2] RIKEN Nishina Center, 2-1 Hirosawa, Wako, Saitama 351-0198, Japan
[3] CEA, Centre de Saclay, IRFU/Service de Physique Nucléaire, 91191 Gif-sur-Yvette, France
[4] Departamento de Física Teórica, Universidad Autónoma de Madrid, 28049, Spain
[5] IPHC, CNRS/IN2P3 et Université de Strasbourg, 67037 Strasbourg, France
[6] State Key Laboratory of Nuclear Physics and Technology, Peking University, Beijing 100871, P.R. China
[7] Department of Physics, University of Tokyo, 7-3-1 Hongo, Bunkyo, Tokyo 113-0033, Japan
[8] GSI Helmholtzzentrum für Schwerionenforschung GmbH, 64291 Darmstadt, Germany
[9] Institut für Kernphysik, Universität zu Köln, 50937 Köln, Germany
[10] School of Computing Engineering and Mathematics, University of Brighton, Brighton BN2 4GJ, United Kingdom
[11] Department of Physics, University of Surrey, Guildford GU2 7XH, United Kingdom
[12] Institute for Nuclear Science & Technology, VINATOM, P.O. Box 5T-160, Nghia Do, Hanoi, Vietnam
[13] Institute of Modern Physics, Chinese Academy of Sciences, Lanzhou 730000, P.R. China
[14] Institut de Physique Nucléaire Orsay, IN2P3-CNRS, 91406 Orsay Cedex, France
[15] Instituto de Estructura de la Materia, CSIC, 28006 Madrid, Spain
[16] Department of Physics, The University of Hong Kong, Pokfulam, Hong Kong
[17] Department of Physics, Faculty of Science, Kyoto University, Kyoto 606-8502, Japan
[18] Horia Hulubei National Institute of Physics and Nuclear Engineering (IFIN-HH), 077125 Bucharest, Romania
[19] KU Leuven, Instituut voor Kern- en Stralingsfysica, 3001 Leuven, Belgium

E-mail: `mlettmann@ikp.tu-darmstadt.de`

Abstract. Low-spin states of neutron-rich ^{84,86,88}Ge were measured by in-flight γ-ray spectroscopy at 270 MeV/u at the RIKEN-RIBF facility. The exotic beams have been produced by primary ^{238}U in-flight fission reactions and impinged on the MINOS device. MINOS combines a 10-cm long LH_2 target with a Time Projection Chamber (TPC) to reconstruct the reaction vertices. The reactions were selected by the BigRIPS and the ZeroDegree spectrometers for

IOP Conf. Series: Journal of Physics: Conf. Series **1023** (2018) 012023 doi:10.1088/1742-6596/1023/1/012023

the incoming and outgoing channels, respectively. Emitted γ radiation was detected by the NaI-array DALI2. De-excitations from the 6_1^+, $4_{1,2}^+$, and $2_{1,2}^+$ states of ^{84,86}Ge and 4_1^+ and $2_{1,2}^+$ states of ^{88}Ge were observed. The data are compared to state-of-the-art shell model and beyond-mean-field calculations. Furthermore, a candidate for a 3_1^+ state of ^{86}Ge was identified. This state plays a key role in the discussion of ground-state triaxiality of ^{86}Ge, along with other features of the low-energy level scheme. This work was published in [1].

1. Introduction

For many years triaxial nuclei have been of high interest in nuclear structure physics. More than 60 years ago two elementary models, introducing a breaking of the axial symmetry of the Bohr Hamiltonian [2], were derived. In these models the deformation is described by the triaxial deformation parameter γ, ranging from 0° (prolate shape) to 60° (oblate shape), and the axial elongation β. A deformation parameter $\gamma = 30°$ reflects the maximum of triaxiality. On one hand, the model by Wilets and Jean [3] assumes a potential energy surface independent of γ, which is referred to as the γ-soft case. On the other hand the rigid triaxial rotor model, by Davydov and Filippov [4], describes the potential energy surface with a distinct minimum for a certain value of γ.

The low-spin spectra of a rigid triaxial rotor and a γ-soft nucleus show strong similarities. Of particular importance is the (quasi-)γ-band head positioned at low energy, typically below the yrast-4^+ state. This unique feature clearly differentiates them from axially symmetric rotors, which show high energies of the γ band. In order to distinguish between the spectra of a rigid triaxial rotor and a γ-soft nucleus the energy difference of the odd and even members is a key part, especially the difference between the 3_γ^+ and 2_γ^+ states compared to the difference of the 3_γ^+ and 4_γ^+ states. While in case of a rigid triaxial rotor the odd-spin states are closer to the lower-lying even-spin states, in a γ-soft nucleus they are closer to the higher-lying states. This fact is usually referred to as staggering $S(J)$ [5, 6] and is described by

$$S(J) = \frac{[E(J) - E(J-1)] - [E(J-1) - E(J-2)]}{E(2_1^+)}, \tag{1}$$

with the spin J. For a rigid triaxial rotor the $S(4)$ is positive. So far only ^{76}Ge [7] has an experimentally-derived positive $S(4)$ value and shows a significant degree of rigid triaxiality in the medium mass region $A < 100$.

2. Experimental setup

The experiment was conducted at the Radioactive Isotope Beam Factory (RIBF). A ^{238}U beam was accelerated to an energy of 345 MeV/u and impinged on a 3-mm-thick ^{9}Be production target at the entrance of the BigRIPS fragment separator [8]. This lead to a cocktail of different nuclei produced by in flight-fission. The isotopes of interest were selected by the $B\rho$-ΔE-$B\rho$ method in the BigRIPS fragment separator, while the particle identification was performed on an event-by-event basis by the time of flight-$B\rho$-ΔE method [9]. ^{87}As and ^{85}Ge were produced in one setting with a measurement time of $\sim$ 22 h, with rates of 2059 and 731 s^{-1}, respectively. An additional setting was utilized to provide ^{89}As with rates of 140 s^{-1} for $\sim$ 10.5 h. At the end of the BigRIPS fragment separator, the selected nuclei impinged on the 99(1)-mm-thick liquid-hydrogen reaction target of MINOS [10] with an kinetic energy of $\sim$ 270 MeV/u. While passing the target cell the ion's kinetic energy was reduced by $\sim$ 70 MeV/u. In the following, we focus on (p, pn) and $(p, 2p)$ reaction channels. The resulting reaction products were identified by the $B\rho$-ΔE-$B\rho$ method in the ZeroDegree spectrometer [8]. The reaction vertices of the protons

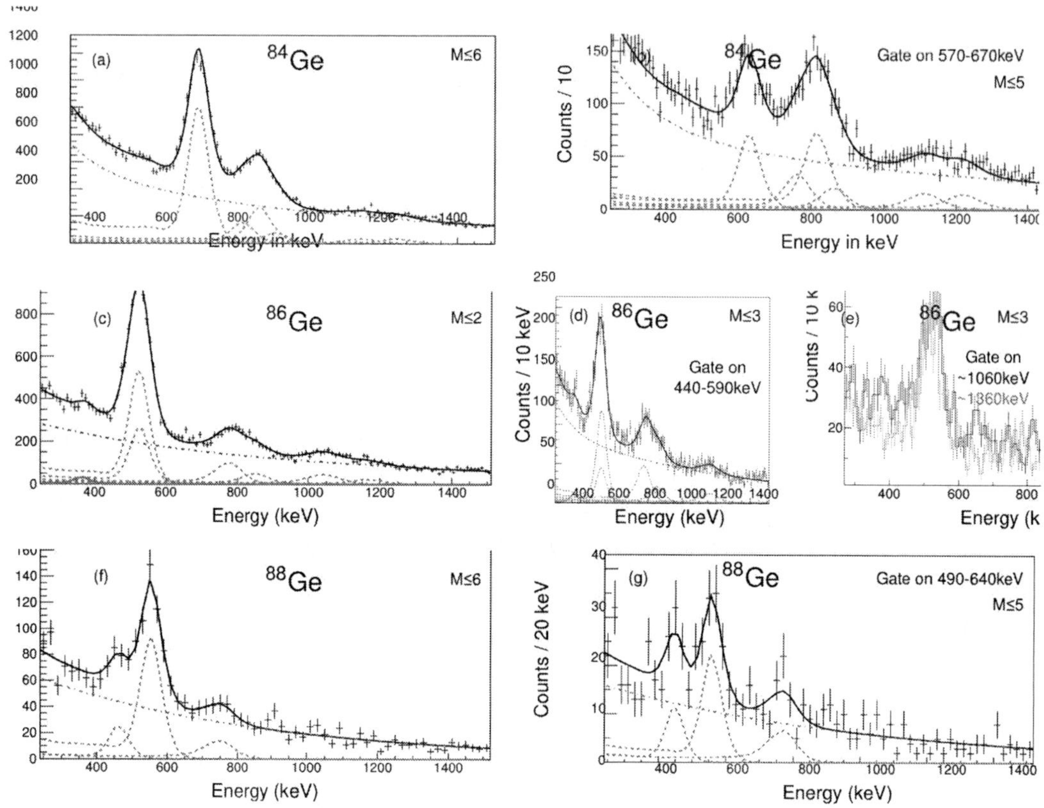

Figure 1. Doppler-corrected DALI2 spectra obtained in the reactions (a) ^{85}Ge(p,pn)^{84}Ge, (c) ^{87}As(p,2p)^{86}Ge, (f) ^{89}As(p,2p)^{88}Ge are shown as black data points, superimposed with the fit of the whole spectrum (black solid line). The respective background (blue dash dotted line) and simulated response functions for each transition (red dotted and solid lines) are shown. Spectra (b), (d), (g) show gates on the 2_1^+ regions. In (e) also gates on the $2_2^+ \rightarrow 0_1^+$ (black), and a neighboring region (red) are shown. The multiplicity cutoff M for ^{86}Ge is chosen to optimize the background-to-peak ratio according to available statistics. For ^{84}Ge higher M is chosen to enhance higher-lying transitions and for ^{88}Ge because of the low statistics. An addback distance of 15 cm was applied to produce the shown spectra.

emitted during (p, pn) or $(p, 2p)$ reactions were reconstructed by the time projection chamber (TPC) of MINOS, which was surrounding the liquid-hydrogen target. The vertex resolution of MINOS is $\lesssim$ 5 mm $FWHM$ and the possibility to detect at least one proton from a $(p, 2p)$ channel is $\gtrsim$ 92% . The de-excitation γ rays were detected by the NaI(Tl) scintillator array DALI2 [12], covering polar angles from 0° to 128° with respect to the center of MINOS. The combination of MINOS and DALI2 was successfully used in the past [13, 14, 15, 16, 17, 18, 19]. A full-energy peak detection efficiency of 35% (23%) was obtained for a 500-keV (1-MeV) γ ray by a simulation within the GEANT4 framework [20] for the DALI2 array. For that, the emission of a γ ray from a nucleus in the center of the MINOS target moving with a kinetic energy of 250 keV, was simulated. Five transitions between 662 keV and 1836 keV of ^{137}Cs, ^{88}Y and ^{60}Co sources were used for the energy calibrations. A calibration error of 1.5 keV was estimated for the energy calibration, while an energy resolution of 9% (6%) FWHM at 662 keV (1.332 MeV) was determined.

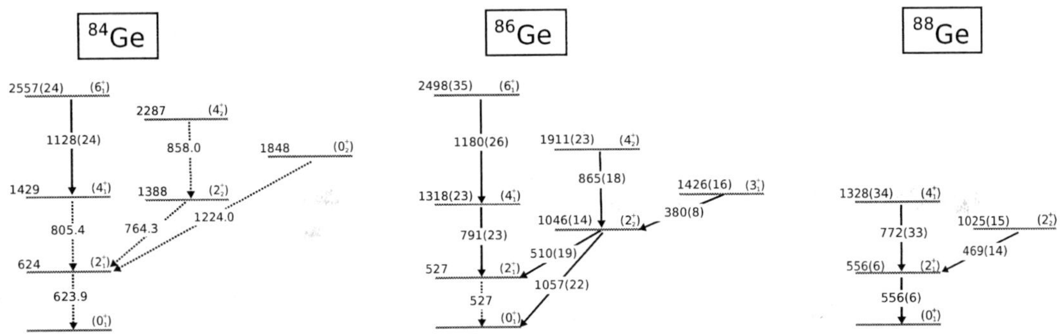

Figure 2. Measured transition energies and proposed level energies of ^{84,86,88}Ge. Dashed arrows denote transition energies taken from literature [21, 22, 23, 24] and solid arrows mark transitions measured for the first time. The tentative spin assignments for ^{84}Ge are taken from [21] (see text).

3. Results

Figure 1 shows the Doppler-corrected γ-ray spectra obtained after the reactions ^{85}Ge(p,pn)^{84}Ge, ^{87}As(p,2p)^{86}Ge, ^{89}As(p,2p)^{88}Ge. The experimental spectra are described by a least-square fit of the DALI2 response function of the γ-ray transitions combined with a two-component experimental background. The DALI2 response function is derived by a Monte Carlo simulation. The uncertainties of the measured transition energies consist of three contributions: a statistical uncertainty due to the fitting procedure, an uncertainty from the energy calibration, and an uncertainty caused by lifetime dependent Doppler broadening and shifts of the observed transitions. An upper limit for the lifetimes is obtained by a χ^2 analysis and is in agreement with theory predictions (see below).

The Doppler-corrected γ spectrum from the reaction ^{85}Ge(p,pn)^{84}Ge is shown in figure 1 (a). Since many γ-ray transition energies are known from β-delayed spectroscopy [21, 22, 23] the reaction serves as a test case. In total six transitions are observed at energies of 629(7) keV, 772(18) keV, 813(10) keV, 867(13) keV, 1128(24) keV and 1229(15) keV. The transition at 1128(24) keV is observed for the first time. The observations are in good agreement with [21] and the proposed level scheme is shown in figure 2 on the left. Since the presented experiment is not sensitive to the spins of the observed states they are tentatively assigned. However, based on systematics of neighboring Ge isotopes the 1128(24)-keV transition is assigned to the $(6_1^+) \rightarrow 4_1^+$ transition. A $\gamma\gamma$-coincidence gate on the energy range of the $2_1^+ \rightarrow 0_1^+$ transition energy is shown in figure 1 (b). Note, that the 629(7)-keV transition still appears in the coincidence condition, due to coincidences with Compton events of higher-energy transitions underneath the 629(7)-keV lineshape, though strongly reduced. ^{86}Ge is produced by a ^{87}As(p,2p) reaction, and the Doppler-corrected spectrum is shown in figure 1 (c). Seven transitions at energies of 380(8) keV, 510(19) keV, 534(8) keV, 791(23) keV, 865(18) keV, 1057(22) keV and 1180(26) keV are identified. The transition at 534(8) keV was observed before via β-decay studies [24] and is in good agreement with the results of this work. The assignment to the decay of the $2_1^+ \rightarrow 0_1^+$ is adopted. The proposed level scheme for ^{86}Ge is shown in the middle of figure 2, based on the following argumentation. The strongest observed γ decay in the spectrum stems from the $2_1^+ \rightarrow 0_1^+$ transition for an even-even nucleus populated via $(p, 2p)$ (see, e.g., [25, 26, 27, 28]). Furthermore, the second strongest transition corresponds to the $(4_1^+) \rightarrow (2_1^+)$ transition. Figure 1 (d) shows a $\gamma\gamma$ coincidence condition on the energy range of the $2_1^+ \rightarrow 0_1^+$. A strong self coincidence is visible at this energy range, much stronger than in the case of ^{84}Ge, which is a hint for the 510(19)/534(8)-keV doublet in this energy range. Furthermore, the transition at

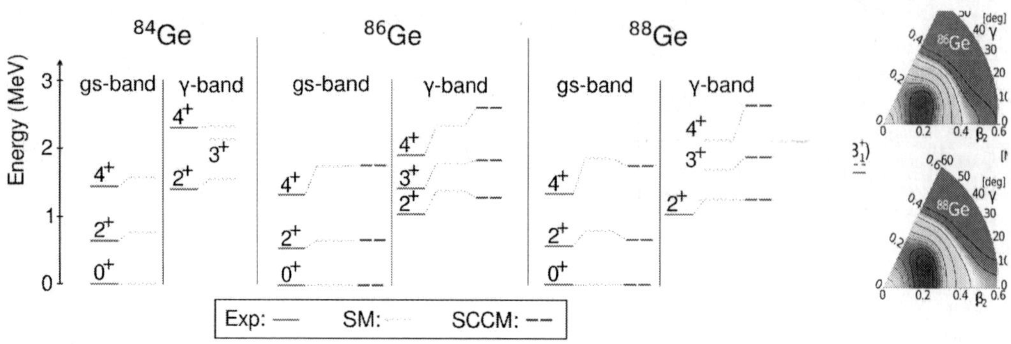

Figure 3. (left) Systematics of the ^{84,86,88}Ge level energies from experiment compared to theoretical predictions from shell model (SM) and SCCM. The shell model calculation for ^{84}Ge is taken from [29] (right) Potential energy surfaces in the particle number variation after projection (PN-VAP) approach [30] for ^{86}Ge (top) and ^{88}Ge (bottom). The spacing between solid contour lines is 2 MeV with intermediate dashed lines for 0.5 MeV steps. The minima are located at about $\beta_2 = 0.2$.

1057(22) keV is absent in the $\gamma\gamma$ coincidence. Therefore, the 1057(22)-keV-γ ray is assigned to the $(2_2^+) \rightarrow (0_1^+)$ transition and the doublet to the competitive $(2_2^+) \rightarrow (2_1^+) \rightarrow (0_1^+)$ decay path. Based on comparison to ^{84}Ge, the transitions at 865(18) keV and 1180(26) keV are assigned to the $(4_2^+) \rightarrow (2_2^+)$ and $(6_1^+) \rightarrow (4_1^+)$ decays, respectively. The transition at 380(8) keV is visible in the $\gamma\gamma$ coincidence gate of the $(2_1^+)/(2_2^+)$ doublet and appears in a gate on the $(2_2^+) \rightarrow (0_1^+)$ transition shown in figure 1 (e) as well. Gating in the neighboring region of $\sim$ 1360 keV the 380(8)-keV transition is absent. This transition is assigned to originate from the $(3_1^+) \rightarrow (2_2^+)$ decay (for details, see [1]).
The Doppler-corrected γ-ray spectrum of ^{88}Ge after a ^{89}As(p,2p) reaction is shown in figure 1 (f). It is the first spectroscopy of ^{88}Ge, thus the three transitions at energies of 469(14) keV, 556(6) keV and 772(33) keV are observed for the first time. The suggested level scheme for ^{88}Ge is shown in figure 2 on the right. According to the assignments before, the strongest transition in the spectrum is assigned to be the $2_1^+ \rightarrow 0_1^+$ transition. A $\gamma\gamma$ coincidence gate on its energy range is shown in figure 1 (g), yielding both transitions. In comparison to ^{86}Ge the γ-ray at 772(33) keV is assigned to the $(4_1^+) \rightarrow (2_1^+)$ transition, and the γ-ray at 469(14) keV is assigned to the $(2_2^+) \rightarrow (2_1^+)$ transition.

4. Discussion

In the following, the obtained results are compared to a shell-model calculation and to a symmetry-conserving configuration mixing Gogny (SCCM) calculation. The left side of figure 3 shows the comparison of the constructed level schemes and the predictions from both theories. Both theoretical predictions of the level sequences are in good agreement with the experiment, whereas the theories overestimate the level energies systematically. The low-lying γ band predicted from both theories, which is pointing to triaxial features of the isotopes, is reproduced by the measured data. Furthermore, both theories predict a 3_1^+ state in ^{86}Ge which is closer to the 2_2^+ state than to the 4_2^+ state, in contrast to ^{84}Ge and ^{88}Ge, where it is predicted to be closer to the 4_2^+ state than to the 2_2^+ state or rater central. A promising candidate for this state is observed through the 380(8)-keV transition in this experiment. This state at 1426(16) keV would dominantly decay to the 2_2^+ state with such a 380(8)-keV transition. With this assignment an $S(4) = 0.20(4)$ staggering parameter is obtained for ^{86}Ge, even larger than for ^{76}Ge where a staggering of $S(4) = 0.091(2)$ [7] has been observed. This reflects an even

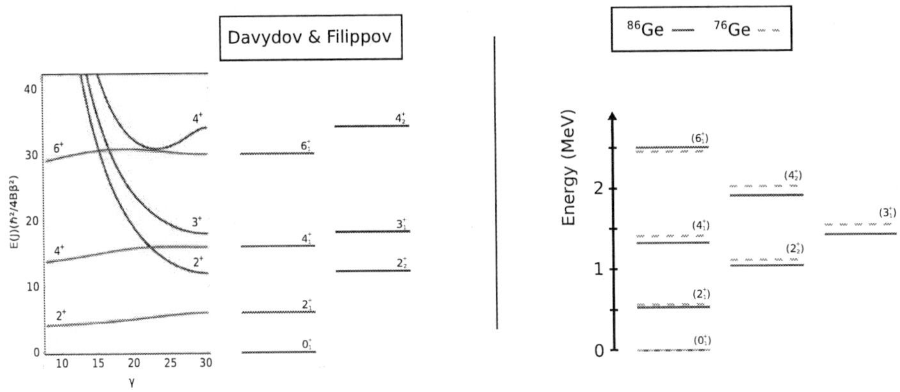

Figure 4. (left) Evolution of the ground-state band and γ-band energies with increasing deformation parameter γ calculated with the model of Davydov and Filippov [4]. (middle) Low-energy level scheme calculated by the model of Davydov and Filippov for $\gamma = 30°$. (right) Comparison of the low-spin spectra of ^{76}Ge and ^{86}Ge.

larger degree of triaxiality in ^{86}Ge than in ^{76}Ge. The similarities of ^{76}Ge and ^{86}Ge are shown in figure 4 in the right. The states in the ground-state and γ band match within 100 keV in both nuclei. In addition, figure 4 (left) shows the behavior of the ground-state band and the γ-band for an increasing deformation parameter γ calculated by the model of Davydov and Filippov. The resulting spectrum for $\gamma = 30°$ is shown in the center of the figure. Besides the distinct reproduction of the ground-state band, the position of the γ-band head is in good agreement with experiment. From the evolution of the 2_2^+ level energy with increasing γ it can be seen that its energy drops below the 4_1^+ level energy for $\gamma > 22°$. Within the model of Davydov and Filippov the γ of ^{86}Ge can be calculated by [31]

$$\frac{E_{2_2^+}}{E_{2_1^+}} = \frac{1 + \sqrt{1 - \frac{8}{9}\sin^2 3\gamma}}{1 - \sqrt{1 - \frac{8}{9}\sin^2 3\gamma}}. \qquad (2)$$

From data we obtain $E_{2_2^+}/E_{2_1^+} = 1.98(3)$. This yields $\gamma = 30(1)°$, as expected for a pronounced triaxial shape. The observation of rigid triaxiality is further extended by the energy surfaces from the SCCM calculation shown in figure 3. A pronounced triaxial minimum is predicted for ^{86}Ge, while more γ softness is predicted for ^{88}Ge.

5. Summary

For the first time, γ-ray spectroscopy of ^{88}Ge has been performed. By means of in-flight γ-ray spectroscopy of ^{84,86,88}Ge 16 transitions have been studied, 10 of which have been observed for the first time. Based on informations of neighboring Ge isotopes and the knowledge from former $(p, 2p)$ studies in this region of the nuclear chart new level schemes for ^{86}Ge and ^{88}Ge are proposed. For ^{86}Ge a candidate for a 3_1^+ state was observed. With this assignment a staggering parameter $S(4) = 0.20(4)$ was obtained for ^{86}Ge, which is even bigger than for the so far known case of ^{76}Ge. Interpreted with the model from Davydov and Filippov, ^{86}Ge is assigned as an example for a pronounced triaxial shape of a nucleus. This is in good agreement with state-of-the-art theoretical predictions in this region.

Acknowledgements
The authors thank the RIBF and BigRIPS teams for providing a stable, high-intensity uranium beam and operating the secondary beams. We acknowledge support from the German BMBF Grant Nos. 05P12RDFN1, 05P12RDFN8 and 05P15PKFNA, the ERC Grant No. MINOS-258567, the Spanish Ministerio de Economía y Competitividad under contracts FPA2014-57196-C5-4-P and FIS−2014−53434, the Vietnam Ministry of Science and Technology, as well as from the Science and Technology Facilities Council (STFC). We further thank GSI for providing computing facilities.

References
[1] M. Lettmann *et al.*, Phys. Rev. C **96**, 011301(R) (2017)
[2] A. Bohr and B. Mottelson, Mat. Fys. Medd. Dan. Vid. Selsk. **27** (1958)
[3] L. Wilets and M. Jean, Phys. Rev. **102**, 788 (1956)
[4] A. S. Davydov and G. F. Filippov, Nucl. Phys. **8**, 237 (1958)
[5] N. V. Zamfir and R. F. Casten, Phys. Lett. B **260**, 265 (1991)
[6] E. A. McCutchan, D. Bonatsos, N. V. Zamfir, R. F. Casten, Phys. Rev. C **76** 024306(R) (2007)
[7] Y. Toh *et al.*, Phys. Rev. C **87** 041304(R) (2013)
[8] T. Kubo *et al.*, Prog. Theor. Exp. Phys. **2012** 03C003 (2012)
[9] N. Fukuda, *et al.*, Nucl. Instrum. Methods Phys. Res. B **317** 323-332 (2013)
[10] A. Obertelli, *et al.*, Eur. Phys. J. A **50** 8 (2014)
[11] C. Santamaria, *et al.*, in preparation (2017)
[12] S. Takeuchi, *et al.*, Nucl. Instrum. Methods Phys. Res. A **763** 8 (2014)
[13] P. Doornenbal, Prog. Theor. Exp. Phys. **2012** 03C004 (2012)
[14] C. Santamaria, *et al.*, Phys. Rev. Lett. **115** 192501 (2015)
[15] S. Chen, *et al.*, Phys. Rev. C **95** 041302(R) (2017)
[16] F. Flavigny, *et al.*, Phys. Rev. Lett. **118** 242501 (2017)
[17] N. Paul, *et al.*, Phys. Rev. Lett. **118** 032501 (2017)
[18] C. M. Shand, *et al.*, Phys. Lett. B **773** 492-497 (2017)
[19] L. Olivier, *et al.*, Phys. Rev. Lett., accepted (2017)
[20] S. Agostinelli, *et al.*, Nucl. Instrum. Methods Phys. Res. A **506** 250-303 (2003)
[21] A. Korgul *et al.*, Phys. Rev. C **88** 044330 (2013)
[22] K. Kolos *et al.*, Phys. Rev. C **88** 047301 (2013)
[23] J. A. Winger *et al.*, Phys. Rev. C **81** 044303 (2010)
[24] K. Miernik *et al.*, Phys. Rev. Lett. **111** 132502 (2013)
[25] P. Doornenbal *et al.*, Phys. Rev. Lett. **111** 212502 (2013)
[26] H. Iwasaki *et al.*, Phys. Rev. Lett. **112** 142502 (2014)
[27] D. Bazin *et al.*, Phys. Rev. Lett. **91** 012501 (2003)
[28] P. Fallon *et al.*, Phys. Rev. C. **81** 041302 (2010)
[29] K. Sieja, T. R. Rodríguez, K. Kolos, D. Verney, Phys. Rev. C. **88** 034327(R) (2013)
[30] M. Anguiano, J. L. Egido, L. M. Robledo, Phys. Lett. B **545** 62-72 (2002)
[31] A. S. Davydov and V. S. Rostovsky, Nucl. Phys. **12** 58-68 (1959)

IOP Conf. Series: Journal of Physics: Conf. Series **1023** (2018) 012024 doi:10.1088/1742-6596/1023/1/012024

Signatures for a nuclear quantum phase transition from $E0$ and $E2$ observables in Gd isotopes

J. Wiederhold[1], R. Kern[1], C. Lizarazo[1], N. Pietralla[1], V. Werner[1], R.V. Jolos[2,3], D. Bucurescu[4], N. Florea[4], D. Ghita[4], T. Glodariu[4], R. Lica[4], N. Marginean[4], R. Marginean[4], C. Mihai[4], R. Mihai[4], I.O. Mitu[4], A. Negret[4], C. Nita[4], A. Olacel[4], S. Pascu[4], L. Stroe[4], S. Toma[4], A. Turturica[4]

[1] Institut für Kernphysik, Technische Universität Darmstadt, Schlossgartenstr. 9, 64289 Darmstadt, Germany

[2] Joint Institute for Nuclear Research, Dubna, Moscow oblast, 141980 Russia

[3] Dubna State University, Dubna, Moscow oblast, 141980 Russia

[4] "Horia Hulubei" National Institute for Physics and Nuclear Engineering, 077125, Bucharest-Magurele, Romania

E-mail: `jwiederhold@ikp.tu-darmstadt.de`

Abstract. Nuclei are complex quantum objects due to complex nucleon-nucleon interactions. They can undergo rather rapid changes in structure as a function of nucleon number. A well known region of such a shape transition is the rare-earth region around $N = 90$, where accessible nuclei range from spherical nuclei at the closed neutron shell at $N = 82$ to deformed nuclei. For a better understanding of this phenomenon, it is of interest to study empirical signatures like the $E2$ transition strength $B(E2; 2_1^+ \rightarrow 0_1^+)$ or the $E0$ excitation strength $\rho^2(E0; 0_1^+ \rightarrow 0_2^+)$. The nuclide ^{152}Gd with 88 neutrons is located close to the quantum phase transition at $N = 90$. The lifetime $\tau(0_2^+)$ of ^{152}Gd has been measured using fast electronic scintillation timing (FEST) with an array of HPGe- and LaBr$_3$- detectors. Excited states of ^{152}Gd were populated via an (α,n)-reaction on a gold-backed ^{149}Sm target. The measured lifetime of $\tau(0_2^+) = 96(6)$ ps corresponds to a reduced transition strength of $B(E2; 0_2^+ \rightarrow 2_1^+) = 111(7)$ W.u. and an $E0$ transition strength of $\rho^2(E0) = 39(3) \cdot 10^{-3}$ to the ground state. This result provides experimental support for the validity of a correlation between $E0$ and $E2$ strengths that is a novel indicator for a quantum phase transition. This work was published as J. Wiederhold *et al.*, Phys. Rev. C **94**, 044302 (2016).

1. Introduction

The rare-earth region around the neutron number $N = 90$ is a well known region where nuclei can undergo a quantum shape phase transition (QPT) [1, 2, 3], a rather rapid change of structure as a function of neutron number. Well known empirical signatures of a shape change in a sequence of even-even nuclei are, e.g., a sudden change in the energy ratio $R_{4/2}$ of the 2_1^+ and 4_1^+ level energies $E(4_1^+)$ and $E(2_1^+)$ or a change in the easier to obtain inverse of the energy $E(2_1^+)$, a sharp change of the $E2$-transition strength connecting the ground state to the first-excited 2_1^+ state $B(E2; 0_1^+ \rightarrow 2_1^+)$, the occurrence of a local extreme point of the two-neutron separation energy S_{2N}, and a maximum of the $E0$-transition strength $\rho^2(E0)$ at the critical point [4, 5, 6]. Studying the correlation of these observables is essential for the understanding of the phenomenon

IOP Conf. Series: Journal of Physics: Conf. Series **1023** (2018) 012024 doi:10.1088/1742-6596/1023/1/012024

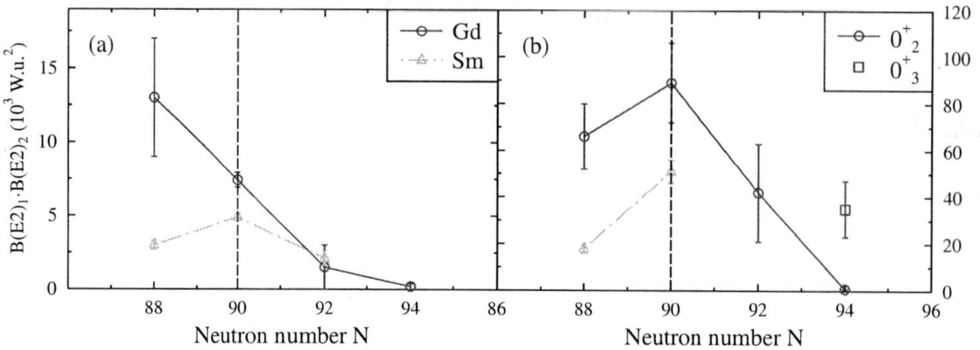

Figure 1. (a) Product $B(E2)_1 \cdot B(E2)_2$ (see text) over neutron number for the Gd- (blue) and Sm- (orange) isotopes. (b) The $E0$ transition rate $\rho^2(E0)$ for the Gd- and Sm-isotopes around N=90. For ^{158}Gd also the $\rho^2(E0; 0_3^+ \rightarrow 0_{gs}^+)$ transition strength is given. Data taken from the Nuclear Data Sheets [7] and [8, 9].

of nuclear shape phase transitions [10, 11, 12].
In a geometrical picture it has been shown [13] that the $E0$-transition strength $\rho^2(E0; 0_2^+ \rightarrow 0_1^+)$ can be connected to a product of $B(E2)$-values. Applying the Q-phonon scheme [14, 15, 16, 17, 18, 19], this results to:

$$\rho^2(E0) \approx \left(\frac{4\pi}{3Z}\right)^2 \frac{5}{(eR^2)^4} B(E2)_2 B(E2)_1, \tag{1}$$

with the proton number Z, the radius R of the atomic nucleus, $B(E2)_1 = B(E2; 2_1^+ \rightarrow 0_1^+)$ and $B(E2)_2 = B(E2; 0_2^+ \rightarrow 2_1^+)$. The left-hand side of Figure 1 shows the available data for the product of $B(E2)$-values, while on the right-hand side the available $\rho^2(E0; 0_2^+ \rightarrow 0_1^+)$-data of the Gd- and Sm-isotopes around $N = 90$ are depicted. The available data partly supports the relation of Equation (1). Both, the $\rho^2(E0; 0_2^+ \rightarrow 0_1^+)$ and the product of $B(E2)$-values, decrease beyond $N = 90$ and the $\rho^2(E0; 0_2^+ \rightarrow 0_1^+)$ maximizes for both isotopic chains at $N = 90$. Moreover, the product of $B(E2)$-values maximizes at $N = 90$ for the Sm-chain. But, in the case of ^{152}Gd the correlation seems to be invalid. The value of the product of $B(E2)$-values seems to increase, not decrease, towards ^{152}Gd with $N = 88$ with an extraordinarily large $B(E2; 0_2^+ \rightarrow 2_1^+)$ value of 178(54) W.u. in this vibrational nucleus, however, with a large uncertainty. The lifetime of the second 0^+ state of ^{152}Gd stems mainly from one recoil distance Doppler shift measurement of this lifetime [20]. But the data are dubious. It is in fact mentioned in that work that the statistics for the measurement were not adequate for an accurate lifetime determination by the standard analysis and it was necessary to sum up data from different target-stopper distances. To validate or eventually improve this value a FEST experiment at the 9 MV Tandem accelerator at the National Institute of Physics and Nuclear Engineering (IFIN-HH) in Bucharest has been carried out.

2. Experimental setup

An α-beam impinging on a 1.3 mg/cm^2 thick ^{149}Sm-target with a 4.3 mg/cm^2 gold backing populated excited states of ^{152}Gd via the ^{149}Sm(α,n)^{152}Gd fusion-evaporation reaction. The beam energy was set to 17.5 MeV, slightly above the Coulomb barrier for this reaction, to select the one-neutron evaporation channel. The target consisted of 93.2% of ^{149}Sm and some small

IOP Conf. Series: Journal of Physics: Conf. Series **1023** (2018) 012024 doi:10.1088/1742-6596/1023/1/012024

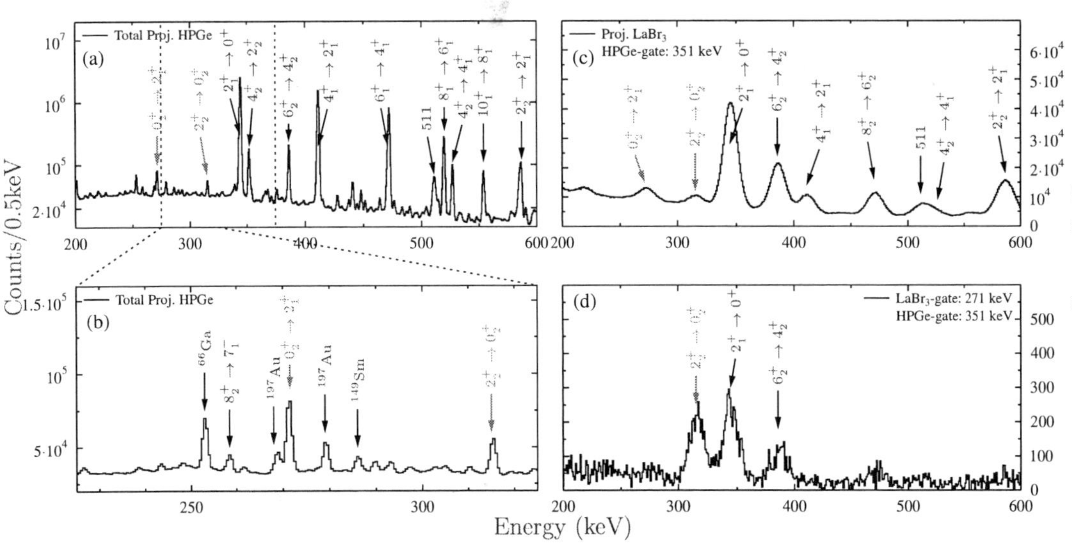

Figure 2. (a) Total Projection of the E_{HPGe}-E_{HPGe}-matrix. (b) Zoomed in on the region of interest between 225 and 325 keV. (c) Energy projection of a E_{LaBr_3}-E_{LaBr_3}-Δt-cube in the energy range from 200 to 600 keV. An additional gate was set in the HPGe-detectors on the $4_2^+ \rightarrow 2_2^+$ transition at 351 keV. Marked in red are the important transitions to determine $\tau(0_2^+)$. (d) Coincidence spectrum gated on the 271 keV $0_2^+ \rightarrow 2_1^+$-transition in spectrum (a).

contaminants from other stable Sm-isotopes, i.e. ^{150}Sm (3.3%), ^{148}Sm(1.5%), ^{152}Sm(1.2%), and 0.8% of other elements. The ROSPHERE detector array was used in a configuration of 11 $LaBr_3$- and 14 HPGe-detectors [21] to detect de-excitation γ-rays of the isotope of interest. The energy calibration and the energy dependent time-walk calibration of the experimental setup were performed using a ^{152}Eu-source with its well known γ-ray transitions from ^{152}Gd and ^{152}Sm ranging from 244 keV to 1299 keV.

3. Analysis and Corrections

3.1. Spectroscopy

Due to the beam energy chosen close to the Coulomb barrier, the reaction channel was clean. The γ-ray transitions of ^{152}Gd were clearly identified, by using the HPGe-detectors with an energy resolution of about 3 keV at 1.4 MeV and 2.5 keV at 300 keV. The γ-ray transitions up to the $10_1^+ \rightarrow 8_1^+$-transition of the yrast band and other bands were identified, including the γ-transitions necessary to determine the lifetime of the 0_2^+ state with FEST, i.e. the $2_2^+ \rightarrow 0_2^+$-transition at 315 keV and the $0_2^+ \rightarrow 2_1^+$-transition at 271 keV (See Figure 2). Only a few contaminants from Coulomb excitation of the gold backing of the target and from fusion-evaporation reactions of α-particles with Cu parts of the target chamber were found in the energy spectrum. These contaminants were eliminated by setting energy gates in the HPGe-detectors, since they can't be separated from the transitions of interest in the $LaBr_3$-detectors. Figure 2(a) shows the total projection of the HPGe-detector-coincidence matrix between 200 and 600 keV and Figure 2(b) shows the zoomed energy spectra in the energy region 225-325 keV around the relevant transitions for determining the lifetime of the 0_2^+ state (marked in red). A few small contaminants close to the $0_2^+ \rightarrow 2_1^+$-transition were identified. Figure 2(c) shows the energy projection of the $LaBr_3$-detector-coincidence matrix in coincidence with the 351 keV ($4_2^+ \rightarrow 2_2^+$) transition observed in the HPGe-detectors. After setting an additional gate on the

IOP Conf. Series: Journal of Physics: Conf. Series **1023** (2018) 012024 doi:10.1088/1742-6596/1023/1/012024

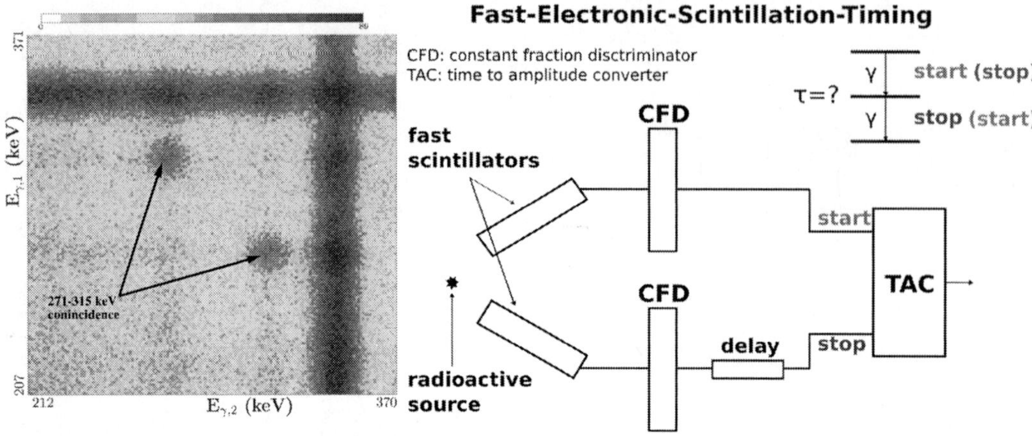

Figure 3. (Left) Part of the $E_{\gamma,LaBr_3} - E_{\gamma,LaBr_3}$ matrix obtained after projection of the $E_{\gamma,LaBr_3} - E_{\gamma,LaBr_3} - \Delta t$ cube. Marked are the coincidence areas of the γ-ray pair 271 and 315 keV, corresponding to the transitions $0_2^+ \rightarrow 2_1^+$ and $2_2^+ \rightarrow 0_2^+$ of ^{152}Gd. (Right) Simplified FEST setup with two detectors and a source.

$0_2^+ \rightarrow 2_1^+$-transition at 271 keV detected in the $LaBr_3$-detectors, the contaminants in the resulting $LaBr_3$-energy spectra vanish (See Figure 2(d)).

3.2. Time-Walk

The lifetime of an excited nuclear state can be determined in a FEST measurement by measuring the time-difference between two signals, e.g. from two γ-ray transitions. A simplified setup of a FEST measurement is shown in Figure 3 on the right-hand side. These two signals can stem from two γ-ray transitions, one populating the state of interest and one depopulating it. The measured time-difference is equal to the effective lifetime of the nuclear states between the two γ-transitions and the energy dependent time-walk of both signals. More details can be found in [22, 23, 24]. Using a ^{152}Eu-source, a time-walk correction can be applied to the data, as described in [25]. After correcting for the time-walk to measure the lifetimes of excited states, an Energy-Energy-Δt cube was sorted for the $LaBr_3$ detectors. By selecting the region of the pair of coincident γ-rays (e.g. 271 keV and 315 keV for the 0_2^+ state) in the Energy-Energy plane of this cube (see Figure 3,(left)) a time-difference spectrum can be extracted. A correction for random coincidences and the Compton background was done, by selecting an area near the peak in the Energy-Energy plane. A time-difference spectrum was also established for the surrounding background. It was normalized to the amount of background underneath the coincidence peaks and subtracted from the time-difference spectrum gated on the coincidence peaks. This way, a background-subtracted time difference spectrum for the pair of coincidences was established.
The typical delayed time distribution, assuming no remaining background contributions, is a convolution of the prompt response function (PRF) of the setup, $P(t')$, and the exponential decay [26]:

$$D_\lambda(t) = n\lambda \int_{-\infty}^{t} P(t')e^{-\lambda(t-t')}dt', \qquad (2)$$

with the transition rate $\lambda = 1/\tau$ and the normalization n.
Depending on the lifetime of the nuclear state of interest, there are different approaches to determine the lifetime, e.g. the slope method [26, 27] or the centroid-shift method [28].
The lifetime of the 0_2^+ state of ^{152}Gd was determined with the centroid-shift method [13]. As mentioned above an additional energy gate had to be set in the HPGe-detectors to exclude

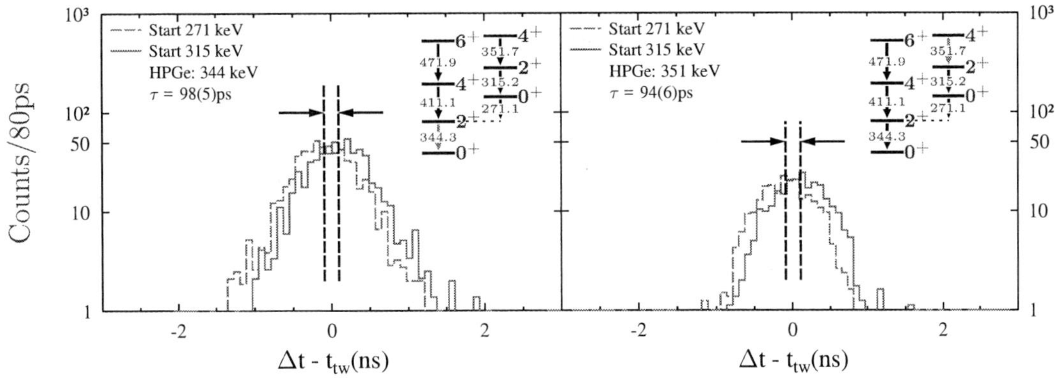

Figure 4. (Left) Time-difference-spectra for the combination of start-gate on 315 keV ($2_2^+ \rightarrow 0_2^+$) and stop-gate on 271 keV ($0_2^+ \rightarrow 2_1^+$) in green and vice versa in red (dashed line) with an additional gate on 344 keV ($2_1^+ \rightarrow 0_{gs}^+$) in the HPGe-detectors. (Right) The same with a different gate for the HPGe-detectors ($4_2^+ \rightarrow 2_2^+$ at 351 keV). The right top corner of every time-difference plot shows a partial level scheme of ^{152}Gd. Marked in blue is the HPGe-gate. The dashed vertical lines mark the centroids of the time-difference spectra. The difference between the centroids is equal to two times the lifetime $\tau(0_2^+)$ of ^{152}Gd [13].

contributions from any contaminants close to the $0_2^+ \rightarrow 2_1^+$-transition at 271 keV. Figure 4 shows time-difference spectra, one with an additional gate above the state of interest, i.e. the $4_2^+ \rightarrow 2_2^+$-transition at 351 keV (Figure 4,(b)), and one below the 0_2^+ state on the $2_1^+ \rightarrow 0_{gs}^+$-transition at 344 keV (Figure 4,(a)). Both obtained lifetimes agree within their uncertainties and the lifetime of the 0_2^+ state of ^{152}Gd was determined to $\tau = 96(6)$ ps or $T_{1/2} = 67(4)$ ps [13].

4. Results and Discussion

From the obtained value for the lifetime $\tau(0_2^+)$ the corresponding $E2$ transition strength with the conversion coefficient $\alpha = 0.0826(12)$ [29] can be calculated to

$$B(E2; 0_2^+ \rightarrow 2_1^+) = 0.536(33)e^2b^2 = 111(7)\text{W.u.} \tag{3}$$

The quoted error represents the statistical uncertainty in determining the centroids and the contributions to the error from the uncertainty of the energy-dependent time-walk. Additionally, a new value for the $\rho^2(E0)$ transition strength can be determined from the $B(E2; 0_2^+ \rightarrow 2_1^+)$-strength using the dimensionless ratio of the $E0$- and $E2$-transitions defined by Rasmussen [30]:

$$X(E0/E2) \equiv \frac{B(E0)}{B(E2)} = \frac{\rho^2(E0)e^2R^4}{B(E2)}. \tag{4}$$

With the literature value of $X(E0/E2) = 0.0122(5)$ [9] one obtains the fairly large value of

$$\rho^2(E0) = 39(3) \cdot 10^{-3}. \tag{5}$$

Figure 5 shows the $\rho^2(E0)$-strength and the product $B(E2)_1 \cdot B(E2)_2$ as a function of the neutron number for the Sm- and Gd-isotopes including the new values for ^{152}Gd. The maximum for the product of $B(E2)$-values lies between N=88 and N=90 and it does not exactly coincide with the location of the maximum of the $E0$ strength. This may be traced back to the approximation of the Q-phonon scheme applied.

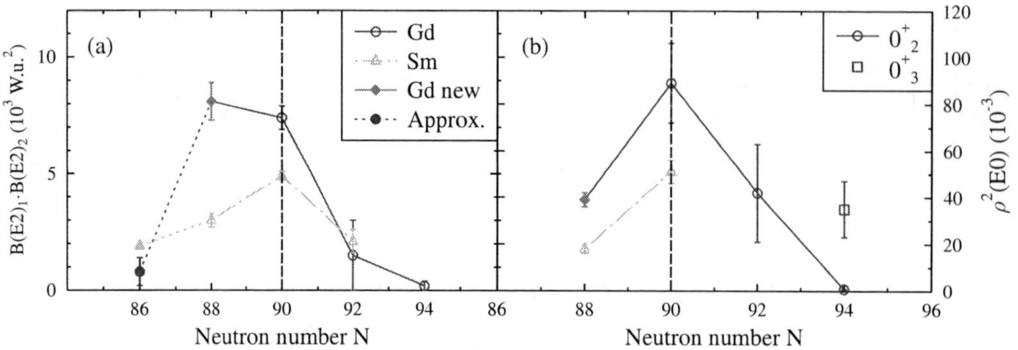

Figure 5. (a) Product $B(E2)_1B(E2)_2$ of Gd- and Sm-isotopes. Red (filled diamond) value from [13]. For N=86 isotopes, there is no lifetime data of the 0_2^+ state. Product $B(E2)_1B(E2)_2$ was approximated as $2 \cdot B(E2)^2$. (b) Experimental values of $\rho^2(E0)$ values for the $0_2^+ \rightarrow 0_1^+$-transition for Gd- (blue) and Sm- (orange) isotopes. Red (filled diamond) value from [13]. For ^{158}Gd the $\rho^2(E0; 0_3^+ \rightarrow 0_1^+)$ value is shown, too. Note the difference to Figure 1 for $N = 88$.

5. Summary

The lifetime of the 0_2^+ state of ^{152}Gd was measured using the fast-timing technique. The lifetime of the 0_2^+ state, 96(6) ps [13] is nearly two times the previously measured value of 53(12) ps [20]. With the new value for $\tau(0_2^+)$ of ^{152}Gd the $\rho^2(E0)$-transition strength were determined to $39(3) \cdot 10^{-3}$ and the $B(E2; 0_2^+ \rightarrow 2_1^+)$-transition strength to 0.536(33) e^2b^2 or 111(7) W.u.
The $\rho^2(E0)$ values for the 0_2^+ state and the absolute strength product of the $E2$-$E2$ cascade, $B(E2)_2B(E2)_1$ peak at N=90 or between N=88 and N=90 for the Sm- and the Gd-isotopic chains, respectively indicating the presence of a quantum shape phase transition.

Acknowledgments

The authors thank the Bucharest Tandem accelerator crew for their support during the experiment. This work was supported by the Deutsche Forschungsgemeinschaft under the Grant No. SFB 634 and the BMBF under the grants 05P15RDFN1/05P15RDFN9 within the collaboration 05P15 NuSTAR R&D.

References

[1] F. Iachello, Phys. Rev. Lett. **87**, 052502 (2001).
[2] R. F. Casten and N. V. Zamfir, Phys. Rev. Lett. **87**, 052503 (2001).
[3] R. Krücken *et al.*, Phys. Rev. Lett. **88**, 232501 (2002).
[4] P. von Brentano *et al.*, Phys. Rev. Lett. **93**, 152502 (2004).
[5] N. Pietralla and O. M. Gorbachenko, Phys. Rev. C **70**, 0113042 (2004).
[6] J. Bonnet *et al.*, Phys. Rev. C **79**, 034307 (2009).
[7] Nuclear Data Sheets, http://www.nndc.bnl.gov/ensdf.
[8] J. L. Wood *et al.*, Nucl. Phys. A **651**, 323 (1999).
[9] T. Kibédi and R. H. Spear, At. Data Nucl. Data Tables **89**, 77 (2005).
[10] R. F. Casten, Prog. Part. Nucl. Phys. **62**, 183 (2009).
[11] P. Cejnar and J. Jolie, Prog. Part. Nucl. Phys. **62**, 210 (2009).
[12] E. A. McCutchan, N. V. Zamfir and R. F. Casten, Phys. Rev. C **69**, 064306 (2004).
[13] J. Wiederhold *et al.*, Phys. Rev. C **94**, 044302 (2016).
[14] G. Siems *et al.*, Phys. Lett. B **320**, 1 (1994).
[15] T. Otsuka and K.-H. Kim, Phys. Rev. C **50**, R1768 (1994).
[16] N. Pietralla *et al.*, Phys. Rev. Lett. **73**, 2962 (1994).

IOP Conf. Series: Journal of Physics: Conf. Series **1023** (2018) 012024 doi:10.1088/1742-6596/1023/1/012024

[17] N. Pietralla *et al.*, Phys. Lett. B **349**, 1 (1995).
[18] N. Pietralla *et al.*, Phys. Rev. C **57**, 150 (1998).
[19] Y. V. Palchikov, P. von Brentano and R. V. Jolos, Phys. Rev. C **57**, 3026 (1998).
[20] N. R. Johnson *et al.*, Phys. Rev. C **26**, 1004 (1982).
[21] D. Bucurescu *et al.*, Nucl. Instrum. Methods Phys. Res. A **837**, 1 (2016).
[22] H. Mach, R. L. Gill and M. Moszyński, Nucl. Instrum. Methods Phys. Res. A **280**, 49 (1989).
[23] M. Moszyński H. Mach, Nucl. Instrum. Methods Phys. Res. A **277**, 407 (1989).
[24] J.-M. Régis *et al.*, Nucl. Instrum. Methods Phys. Res. A **726**, 191 (2013).
[25] N. Marginean *et al.*, Eur. Phys. J. A **46**, 329 (2010).
[26] L. Boström *et al.*, Nucl. Instrum. Methods. **44**, 61 (1966).
[27] V. Werner *et al.*, Phys. Rev. C **93** 034323 (2016).
[28] Z. Bay, Phys. Rev. **77**, 419 (1950).
[29] T. Kíbedi *et al.*, Nucl. Instrum. Methods. Phys. Res. A **589**, 202 (2008).
[30] J. O. Rasmussen, Nucl. Phys. **19**, 85 (1960).

IOP Conf. Series: Journal of Physics: Conf. Series **1023** (2018) 012025 doi:10.1088/1742-6596/1023/1/012025

Exotic states of ^{13}C nuclei

N Burtebayev[1], D M Janseitov[1,2], Zh Kerimkulov[1], Marzhan Nassurlla[1], Y S Mukhamejanov[1], A S Aimaganbetov[3] and D S Valiolda[4]

[1] Institute of Nuclear Physics, Almaty, Kazakhstan
[2] Joint Institute for Nuclear Research, Dubna, Russian Federation
[3] L.N. Gumilyov Eurasian National University, Astana, Kazakhstan
[4] Al-Farabi Kazakh National University, Almaty, Kazakhstan

E-mail: janseit.daniar@gmail.com

Abstract. The new study of differential cross sections of ^{13}C$(\alpha,\alpha')^{13}$C inelastic scattering at E_α=29 MeV with the excitation of the 9.90 MeV state of the initial nucleus has been carried out. Modified diffraction model analysis showed that the root-mean-square (RMS) radius of the nucleus at this level might be approximately 20% less than the one of the ground state. We have showed the possibility of the existence of various states in the structure of the ^{13}C nucleus.

1. Introduction

Recently, data on the existence of excited states of light nuclei with anomalously large radius has been obtained. Firstly, these excited states have α-cluster structure, such as the famous Hoyle state (0^+, E=7.65 MeV) of the ^{12}C nucleus, which plays a significant role in stellar nucleosynthesis, and their analogs in ^{11}B and ^{13}C [1-4]. Secondly, these excited states have a neutron halo as those in ^{13}C ($1/2^+$, 3.09 MeV) [5,6] and in ^{9}Be ($1/2^+$, 1.68 MeV)[7].

Anomalously large radii of nuclei in excited states were predicted in various theoretical works, especially after the appearance of a hypothesis about the possible existence of an α-particle condensate [8]. The specific radii were obtained as a result of the experiments and enabled us to make a choice between different theoretical models. At the same time, the fact of observing dilute states agrees with existing concepts about the nucleus. But, is it possible for the existence of other, exotic states in stable nuclei?

In current work we continue the investigation of the nature of ^{13}C excited states at low energies.

2. Experimental Details

The experimental angular distributions of α-particles scattering by ^{13}C nuclei at energies E_α=29 MeV has been measured on the extracted beam of the U-150M isochronous cyclotron of the INP (Almaty, Kazakhstan). In the experiment we used sets of ΔE-E telescopes with silicon semiconductor detectors. Three telescopes of ΔE (with a thickness of 100 μm) and E (with a thickness of 2 mm) detectors were used for registration of scattered α-particles. The target was a self-sustaining film made of ^{13}C isotope (86% enrichment) with a thickness of $\sim$0.35 mg/cm^2. Targets contained some admixtures of the ^{12}C and ^{16}O nuclei. Figure 1 shows a typical spectrum of ^{13}C nucleus.

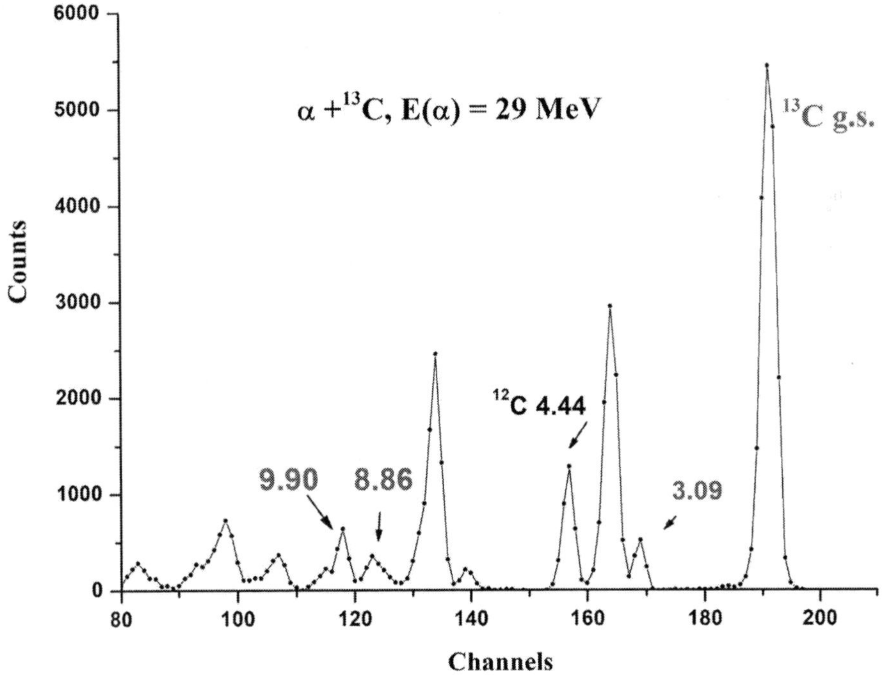

Figure 1. Typical spectrum of $^{13}C(\alpha,\alpha)^{13}C$ at 32 degree, E_α=29 MeV. Red color - exotic excited states of ^{13}C nucleus.

3. Results and discussion

The analysis of the experimental data was made in several stages. At the first stage, the data on elastic scattering were analyzed within the framework of the standard optical model of the nucleus and the double folding (DF) model. Optical potential (OP) parameters have been chosen for achieving the best agreement between the theoretical and experimental angular distributions. This is described in detail in our previous papers [5,9]. The obtained values of the optical potential for a wide range of energies [5,9] are given in Table 1.

Table 1. Parameters of optical and folding potentials for the α+^{13}C system.

E, MeV	Set	V, MeV	r, fm	a, fm	N_r	W_V, MeV	r_V, fm	a_V, fm	r_C, fm
29	OM	147.22	1.112	0.736		12.844	1.6	0.731	1.28
29	DF				0.98	12.844	1.6	0.731	1.28
65	OM	123.07	1.112	0.8		14.97	1.6	0.76	1.28
65	DF				0.98	14.97	1.6	0.76	1.28
90	OM	115.5	1.112	0.645		20.21	1.6	0.6	1.28
90	DF				0.96	20.21	1.6	0.6	1.28

IOP Conf. Series: Journal of Physics: Conf. Series **1023** (2018) 012025 doi:10.1088/1742-6596/1023/1/012025

The task of determining the parameters of the OP from the analysis of elastic scattering can not be solved uniquely. At the second stage, in order to verify the adequacy of the set of potential parameters for a "true", OP were tested with corresponding data on inelastic scattering of α-particles by ^{13}C nuclei, within the framework of CC (couple channels) and DF methods [9].

All our calculations for determining radii of the nuclei in the excited states were carried out within the framework of the modified diffraction model (MDM), which is described in detail in [2].

The ^{13}C nucleus seems to be unique, because several different structures coexist in its spectrum. In addition to the usual levels of the shell model, there are two "dilute" states of various types: one of them has a neutron halo (3.09 MeV ($1/2^+$)), and the other one is an analogue of the Hoyle state (8.86 MeV ($1/2^-$)). In our several earlier works [3,5,10,11], we provided calculations made within the framework of the MDM, which confirm that these states actually exist in the structure of the ^{13}C nucleus.

We can not exclude existence of structures (compact states), for example, even more "exotic" with the reduced radii [12]. It is interesting to note that the 9.90 MeV state and other parts of its rotational band are strongly excited in the α-cluster transfer nuclear reactions: (^{6}Li,d) and (^{7}Li,t)^{9}Be nucleus [13]. This means that the α-cluster structure of the 9.90 MeV state has a strong α+^{9}Be component, which can eventually lead to an enhancement of the radius of this state. Indeed, in [14], authors assume that the RMS radius of a given state might be large.

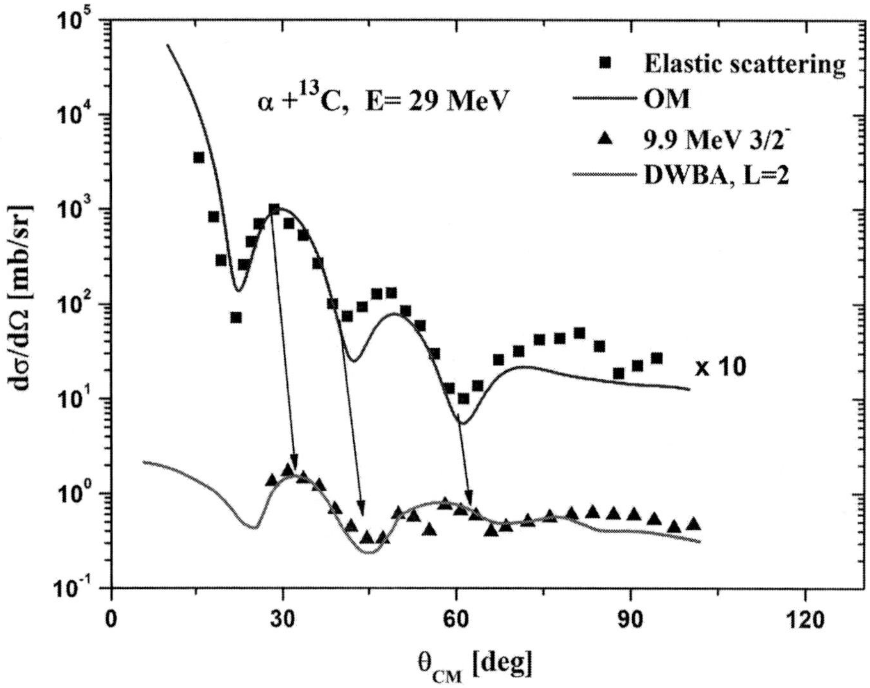

Figure 2. Differential cross sections of elastic α+^{13}C and inelastic scattering for the 9.90 MeV state of ^{13}C nucleus at energies of α-particles 29 MeV. Solid curves are calculations within the OM and the distorted-wave Born approximation (DWBA), respectively. Straight lines connect extrema, which, when the diffraction radii are equal, must be in antiphase.

According to Blair's rules [15], the angular distributions of the elastic and inelastic scattering must coincide in phase. The shift of the minima and maxima of the inelastic scattering cross sections toward smaller or larger angles in comparison with the elastic scattering curve, is an indicator, of increase or decrease, respectively, of the diffraction radius of a certain excited state. In our case, as Figure 2 shows, we observed a clear systematic shift of the minima and maxima of the inelastic scattering cross section toward large angles in comparison with the elastic scattering curve. This character of the shift, as we noted above, is an indicator of a decrease in diffraction radius of the 9.90 MeV excited state.

Indeed, analysis of the differential cross section of the 9.90 MeV state within MDM showed that the RMS radius of this state is smaller by 20% than the ground state Table 2. For comparison, Table 2 also shows the results for other incident particles and energies [5,12]. Within the limits of error, the obtained results are similar.

Table 2. Diffraction and RMS radii for the state of the ^{13}C nucleus at different energies

E*,MeV, I^{π}	R_{dif},fm	R_{rms},fm	E_{α}, MeV
$0.00, 1/2^-$	5.31 ± 0.07	2.33	65
$9.90, 3/2^-$	4.99 ± 0.07	1.97 ± 0.07	29
$9.90, 3/2^-$	5.00 ± 0.12	2.02 ± 0.14	65
$9.90, 3/2^-$	4.80 ± 0.20	1.76 ± 0.23	90

In this paper, we considered a wider range of experimental data [5,9,12]. From this analysis, we can conclude that the predicted enhancement of the radius [14] of the 9.90 MeV state of the ^{13}C nucleus is not observed, but, on the contrary, the ^{13}C nucleus has a more compact size in this excited state.

4. Conclusion

In this paper we have obtained data indicating the possible existence of nuclear states with anomalously small dimensions. Joint analysis of the results of experiments within the study of inelastic scattering of α particles by ^{13}C at the energies of 29, 65 [5] and 90 MeV [12] with excitation of the $3/2^-$, E=9.90 MeV state showed that another phenomenon was discovered in the structure of the ^{13}C nucleus a "supercompact" size. The 9.9 MeV ($3/2^-$) excited state has a smaller radius than the ground state of the ^{13}C nucleus. So far, such "condensed" excited states of nuclei have not yet been observed.

5. Acknowledgements

The work was supported by grants AP05133606 and AP05131597 of Ministry of Education of Republic of Kazakhstan.

6. References

[1] Demyanova A S, Glukhov Y A, Belyaeva T L, Dmitriev S V, Goncharov S A, Khlebnikov S V, Maslov V A, Molchanov Yu D, Penionzhkevich Yu E, Revenko R V, Safonenko M V, Sobolev Yu G, Trzaska W, Tyurin G P, Zherebchevski V I and Ogloblin A A 2008 *Nuclear Phys. A.* **805** 489
[2] Danilov A N, Belyaeva T L, Demyanova A S, Goncharov S A and Ogloblin A A 2009 *Phys. Rev. C.* **80** 054603
[3] Demyanova A S, Ogloblin A A, Danilov A N, Belyaeva T L and Goncharov S A 2011 *Int. J. Mod. Phys. E* **20** 915

[4] Danilov A N, Demyanova A S, Ogloblin A A, Dmitriev S V, Belyaeva T L, Goncharov S A, Gurov Yu B, Maslov V A, Sobolev Yu G, Trzaska W, Khlebnikov S V, Burtebaev N, Zholdybayev T, Saduyev N, Heikkinen P, Julin R and Tyurin G P 2014 *EPJ Web Conf.* **66** 03007

[5] Demyanova A S, Danilov A N, Dmitriev S V, Ogloblin A A, Belyaeva T L, Burtebaev N, Drobyshev P, Goncharov S A, Gurov Yu B, Heikkinen P, Julin R, Khlebnikov S V, Maslov V A, Nassurlla M, Penionzhkevich Yu E, Sobolev Yu G, Trzaska W, Tyurin G P and Zherebchevskii V I 2014 *EPJ Web Conf.* **66** 02027

[6] Liu Z H, Lin C J, Zhang H Q, Li Z C, Zhang J S, Wu Y W, Yang F, Ruan M, Liu J C, Li S Y and Peng Z H 2001 *Phys. Rev. C* **64** 034312

[7] Ogloblin A A, Danilov A N, Belyaeva T L, Demyanova A S, Goncharov S A and Trzaska W 2011 *Phys. Rev. C* **84** 054601

[8] Tohsaki A, Horiuchi H, Schuck P and Ropke G 2001 *Phys. Rev. Lett.* **87** 192501

[9] Burtebayev N, Sakhiyev S K, Janseitov D M, Kerimkulov Zh, Alimov D and Danilov A N 2016 *Int. J. Mod. Phys. E* **25** 1650078

[10] Ogloblin A A, Demyanova A S, Danilov A N, Belyaeva T L, Goncharov S A and Trzaska W 2016 *Physics of Atomic Nuclei* **79** 514

[11] Demyanova A S, Ogloblin A A, Danilov A N, Goncharov S A, Belyaeva T L, Sobolev Yu G, Khlebnikov S V, Burtebaev N, Trzaska W, Heikkinen P, Tyurin G P, Janseitov D and Gurov Yu B 2016 *EPJ Web Conf.* **117** 0401

[12] Ogloblin A A, Demyanova A S, Danilov A N, Goncharov S A, Belyaeva T L, Trzaska W and Sobolev Yu G 2015 *JETP Letters* **102** 227

[13] Goldberg V Z, Davidov V V, Ogloblin A A, Sakuta S B and Tshuev V I 1971 *Proc. of Soviet Academy of Science, Ser. Fiz.* **35** 1663

[14] Furutachi N and Kimura M 2011 *Phys. Rev. C* **83** 021303

[15] Austern N and Blair J S 1965 *Annals of Physics* **33** 15

Calculation of total reaction cross sections of ^{6}He, ^{8,9}Li on nuclei within the microscopic optical potential model

K.V.Lukyanov, Yu.G.Sobolev, Yu.E.Penionzhkevich, E.V.Zemlyanaya
Joint Institute for Nuclear Research, Dubna 141980, Russia
E-mail: luky@jinr.ru

Abstract. Experimental data on total reaction cross sections of exotic nuclei ^{6}He, ^{8,9}Li interaction with the ^{181}Ta, ^{59}Co, ^{28}Si and ^{9}Be targets in a wide energy range are analyzed in the framework of a hybrid model of microscopic optical potential (OP). The real part of OP is obtained on the basis of the double folding model, while the imaginary part of OP is calculated within the high energy approximation. This theoretical approach demonstrates the fairly well agreement with experimental data at the wide energy regions.

1. Introduction

In last few decades the nuclear structure of light exotic nuclei is intensively investigated both theoretically and experimentally. In particular, experimental data of the total reaction cross section σ_R of interaction of stable nuclei with the exotic helium and lithium isotopes in a wide energy range are of interest for their physical properties study.

Current calculations continues the previous one [1], made for ^{4,6}He and ^{6,7}Li reactions with the ^{28}Si target, where we considered a possibility of the microscopic folding potentials [2] to obtain the total reaction cross sections. It was shown, that such optical potentials based on the density distributions of the nuclei-participants give rather good description of experimental data. In this work we apply the same microscopical optical potentials to study the total reaction cross sections of the ^{6}He and ^{8,9}Li on ^{9}Be, ^{28}Si, ^{59}Co and ^{181}Ta targets. The motivation is that the microscopic models in their origin do not contain free parameters and thus provide the possibility to test the models of nuclear structure of exotic nuclei.

The experimental data used in this calculations were obtained at the cyclotron U-400M FLNR JINR with ACCULINA beam-line. The energy of ^{6}He and ^{8,9}Li particles has been changed in an energy range $\sim 10 - 25$ AMeV [3]-[6].

2. Calculation technique

2.1. Microscopic optical potential

In our calculations we use the microscopic optical potential (OP) of the form

$$U_{opt}(r) = \left[N_r V^{DF} - N_r^{(1)} r \frac{dV^{DF}}{dr}\right] + i\left[N_{im} W^H - N_{im}^{(1)} r \frac{dW^H}{dr}\right], \quad (1)$$

where V^{DF} is the double-folding potential with exchange effects included, and W^H is obtained within the so-called high-energy approximation (HEA) model, $N_{r,im}$ are fitted renormalization coefficients. The "surface terms" of OP dV^{DF}/dr and dW^H/dr describe effect of nuclear collective modes on the mechanism of nucleus-nucleus scattering. It was shown in [1] that this form of OP allows one to get the fairly well agreement with the experimental data comparing to the other potential forms of OP.

2.2. Double-folding potential

The real part V^{DF} of constructed OP is the double-folding nucleus-nucleus potential (see [7],[2]) consists of the direct and exchange parts:

$$V^{DF} = V^D + V^{EX}, \tag{2}$$

$$V^D(r) = \int d^3r_p d^3r_t\, \rho_p(\mathbf{r}_p)\, \rho_t(\mathbf{r}_t)\, v^D_{NN}(s), \quad \mathbf{s} = \mathbf{r} + \mathbf{r}_t - \mathbf{r}_p, \tag{3}$$

$$V^{EX}(r) = \int d^3r_p\, d^3r_t\, \rho_p(\mathbf{r}_p,\, \mathbf{r}_p + \mathbf{s})\, \rho_t(\mathbf{r}_t, \mathbf{r}_t - \mathbf{s}) v^{EX}_{NN}(s)\, \exp\left[\frac{i\mathbf{K}(r)\cdot s}{M}\right], \tag{4}$$

where $\rho_{p,t}$ are the one-particle nuclear densities of the projectile (p) and target (t) nulcei. In our calculations we apply the effective Paris nucleon-nucleon CDM3Y6 potential v_{NN} having the form

$$v_{NN}(E, \rho, s) = g(E)\, F(\rho)\, v(s), \tag{5}$$

where

$$g(E) = 1 - 0.003E/A_p, \tag{6}$$

$$F(\rho) = C\Big[1 + \alpha \exp(-\beta\rho) - \gamma\rho\Big], \tag{7}$$

$$v(s) = \sum_{i=1,2,3} N_i \frac{\exp(-\mu_i s)}{\mu_i s}, \tag{8}$$

$$\rho = \rho_p + \rho_t. \tag{9}$$

The parameters C, α, γ, N_i and μ_i are known and presented, for example, in [7].

The energy dependence of V^{EX} arises primarily from the exponential function in the integrand, where

$$K(r) = \{2Mm/\hbar^2[E - V^{DF}(r) - V_c(r)]\}^{1/2} \tag{10}$$

is the local nuclear collision momentum, $M = A_pA_t/(A_p + A_t)$, and m is the nucleon mass. One can see that $K(r)$ depends on V^{DF}, and thus reveals the typical non-linear problem. The iteration procedure is described in [8] and implemented here.

2.3. HEA microscopical potential

When constructing microscopic OP, many authors usually apply only the real double-folding potential V^{DF} without accounting for the exchange term (4), and the imaginary part is taken in a phenomenological form with free parameters fitted to experimental data for each specific energy. In some cases they also use their form to be proportional to the form of V^{DF}. In our calculations, we use the imaginary part W^H obtained in [2] within the high-energy theory of nuclear scattering [9, 10]:

$$U^H_{opt} = V^H + W^H, \tag{11}$$

$$V^H = -\frac{2E}{k(2\pi)^2}\sigma_{NN}\alpha_{NN}\int_0^\infty dq q^2 j_0(qr)\rho_p(q)\rho_t(q)f_N(q), \tag{12}$$

$$W^H = -\frac{2E}{k(2\pi)^2}\sigma_{NN}\int_0^\infty dq q^2 j_0(qr)\rho_p(q)\rho_t(q)f_N(q). \tag{13}$$

Here $\rho(q) = \int d^3r\, e^{iqr} \rho(r)$ is the form factor of a point like nuclear density; σ_{NN} and α_{NN} are the total NN cross section and the ratio of real to imaginary part of NN-amplitude of scattering respectively; $f_N(q)$ corresponds to the form of NN-amplitude of scattering.

2.4. Density models

It was shown in [1] that the behavior of total cross sections are sensitive to the form of densities of interacting nuclei in their peripheral region. Generally, microscopical optical potentials do not contain free parameters and thus provide the possibility to test the models of nuclear structure. Figure 1 demonstrates how the density models of the ^{6}He projectile affects the resulting ^{6}He+^{28}Si total reaction cross section. As a result of studies in [1] it was concluded, that the densities with the realistic exponential behavior at large distances are preferable.

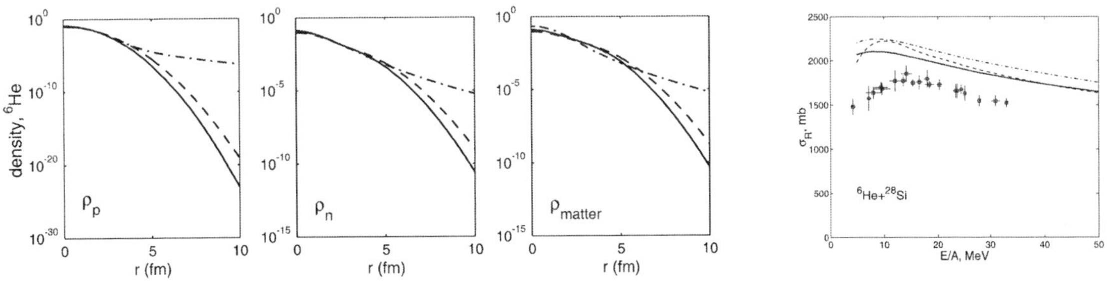

Figure 1. Different models of density distributions of ^{6}He (solid curves: Tanihata model [11]; dash-dotted: LSSM [12]; dashed - COSMA [13]) and the resulting total reaction cross section calculations for ^{6}He+^{28}Si (solid curves: Tanihata model; dashed curves: LSSM; dash-dotted curves: COSMA). Reproduced from [1].

Therefore in our calculations we used the large-scale shell-model (LSSM) densities [12] for the helium projectile and Table densities [14] for other nuclei.

3. Results of calculations

Numerical calculations of cross sections were made basing on the given potential (1) by using the DWUCK4 code [15]. The renormalization coefficients $N_{r,im}$ of microscopic OPs (1) were introduced to get a better agreement to the data at larger energies by fitting their strengths V^{DF} and W^H. Two other parameters $N^{(1)}_{r,im}$ are responsible to the contribution of collective terms.

Table 1. Renormalization coefficients of the OP (1)

	^{6}He				^{8}Li				^{9}Li			
	^{9}Be	^{28}Si	^{59}Co	^{181}Ta	^{9}Be	^{28}Si	^{59}Co	^{181}Ta	^{9}Be	^{28}Si	^{59}Co	^{181}Ta
N_r	0.5	0.7	0.7	0.9	0.1	0.85	1.0	1.0	0.7	1.0	1.0	1.0
N_{im}	0.3	0.5	0.5	0.9	0.1	0.75	1.0	1.0	0.5	1.0	1.0	1.0
$N_r^{(1)}$	0.4	0.4	0.7	0.9	0.1	0.5	0.4	0.2	0.4	0.6	0.6	0.2
$N_{im}^{(1)}$	0.03	0.03	0.035	0.03	0.01	0.07	0.08	0.35	0.03	0.12	0.13	0.35

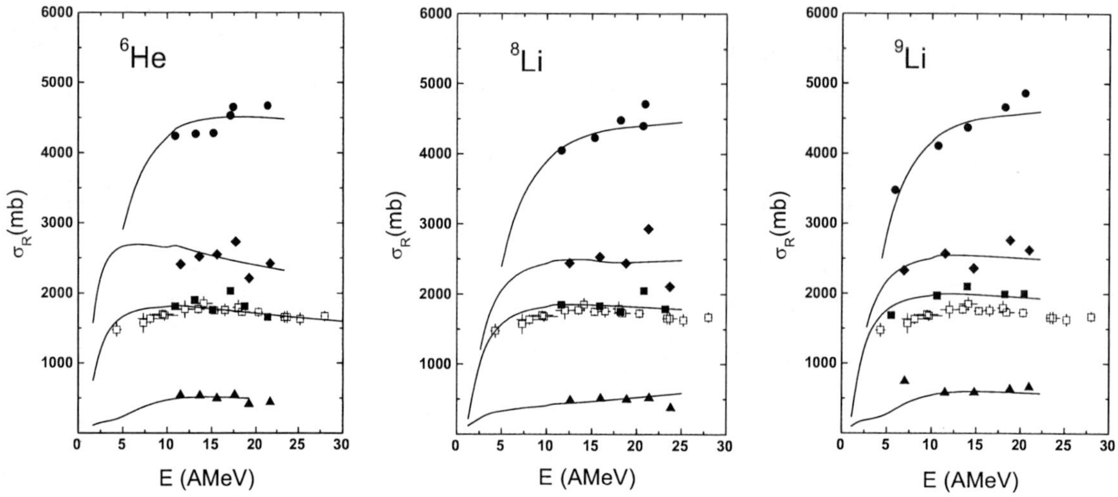

Figure 2. Calculations of total reaction cross sections of ^{6}He, ^{8,9}Li with different targets. Projectile is in left top corner of each figure. Lines show results of calculations, points - experimental data (black triangles: ^{9}Be target; black squares: ^{28}Si target; black diamonds: ^{59}Co target; black disks: ^{181}Ta target; empty squares are the ^{6}He+^{28}Si reaction experimental data from [1] for comparison).

Basically we obtain qualitative agreement with experimental data for ^{6}He and ^{8,9}Li reactions on different targets, shown on Figure 2. The corresponding fitted coefficients of the depths of OPs are represented in Table 1. We mention that there exist some kind of ambiguity when comparing calculated cross sections with the experimental data. Usually one performs the χ^2 fitting procedure to obtain the best set of adjusted parameters, and such a procedure assumes enough experimental points in a wide energy range. Unfortunately, in our case one sees that the number of experimental data is very small. It means that one can fit several lines with similar χ^2 values, but with difference in their behavior out of the energy range $\sim 10-25$ AMeV. To exclude this uncertainty more experimental data are needed.

According to Table 1, one sees that the N_r and N_{im} coefficients for the volume parts of OPs are close to 1 for reactions with ^{28}Si, ^{59}Co and ^{181}Ta. Also, influence of renormalization on the surface terms (derivatives) is less significant, especially of its imaginary part, represented by $N_{im}^{(1)}$. On the other hand, one can see very low N-values for reactions with ^{9}Be, especially in case of the ^{8}Li projectile. The possible reason is the structure of ^{9}Be, which could be represented by cluster model, for example of two α-particles and neutron in periphery. From this point of view, it can be also considered as an exotic nuclei. In this case, the used gaussian form of density distribution doesn't corresponds to it real form. Therefore the other density models need to be checked, and interaction between two exotic nuclei needs in more detailed investigations too.

4. Conclusion

Microscopic model of the nucleus-nucleus OP has been constructed by using physical characteristics of structure of densities of colliding nuclei, and the effective nucleon-nucleon forces in nuclear medium. It was shown that a renormalization of strengths of this potentials by introducing the respective renormalization parameters allow us to explain the experimental data. We obtain the qualitative agreement to the most energy range.

Simultaneously, there are only few experimental points for each reaction (except ^{6}He on ^{28}Si target), which sets the ambiguity of parameters at low and high energy regions. More experimental data (also data on differential cross sections) are required to reduce this ambiguity. We need the further developing of the theory and model to understand the inconsistency with experimental data.

5. Acknowledgement

K.V.Lukyanov and E.V.Zemlyanaya thanks the RFBR (grant 17-52-18057) for the partial support of this work.

References

[1] K.V.Lukyanov et al. 2008 *Bulletin of the Russian Academy of Sciences. Physics.* Vol.**72** No.3 pp.356-360
[2] V.K.Lukyanov, E.V.Zemlyanaya, K.V.Lukyanov 2006 *Physics of Atomic Nuclei* Vol.**69** N2 p.240
[3] Yu.G.Sobolev et al. 2012 *Bulletin of the Russian Academy of Sciences. Physics* Vol.**76** N.8 pp.952-957
[4] Yu.G.Sobolev et al. 2015 *Exotic Nuclei. Proceedings of the International symposium on Exotic Nuclei EXON 2014* - World Scientific Publishing pp.147-152.
[5] Yu.E.Penionzhkevich, Yu.G.Sobolev, V.V.Samarin, M.A.Naumenko 2017 *Physics of Atomic Nuclei* Vol.**80** N5 pp.51-64
[6] Yu.G.Sobolev et al. 2017 *Physics of Particles and Nuclei* Vol.**48** No.6 pp.922-926
[7] D.T.Khoa and G.R.Satchler 2000 *Nucl. Phys. A* **668** 3
[8] K.V.Lukyanov 2007 Double Folding Model of Nucleus-Nucleus Potential: Formulae, Iteration Method and Computer Code. *Communication of JINR* P11-2007-38 Dubna
[9] R.J.Glauber 1959 *Lectures on Theoretical Physics* (New York: Interscience) p.315
[10] A.G.Sitenko 1959 *Ukr. Fiz. Journal* **4** 152
[11] I.Tanihata et al. 1992 *Phys. Lett. B* **289** 261
[12] S.Karataglidis, P.J.Dortmans, K.Amos, and C.Bennhold 2000 *Phys. Rev. C* **61** 024319
[13] M.V.Zhukov, B.V.Danilin, D.V.Fedorov et al. 1993 *Phys. Rep.* **231** 151
[14] J.D.Patterson and R.J.Peterson 2003 *Nucl. Phys. A* **717** 235
[15] P.D.Kunz and E.Rost 1993 *Computational Nuclear Physics*, V.**2** (Eds: Langanke K.et al., Springer Verlag) p.88

IOP Conf. Series: Journal of Physics: Conf. Series **1023** (2018) 012027 doi:10.1088/1742-6596/1023/1/012027

Cluster Structure of ^{9}Be by Scattering of Deuterons

S.M. Lukyanov[1], A. S. Denikin[1], M. A. Naumenko[1], V. Burjan[2], W.H. Trzaska[3], M. Harakeh[4], D.Etasse[5],V. Kroha[2], J. Mrazek[2], K. Mendibayev[1] I. Sivaček[2], V. Glagolev[2], Š. Piskoř[2], Yu.E. Penionzhkevich[1], N.K. Skobelev[1], I.Stefan[6], D. Verney[6], K. Kuterbekov[7],T. Zholdyvayev [8]

[1] Flerov Laboratory of Nuclear Reactions, Dubna, Russian Federation
[2] Nuclear Physics Institute, Řež, Czech Republic
[3] Department of Physics, University of Jyväskylä, Jyväskylä, Finland
[4] KVI-CART, University of Groningen, Groningen, The Netherlands
[5] LPC-Caen, ENSICAEN, Université de Caen, CNRS/IN2P3-ENSI, Caen, France
[6] Institut de Physique Nucléaire, Univ. Paris-Sud, Université Paris-Saclay, Orsay, France
[7] Eurasian Gumilev University, Astana, Kazakhstan
[6] Institute of Nuclear Physics, Almaty, Kazakhstan

E-mail: lukyan@jinr.ru

Abstract. Angular distributions of protons, deuterons, tritons and alpha particles emitted in the reaction ^{2}H+^{9}Be at E_{lab}=19.5 MeV were measured with an aim to shed light on the internal cluster structure of ^{9}Be and to study possible cluster transfer of ^{5}He. The analyses suggest a significant contribution of five-nucleon transfer in the reaction channel ^{9}Be(d,^{4}He)^{7}Li.

1. Introduction

Due to the Borromean structure, a special attention has been focused on the ^{9}Be nucleus, the breakup of which can occur directly into two alpha particles and a neutron or via one of two unstable intermediate nuclei such as ^{8}Be or ^{5}He [1, 2]. Scattering of a projectile, such as ^{1,2}H or ^{3,4}He, on a target is a standard tool to study the structure of nuclei. This method involves the angular distribution measurement of elastic and inelastic scattering. Energy distribution of the products gives information about internal structure of the interacting nuclei. The angular distributions of the ^{9}Be(^{3}He,^{3}He)^{9}Be, ^{9}Be(^{3}He,^{5}He)^{7}Be, ^{9}Be(^{3}He,^{5}Li)^{7}Li, ^{9}Be(^{3}He,^{6}Be)^{6}He, and ^{9}Be(^{3}He,^{6}Li)^{6}Li reaction channels were measured in Refs. [3, 4] and described within the framework of the optical model, the coupled-channel approach and the distorted-wave Born approximation. The performed analysis of the experimental data shows that the potential parameters are quite sensitive to the exit channel and hence to the cluster structure of the populated states, which allows one to make general observations and conclusions regarding the internal structure of the target and product nuclei. The experiment [3, 4] was designed to study the breakup of ^{9}Be in an attempt to determine the contribution of the ^{8}Be+n and ^{5}He+α channels in the inclusive measurements. We found that the ratio about 2.7:1 may be assigned to the contributions of these two channels, respectively. The determined value justifies that the ^{5}He+α breakup channel plays an important role.

Another aspect is an attempt to find not only the cluster structure (for instance, ^{5}He), but also clarify how the cluster structure is involved into nuclear reaction mechanism. Indeed,

IOP Conf. Series: Journal of Physics: Conf. Series **1023** (2018) 012027 doi:10.1088/1742-6596/1023/1/012027

starting from Detraz [5, 6] multiparticle-multihole structures were expected to occur at rather low excitation energies in nuclei. Four-nucleon transfer reactions are being extensively studied. One may hope that their major features, in spite of the a-priori complexity of such a transfer, can be understood assuming that the nucleons are transferred as a whole, strongly correlated in a cluster which has the internal quantum numbers of a free particle.

2. Experimental Method

The experiments were performed using ^{2}H ion beams at 19.5 MeV, 25 MeV and 35 MeV energies delivering by the accelerator facilities at INP (Řež, Czech Republic), IPNO (Orsay, France) and Department of Physics, University of Jyväskylä, Jyväskylä (Finland), respectively. The average beam current during the experiments was maintained at 10 nA. The self-supporting Be target was prepared from a thin beryllium foil of 99 % purity. To measure (in)elastically scattered ions, a set of four telescopes each consisting of ΔE_0, ΔE, E_r detectors with thicknesses of 12, 100 μm and 3 mm, respectively, were used. The telescopes were mounted at a distance of about 19 cm from the target in a reaction chamber. Particle identification was performed based on the energy-loss measurements of ΔE and residual energy E_r, i.e., by the so-called ΔE-E method. An example of two-dimensional plots (yield vs. energy loss ΔE and residual energy E_r) is shown in Fig. 1.

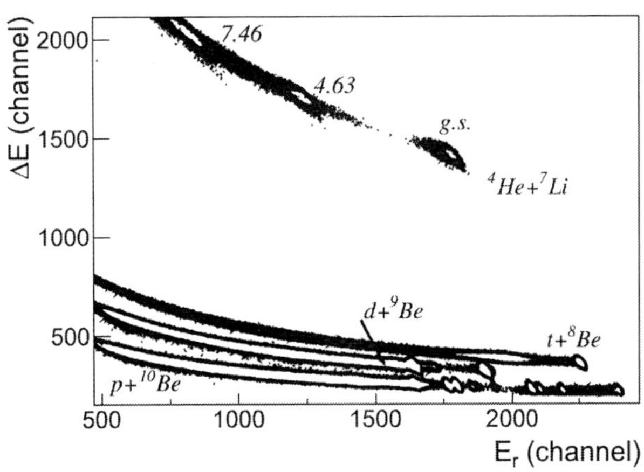

Figure 1. Particle identification plots for the products of the ^{2}H+^{9}Be reaction: p, d, t, and ^{4}He. ΔE is the energy loss and E_r is the residual energy. Excited states for the ^{7}Li reaction channel ^{7}Li+α are indicated.

This experimental technique allows us to identify the particles p, d, t, and ^{4}He and determine their total deposited energies. The spectra of total deposited energy are shown in Fig. 2. All peaks, which can be observed in the histograms in Fig. 2, were identified and found to belong to the ground and the excited states of ^{10}Be, ^{9}Be, ^{8}Be, and ^{7}Li, as the complementary products to the detected particles p, d, t, and ^{4}He, respectively.

3. Results and Data Analysis

The differential cross section for the elastic, inelastic, and transfer reaction channelsof 19.5 MeV deuterons from ^{9}Be target are presented in Fig. 3. All calculations and fitting of these experimental data have been performed using the FRESCO [7] code in the framework of the CC method.

IOP Conf. Series: Journal of Physics: Conf. Series **1023** (2018) 012027 doi:10.1088/1742-6596/1023/1/012027

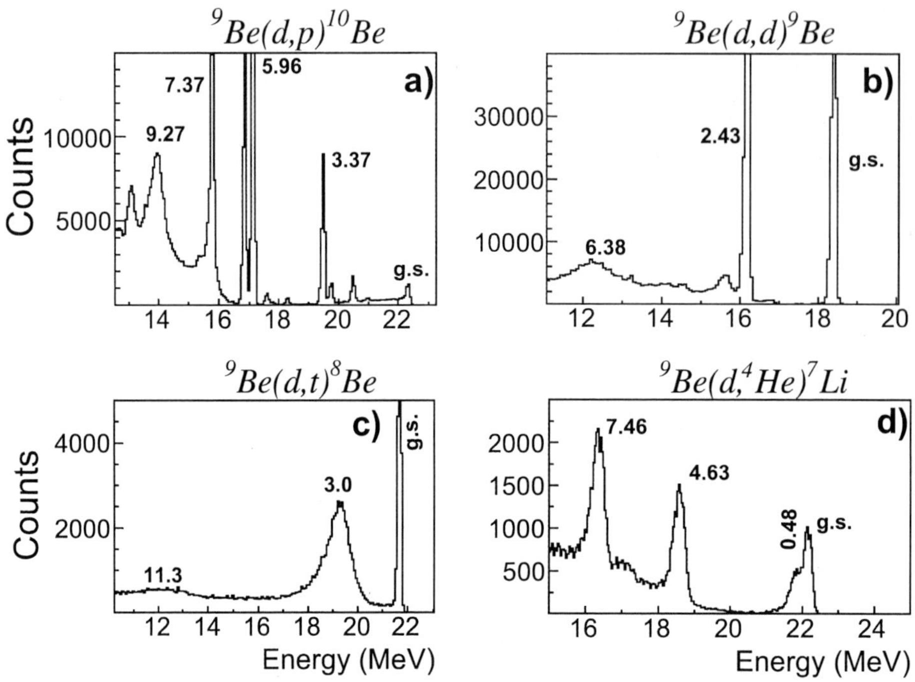

Figure 2. Total deposited energy spectra measured at θ_{lab}=32° for the detected p (panel a), d (panel b), t(panel c), and ^{4}He(panel d). The ground and the excited states of ^{7}Li for the case of detected complementary product ^{4}He as well as the ground states and the excited states for ^{8}Be, ^{9}Be, and ^{10}Be in the case of detected t, d, and p, as complementary products, respectively, were unambiguously identified.

The angular distributions of the ^{9}Be(d,d)^{9}Be, ^{9}Be(d,p)10Be, ^{9}Be(d,t)^{8}Be, ^{9}Be(d,^{3}He)^{8}Li, and ^{9}Be(d,4He)^{7}Li reaction channels were measured at various values of the projectile energies. Data analyze is still in progress. The calculation for (in)elastic scattering (Fig. 3a) at deuteron energy at 19.5 MeV was performed using the method of strong channel coupling in the adiabatic rotational model. It is assumed that the state 2.43 MeV is the first excited state in the rotational band of ^{9}Be. The quadrupole deformation parameter of the target nucleus found from the performed analysis of data is β_2= 0.64, which perfectly coincides with the previous analysis of inelastic scattering of ^{4}He nuclei on the same target [3]. The optical potential parameters used for the calculation are similar to those obtained in Ref. [8]. Good agreement between experimental data and calculations is observed, as shown in Fig. 3a.

A special attention was paid to the ^{4}He+^{7}Li reaction channel, which can proceed through two different mechanisms: either transfer of d (dashed line in Fig. 3b) or transfer of ^{5}He (solid line in Fig. 3b) from the target to the projectile. In the exit channel both these mechanisms are indistinguishable from each other. However, in the former case the ^{4}He nuclei are expected to fly in the forward direction in the center of mass system, whereas in the latter case the alpha particles should preferably fly at the backward angles in the center of mass system. The calculations carried out within the DWBA method for these two mechanisms are shown as the dashed and solid curves in Fig. 3b. Their coherent sum shown as the black curve is in good agreement with the experimental points. It is interesting to note that for the correct description of the data, it is necessary to set the sufficiently large values of the spectroscopic factors for

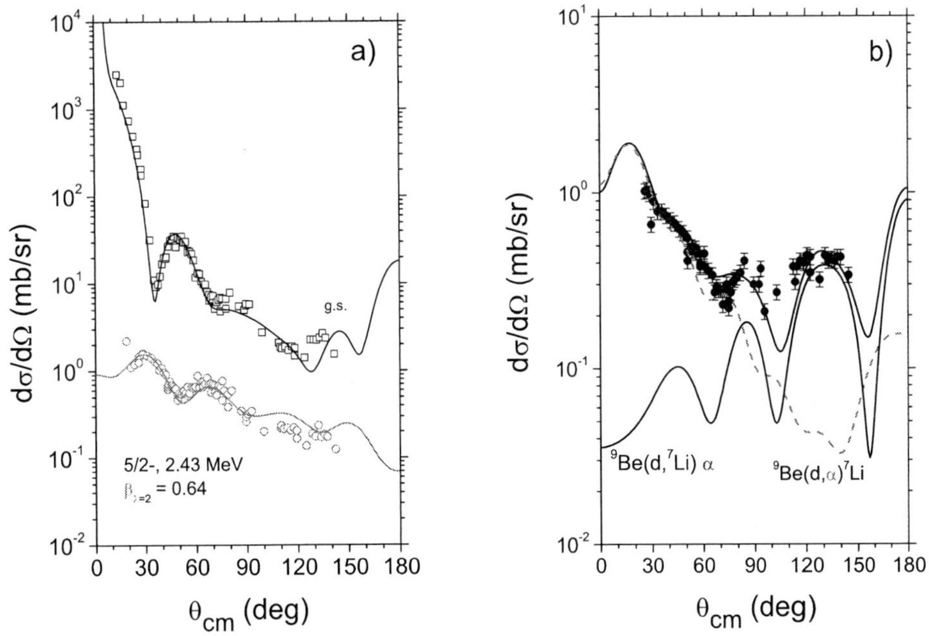

Figure 3. a) Experimental angular distribution for elastic(□) and inelastic scattering(∘) of 19.5 MeV deuterons from ^{9}Be target compared with the calculations. The curves are the results of the optical-model and the coupled-channel calculations. b) Experimental angular distribution (•) for ^{4}He+^{7}Li reaction channel (see details in text).

the systems ^{9}Be $=\alpha+$ ^{5}He (S=1.2) and ^{7}Li = d + ^{5}He (S=1.0). In addition, to describe the structure of the angular distributions it is also necessary to assume a 30% admixture of the *d*-state in the structure of ^{7}Li.

In summary, angular distributions for the ^{9}Be(*d*,*d*)^{9}Be*, ^{9}Be(*d*,*p*)^{10}Be, ^{9}Be(*d*,*t*)^{8}Be, and ^{9}Be(*d*,^{4}He)^{7}Li channels were measured. Experimental angular distributions were described within the optical model, the coupled channel approach, and the distorted wave Born approximation. The optical model provides good agreement with the elastic scattering. The DWBA calculations agree well with the transfer reaction data. The spectroscopic factors for the systems ^{9}Be $=\alpha+$ ^{5}He and ^{7}Li = d + ^{5}He are close to unity, which confirms significant contribution of the considered cluster configurations to the structure of ground states. The analysis shows that the contribution of the compound nucleus mechanism is negligible. In the (*d*, ^{4}He) channel (see Fig. 3), the deuteron transfer provides only a small contribution, whereas a relatively large contribution of ^{5}He transfer was found in agreement with the result [8, 9]. This demonstrates that the specific structure of the ^{9}Be nucleus as a weakly bound system of two alpha particles and a neutron strongly favors the five-nucleon transfer compared to the deuteron transfer.

The authors thank the CANAM project (http://users.canam.ujf.cas.cz), mobility grant from the Academy of Finland, IN2P3-Dubna agreement for providing beam time for the experiments and support. This work was also supported by Russian Science Foundation (17-12-01170).

References

[1] T A D Brown, P Papka, B R Fulton *et al.*, *Phys. Rev. C* **76**, 054605 (2007).
[2] P Papka, T A D Brown, B R Fulton *et al.*, *Phys. Rev. C* **75**, 045803 (2007).
[3] S M Lukyanov, A S Denikin, E I Voskoboynik *et al.*, *J. Phys. G* **41**, 035103 (2014).
[4] S M Lukyanov, M Harakeh, M A Naumenko *et al.*, *World Journal of Nuclear Science and Technology* **5** 265 (2015). doi: 10.4236/wjnst.2015.54026.
[5] C Détraz, H H Duhm and H Hafner, *Nucl. Phys. A* **147**, 488 (1970).
[6] C Détraz, F Pougheon, M Bernas *et al.*, *Nucl. Phys. A* **228**, 39 (1974).
[7] http://www.ianthompson.org/surrey and http://nrv.jinr.ru
[8] A Szczurek, K Bodek, J Krug *et al.*, *Zeitschrift für Physik A Atomic Nuclei*, **333**, 271 (1989).
[9] L. Jarczyk, B. Kamys, Z. Rudy *et al.*, *Zeitschrift für Physik A Atomic Nuclei*, **325**, 303 (1986).

IOP Conf. Series: Journal of Physics: Conf. Series **1023** (2018) 012028 doi:10.1088/1742-6596/1023/1/012028

Nuclear effects in (anti)neutrino charge-current quasielastic scattering at MINERνA kinematics

M.V. Ivanov[1], **A.N. Antonov**[1], **G.D. Megias**[2], **R. González-Jiménez**[3], **M.B. Barbaro**[4], **J.A. Caballero**[2], **T.W. Donnelly**[5], **and J.M. Udías**[6]

[1] Institute for Nuclear Research and Nuclear Energy, Bulgarian Academy of Sciences, Sofia 1784, Bulgaria

[2] Departamento de Física Atómica, Molecular y Nuclear, Universidad de Sevilla, 41080 Sevilla, Spain

[3] Department of Physics and Astronomy, Ghent University, Proeftuinstraat 86, B-9000 Gent, Belgium

[4] Dipartimento di Fisica, Università di Torino and INFN, Sezione di Torino, Via P. Giuria 1, 10125 Torino, Italy

[5] Center for Theoretical Physics, Laboratory for Nuclear Science and Department of Physics, Massachusetts Institute of Technology, Cambridge, Massachusetts 02139, USA

[6] Grupo de Física Nuclear, Departamento de Física Atómica, Molecular y Nuclear, Facultad de Ciencias Físicas, Universidad Complutense de Madrid, Madrid E-28040, Spain

E-mail: martin@inrne.bas.bg

Abstract. We compare the characteristics of the charged-current quasielastic (anti)neutrino scattering obtained in two different nuclear models, the phenomenological SuperScaling Approximation and the model using a realistic spectral function $S(p, \mathcal{E})$ that gives a scaling function in accordance with the (e, e') scattering data, with the recent data published by the MiniBooNE, MINERνA, and NOMAD collaborations. The spectral function accounts for the nucleon-nucleon (NN) correlations by using natural orbitals from the Jastrow correlation method and has a realistic energy dependence. Both models provide a good description of the MINERνA and NOMAD data without the need of an *ad hoc* increase of the value of the mass parameter in the axial-vector dipole form factor. The models considered in this work, based on the the impulse approximation (IA), underpredict the MiniBooNE data for the flux-averaged charged-current quasielastic $\nu_\mu(\overline{\nu}_\mu)$ + ^{12}C differential cross section per nucleon and the total cross sections, although the shape of the cross sections is represented by the approaches. The discrepancy is most likely due to missing of the effects beyond the IA, *e.g.*, those of the 2p–2h meson exchange currents that have contribution in the *transverse responses.*

1. Introduction

The MINERνA Collaboration has recently measured differential cross sections for neutrino and antineutrino charged-current quasielastic (CCQE) scattering on a hydrocarbon target [1, 2]. "Quasielastic" events are defined, in this case, as containing no mesons in the final state. The beam energy goes from 1.5 to 10 GeV and is peaked at $E_\nu \sim 3.5$ GeV. At lower energies $E_\nu \sim 0.8$ GeV the MiniBooNE experiment has reported [3, 4] CCQE cross sections that are higher than most theoretical predictions based on the impulse approximation (IA), leading to the suggestion that non-QE processes induced by two-body currents may play a significant role in this energy domain [5–8]. These effects have sometimes been simulated, in the Relativistic Fermi

IOP Conf. Series: Journal of Physics: Conf. Series **1023** (2018) 012028 doi:10.1088/1742-6596/1023/1/012028

Gas (RFG) framework, by a value of the nucleon axial-vector dipole mass $M_A = 1.35$ GeV [3, 4], which is significantly larger than the standard value $M_A = 1.03$ GeV extracted from neutrino-deuterium quasielastic scattering. On the other hand, higher-energy data from the NOMAD experiment ($E_\nu \sim 3 - 100$ GeV) [9] are well accounted for by IA models [10]. The MINERνA experiment is situated in between these two energy regions and its interpretation can therefore provide valuable information on the longstanding problem of assessing the role of correlations and meson exchange currents (MEC) in the nuclear dynamics [11–13].

In this paper we present results corresponding to two different nuclear models: the SuSA (SuperScaling Approximation) and the model using a realistic spectral function $S(p, \mathcal{E})$. Both have been extensively tested against existing QE electron scattering data over a wide energy range. The detailed description of these models can be found in our previous work (see, *e.g.*, [14] and [15–17]). Here we just summarize their main features.

2. Theoretical Scheme and Results

SuSA [14] is based on the idea of using electron scattering data to predict CC neutrino cross sections: a phenomenological "superscaling function" $f(\psi)$, depending only on one "scaling variable" $\psi(q, \omega)$ and embodying the essential nuclear dynamics, can be extracted from QE longitudinal (e, e') data within a fully relativistic framework. This function is then multiplied by the appropriate charge-changing $N \to N$ ($n \to p$ for neutrino and $p \to n$ for antineutrino scattering) weak interaction cross sections to obtain the various response functions that contribute to the inclusive neutrino-nucleus cross section [18]. On the one hand, the model gives a good representation of the purely nucleonic contributions to the existing QE electron scattering data, to the extent that the quasielastic scattering can be isolated. On the other hand, it does not account for the inelastic scattering and MEC which are mainly seen in the transverse channel. For the former, the SuSA approach has been successfully extended to higher energies into the non-QE regime where inelastic contributions dominate [19]. The latter have been modeled using extensions of the RFG for two-body operators and typically cause $10-20\%$ scaling violations.

The second model we consider is the model using a realistic spectral function $S(p, \mathcal{E})$ that gives a scaling function in accordance with the (e, e') scattering data [15–17]. Within the PWIA (see, *e.g.*, [15, 20] and details therein) the differential cross section for the $(e, e'N)$ process factorizes in the form

$$\left[\frac{d\sigma}{d\epsilon' d\Omega' dp_N d\Omega_N}\right]^{PWIA}_{(e,e'N)} = K\sigma^{eN}(q, \omega; p, \mathcal{E}, \phi_N) S(p, \mathcal{E}) \,, \tag{1}$$

where σ^{eN} is the electron-nucleon cross section for a moving off-shell nucleon, K is a kinematical factor and $S(p, \mathcal{E})$ is the spectral function giving the probability to find a nucleon of certain momentum and energy in the nucleus. In Eq. (1): p is the missing momentum and $\mathcal{E}$ is the excitation energy of the residual system. The scaling function can be represented in the form:

$$F(q, \omega) \cong \frac{[d\sigma/d\epsilon' d\Omega']_{(e,e')}}{\overline{\sigma}^{eN}(q, \omega; p = |y|, \mathcal{E} = 0)} \,, \tag{2}$$

where the electron-single nucleon cross section $\overline{\sigma}^{eN}$ is taken at $p = |y|$, y being the smallest possible value of p in electron-nucleus scattering for the smallest possible value of the excitation energy ($\mathcal{E} = 0$). The theoretical concept of superscaling has been introduced within the relativistic Fermi gas (RFG) model [21, 22]. In the RFG model the scaling function $f_{\rm RFG}(\psi') = k_F \cdot F$ has the form [23]:

$$f_{\rm RFG}(\psi') \simeq \frac{3}{4}\left(1 - \psi'^2\right)\theta\left(1 - \psi'^2\right) . \tag{3}$$

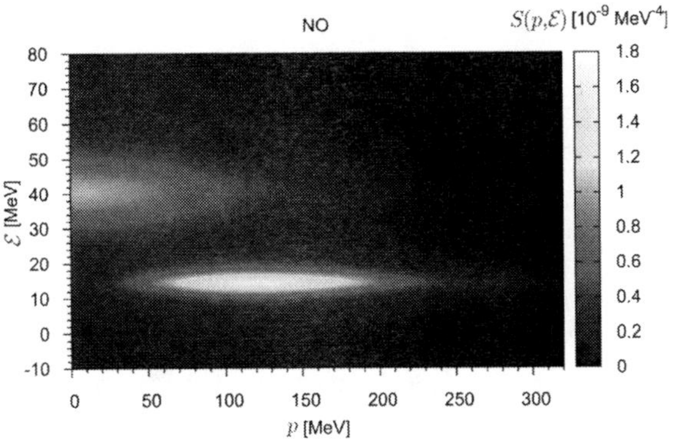

Figure 1. The ^{12}C realistic spectral function $S(p, \mathcal{E})$, which is constructed using natural orbital single-particle momentum distributions from the Jastrow correlation method and Lorentzian function for the energy dependence.

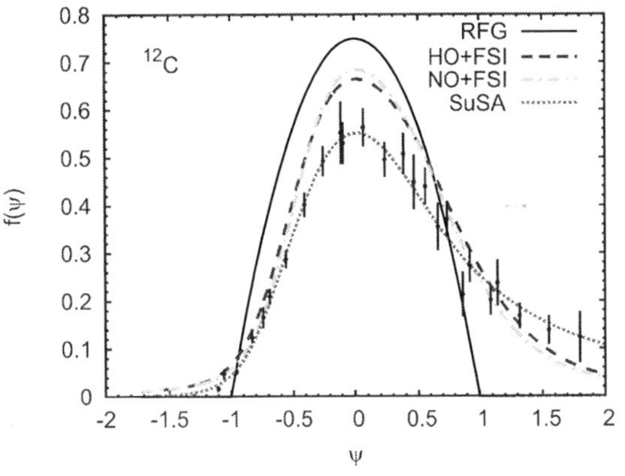

Figure 2. Results for the scaling function $f(\psi)$ for ^{12}C obtained using HO+FSI and NO+FSI approaches are compared with the RFG and SuSA results, as well as with the longitudinal experimental data.

As pointed out in [23], however, the actual dynamical physical reason of the superscaling is more complex than that provided by the RFG model.

In Ref. [15] more information about the spectral function was extracted within PWIA from the experimentally known scaling function. It contains effects beyond the mean-field approximation leading to a realistic energy dependence and accounts for short-range NN correlations. It is written in the form:

$$S(p, \mathcal{E}) = \sum_i 2(2j_i + 1) n_i(p) L_{\Gamma_i}(\mathcal{E} - \mathcal{E}_i), \tag{4}$$

where the Lorentzian function is used:

$$L_{\Gamma_i}(\mathcal{E} - \mathcal{E}_i) = \frac{1}{\pi} \frac{\Gamma_i/2}{(\mathcal{E} - \mathcal{E}_i)^2 + (\Gamma_i/2)^2} \tag{5}$$

with Γ_i being the width of a given s.p. hole state. In the calculations we used the values $\Gamma_{1p} = 6$ MeV and $\Gamma_{1s} = 20$ MeV, which are fixed to the experimental widths of the 1p and 1s states in ^{12}C [24]. In Eq. (4) the s.p. momentum distributions $n_i(p)$ were taken firstly to correspond to harmonic-oscillator (HO) shell-model s.p. wave functions, and second, to natural orbitals (NOs) s.p. wave functions $\varphi_\alpha(\mathbf{r})$ defined in [25] as the complete orthonormal set of s.p. wave functions that diagonalize the one-body density matrix $\rho(\mathbf{r}, \mathbf{r}')$:

$$\rho(\mathbf{r}, \mathbf{r}') = \sum_\alpha N_\alpha \varphi^*_\alpha(\mathbf{r}) \varphi_\alpha(\mathbf{r}'), \tag{6}$$

where the eigenvalues N_α ($0 \leq N_\alpha \leq 1$, $\sum_\alpha N_\alpha = A$) are the natural occupation numbers. In [15] we used $\rho(\mathbf{r}, \mathbf{r}')$ obtained within the lowest-order approximation of the Jastrow correlation methods [26]. The realistic spectral function $S(p, \mathcal{E})$ is presented in figure 1, where the two shells $1p$ and $1s$ are clearly visible.

For accounting for the FSI we follow the approach given in Ref. [27] concerning two types of FSI effects, the Pauli blocking and the interaction of the struck nucleon with the spectator system by means of the time-independent optical potential (OP) $U = V - \imath W$. The latter can be accounted for [28] by the replacing in the PWIA expression for the inclusive electron-nucleus scattering cross section

$$\frac{d\sigma_t}{d\omega d|\mathbf{q}|} = 2\pi\alpha^2 \frac{|\mathbf{q}|}{E_{\mathbf{k}}^2} \int dE\, d^3p\, \frac{S_t(\mathbf{p},E)}{E_{\mathbf{p}}E_{\mathbf{p}'}} \delta(\omega + M - E - E_{\mathbf{p}'}) L_{\mu\nu}^{\text{em}} H_{\text{em},t}^{\mu\nu} \tag{7}$$

the energy-conserving delta-function by

$$\delta(\omega + M - E - E_{\mathbf{p}'}) \to \frac{W/\pi}{W^2 + [\omega + M - E - E_{\mathbf{p}'} - V]^2}. \tag{8}$$

In Eq. (7) the index t denotes the nucleon isospin, $L_{\mu\nu}^{\text{em}}$ and $H_{\text{em},t}^{\mu\nu}$ are the leptonic and hadronic tensors, respectively, and $S_t(\mathbf{p},E)$ is the proton (neutron) spectral function. The real (V) and imaginary (W) parts of the OP in (7) and (8) are obtained in Ref. [29] from the Dirac OP.

The CC (anti)neutrino cross section in the target laboratory frame is given in the form (see for details [14, 30])

$$\left[\frac{d^2\sigma}{d\Omega dk'}\right]_\chi \equiv \sigma_0 \mathcal{F}_\chi^2, \tag{9}$$

where $\chi = +$ for neutrino-induced reaction *e.g.*, $\nu_\ell + n \to \ell^- + p$, where $\ell = e, \mu, \tau$) and $\chi = -$ for antineutrino-induced reactions (*e.g.*, $\overline{\nu}_\ell + p \to \ell^+ + n$). The quantity $\mathcal{F}_\chi^2$ in (9) depends on the nuclear structure and is presented [14] as a generalized Rosenbluth decomposition containing leptonic factors and five nuclear response functions, namely charge-charge (CC), charge-longitudinal (CL), longitudinal-longitudinal (LL), vector-transverse (T) and axial-transverse (T') expressed by the nuclear tensor and the scaling function.

To obtain the scaling function we use the spectral function $S(p,\mathcal{E})$ from (4) with $n_i(p)$ corresponding to HO or NOs s.p. wave functions, and the Lorentzian function (5). We calculate the electron-^{12}C cross section by using Eqs. (7) and (8) and the scaling function $F(q,\omega)$ within the PWIA from Eq. (2). By multiplying $F(q,\omega)$ by k_F the scaling function $f(\psi')$ is obtained. In this way the results (figure 2) for the HO+FSI (dashed line) and NO+FSI (dash-dotted line) are obtained. As a reference are shown also the scaling functions in the cases of SuSA (dotted line) and RFG (solid line). The accounting for FSI leads to a small asymmetry of the scaling function. Also, we found that the asymmetry in the scaling function gets larger by using the Lorentzian function [Eq. (5)] for the energy dependence of the spectral function than by using the Gaussian function [15–17].

The results for the total cross sections obtained in [16] within the HO+FSI and NO+FSI are given in figure 3 and compared with the SuSA and RFG results and the MiniBooNE [3, 4] and NOMAD [9] data (up to 100 GeV). All models give results that agree with the NOMAD data but underpredict the MiniBooNE ones, more seriously in the ν_μ than in $\overline{\nu}_\mu$ cases. The discrepancy with the MiniBooNE data (at energies < 1 GeV) is most likely due to missing effects beyond the IA, *e.g.* those of the 2p-2h excitations that have contributions in the transverse responses. This concerns also the similar disagreement that appears when the phenomenological scaling function in SuSA is used. The latter, being exctracted from the (e,e') data is a purely longitudinal QE response and thus is nearly insensitive to 2p-2h MEC contributions.

In figure 4 we display the flux-folded differential cross section $d\sigma/dQ_{\text{QE}}^2$ for both neutrino (left panel) and antineutrino (right panel) scattering off a hydrocarbon (CH) target as a function of the reconstructed four-momentum transfer squared (Q_{QE}^2), that is obtained in the same way as for the experiment, assuming an initial state nucleon at rest with a constant binding energy, E_b, set to 34 MeV (30 MeV) in the neutrino (antineutrino) case. The cross sections are folded with the MINERνA ν_μ and $\overline{\nu}_\mu$ fluxes [1, 2], and the nucleon's axial mass has the standard value $M_A = 1.03$ GeV. We observe that RFG, SuSA, HO+FSI and NO+FSI approaches yield

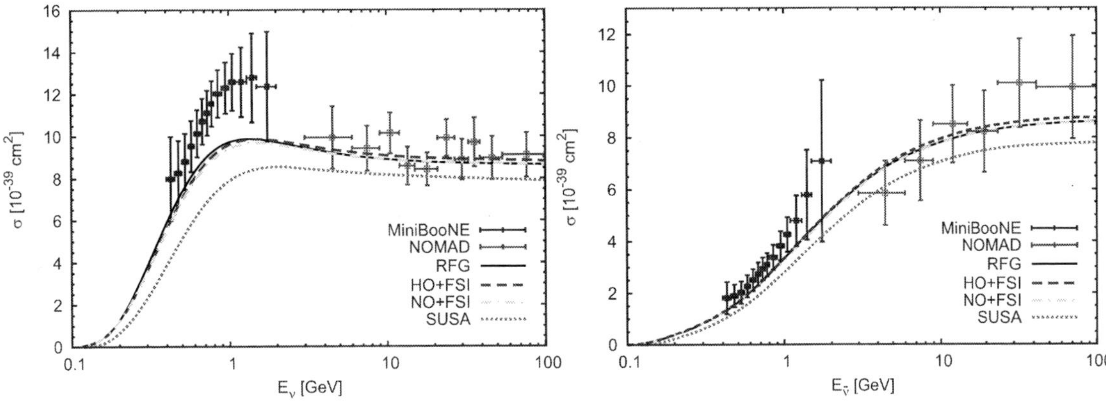

Figure 3. (left panel) CCQE ν_μ+^{12}C total cross sections per nucleon displayed versus neutrino energy E_ν evaluated using the RFG, HO+FSI, NO+FSI, and SuSA approaches with the standard value of the axial-vector dipole mass $M_A = 1.03$ GeV$/c^2$ are compared with the MiniBooNE [3, 4] and NOMAD [9] experimental data; (right panel) CCQE $\overline{\nu}_\mu$+^{12}C total cross section.

predictions in excellent agreement with the experimental data, leaving not much space for large effects of 2p-2h contributions. HO+FSI and NO+FSI results are higher than the SuSA ones and lie closer to the RFG results. In the RFG calculation, we use the formalism of [21], assuming a Fermi momentum of 228 MeV$/c$ and an energy shift of 20 MeV. This is not the same as the RFG modeling of GENIE [31] and NuWRO [32], which could explain the slight difference between our RFG results and the ones reported in [1, 2]. Note that the RFG model with the standard value of the axial mass (black-solid curve) also fits the data, being in very good agreement with the other approaches. Finally, the spread in the curves corresponding to the four models is less than 7% in the case of neutrinos and less than 5% in the case of antineutrinos. The theoretical results presented here include the whole energy range for the neutrino. The experimentalists implement several cuts on the phase space of the data, such as restricting the kinematics to contributions from neutrino energies below 10 GeV. The impact of such a cut on the results we present here is smaller than 0.2%, in the worst case. In the experimental analysis, several cuts were imposed to

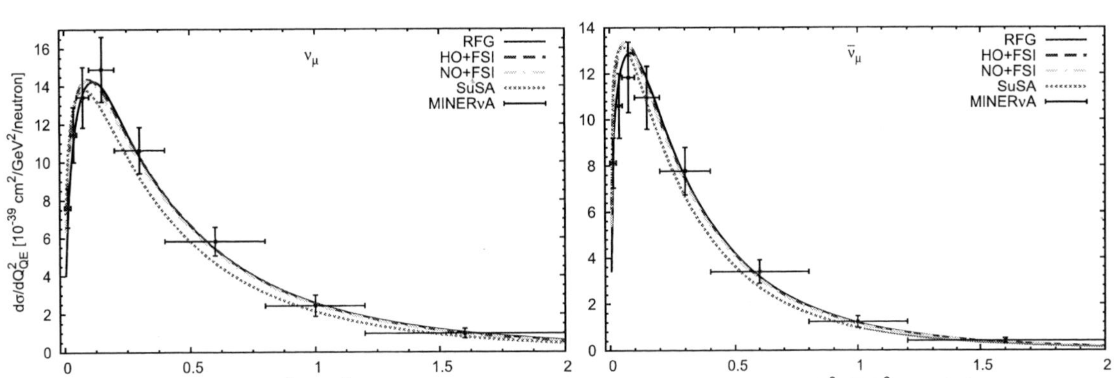

Figure 4. Flux-folded CCQE ν_μ+^{12}C (left panel) and $\overline{\nu}_\mu$+^{12}C (right panel) scattering cross section per target nucleon as a function of $Q^2_{\rm QE}$ and evaluated in the SuSA, RFG, HO+FSI, and NO+FSI models; data [1, 2].

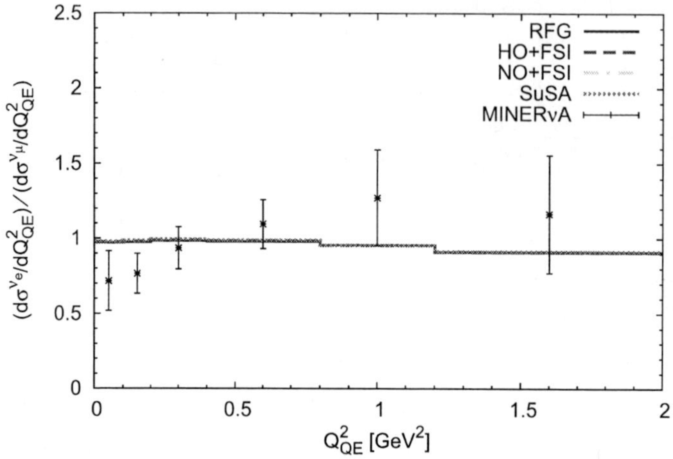

Figure 5. The ratio of the MINERνA ν_e CCQE differential cross section [34] as a function of $Q^2_{\rm QE}$ to the analogous result from MINERνA for ν_μ [1].

the initial data sample to increase the ratio of true quasielastic events in the sample. The effect of these cuts has been incorporated into the efficiency factors of the experiment, and thus, the data have been corrected for them [33]. We apply no cuts to the theoretical results, as the data have been corrected for their effect.

For completeness, we present in the figure 5 the results corresponding the ratio of the MINERνA ν_e CCQE differential cross section [34] as a function of the reconstructed four-momentum $Q^2_{\rm QE}$ to the analogous result from MINERνA for ν_μ [1]. The theoretical results for the ratio obtained within the RFG, HO+FSI, NO+FSI, and SuSA models almost coincide and are in good agreement with the data. For small values of the reconstructed four-momentum (first two bins) our results overpredict the data. It is important to note that the large error bars presented by the data make this particular analysis rather questionable.

3. Conclusions

1) The results with different spectral functions (HO and NO) give quite similar results (within 5–7%) for the CCQE cross sections, signaling that the process is not too sensitive to the specific treatment of the bound state.
2) The FSI leads to an increase of about 2% using spectral functions with HO and NO s.p. wave functions, almost independently of the neutrino energy.
3) All approaches based on IA underestimate MiniBooNE data for the flux-averaged CCQE ($\nu_\mu(\overline{\nu}_\mu)$ + ^{12}C) differential cross sections and the total cross section although the shape of the cross section is represented by NO+FSI and HO+FSI approaches. For $\overline{\nu}$ the agreement is much better.
4) All models give results that are compatible with the Minerνa and NOMAD data. This points to the importance of the evaluation of non-impulsive contributions, like those associated to MEC and their evolution with energy. The 2p-2h contributions may be responsible for the observed discrepancy in our analyses. Similar disagreement is observed for the phenomenological scaling function of SuSA, that is *purely longitudinal* QE response and *2p-2h MEC should not contribute to it when properly extracted from QE electron scattering, but could contribute to QE neutrino scattering because of the axial current.*

IOP Conf. Series: Journal of Physics: Conf. Series **1023** (2018) 012028 doi:10.1088/1742-6596/1023/1/012028

Acknowledgments
This work was partially supported by the Bulgarian National Science Fund under Contracts No. DFNI-T02/19 and No. DFNI-E02/6.

References

[1] Fiorentini G A *et al.* (MINERvA Collaboration) 2013 *Phys. Rev. Lett.* **111**(2) 022502
[2] Fields L *et al.* (MINERvA Collaboration) 2013 *Phys. Rev. Lett.* **111**(2) 022501
[3] Aguilar-Arevalo A A *et al.* (MiniBooNE Collaboration) 2010 *Phys. Rev. D* **81**(9) 092005
[4] Aguilar-Arevalo A A *et al.* (MiniBooNE Collaboration) 2013 *Phys. Rev. D* **88**(3) 032001
[5] Martini M, Ericson M, Chanfray G and Marteau J 2010 *Phys. Rev. C* **81**(4) 045502
[6] Amaro J, Barbaro M, Caballero J, Donnelly T and Williamson C 2011 *Physics Letters B* **696** 151–155
[7] Amaro J E, Barbaro M B, Caballero J A and Donnelly T W 2012 *Phys. Rev. Lett.* **108**(15) 152501
[8] Nieves J, Simo I R and Vacas M V 2012 *Physics Letters B* **707** 72–75
[9] Lyubushkin V *et al.* (NOMAD Collaboration) 2009 *Eur. Phys. J. C* **63** 355–381
[10] Megias G, Amaro J, Barbaro M, Caballero J and Donnelly T 2013 *Physics Letters B* **725** 170–174
[11] Donnelly T, van Orden J, de Forest T and Hermans W 1978 *Physics Letters B* **76** 393–396
[12] De Pace A, Nardi M, Alberico W M, Donnelly T W and Molinari A 2003 *Nucl. Phys. A* **726** 303
[13] Amaro J E, Maieron C, Barbaro M B, Caballero J A and Donnelly T W 2010 *Phys. Rev. C* **82**(4) 044601
[14] Amaro J E, Barbaro M B, Caballero J A, Donnelly T W, Molinari A and Sick I 2005 *Phys. Rev. C* **71** 015501
[15] Antonov A N *et al.* 2011 *Phys. Rev. C* **83**(4) 045504
[16] Ivanov M V *et al.* 2014 *Phys. Rev. C* **89**(1) 014607
[17] Ivanov M V *et al.* 2015 *Phys. Rev. C* **91**(3) 034607
[18] Megias G D *et al.* 2014 *Phys. Rev. D* **89**(9) 093002
[19] Maieron C *et al.* 2009 *Phys. Rev. C* **80** 035504
[20] Caballero J A, Barbaro M B, Antonov A N, Ivanov M V and Donnelly T W 2010 *Phys. Rev. C* **81**(5) 055502
[21] Alberico W M *et al.* 1988 *Phys. Rev. C* **38** 1801
[22] Barbaro M B, Cenni R, Pace A D, Donnelly T W and Molinari A 1998 *Nucl. Phys. A* **643** 137
[23] Donnelly T W and Sick I 1999 *Phys. Rev. C* **60** 065502
[24] Dutta D 1999 *The (e,e′p) Reaction Mechanism in the Quasi-Elastic Region* Ph.D. thesis Northwestern University
[25] Löwdin P O 1955 *Phys. Rev.* **97**(6) 1474–1489
[26] Stoitsov M V, Antonov A N and Dimitrova S S 1993 *Phys. Rev. C* **48**(1) 74–86
[27] Ankowski A M and Sobczyk J T 2008 *Phys. Rev. C* **77**(4) 044311
[28] Horikawa Y, Lenz F and Mukhopadhyay N C 1980 *Phys. Rev. C* **22**(4) 1680–1695
[29] Cooper E D, Hama S and Clark B C 2009 *Phys. Rev. C* **80**(3) 034605
[30] Antonov A N *et al.* 2006 *Phys. Rev. C* **74**(5) 054603
[31] Andreopoulos C 2009 *Acta Phys. Pol. B* **40**(9) 2461
[32] Golan T, Juszczak C and Sobczyk J T 2012 *Phys. Rev. C* **86**(1) 015505
[33] Chvojka J 2012 *Anti-Neutrino Charged Current Quasi-Elastic Scattering in MINERνA* Ph.D. thesis University of Rochester
[34] Wolcott J *et al.* (MINERvA Collaboration) 2016 *Phys. Rev. Lett.* **116**(8) 081802

IOP Conf. Series: Journal of Physics: Conf. Series **1023** (2018) 012029 doi:10.1088/1742-6596/1023/1/012029

Elastic scattering and breakup reactions of the exotic nucleus ^{8}B on nuclear targets

V K Lukyanov[1], D N Kadrev[2], A N Antonov[2], E V Zemlyanaya[1], K V Lukyanov[1], M K Gaidarov[2], and K Spasova[2]

[1] Joint Institute for Nuclear Research, Dubna 141980, Russia

[2] Institute for Nuclear Research and Nuclear Energy, Bulgarian Academy of Sciences, Sofia 1784, Bulgaria

E-mail: nelispas@gmail.com

Abstract.

Microscopic calculations of the optical potentials (OPs) and elastic scattering cross sections of the proton-rich nucleus ^{8}B on ^{12}C, ^{58}Ni and ^{208}Pb targets are presented. The density distributions of ^{8}B obtained within the variational Monte Carlo (VMC) model and the three-cluster model (3CM) are used to construct the optical potentials (OP). The real part of the hybrid OP (ReOP) is calculated using the folding model with the direct and exchange terms included, while the imaginary part (ImOP) is obtained on the base of the high energy approximation (HEA). In addition, the cluster model, in which ^{8}B consists of a proton halo and a ^{7}Be core is applied to calculate the breakup cross sections of ^{8}B on ^{9}Be, ^{12}C and ^{197}Au targets, as well as the momentum distributions of ^{7}Be fragments. A comparison with the available experimental data is made and a good agreement is obtained.

1. Introduction

During the last decades the study of the exotic nuclei is one of the fast developed fields in the nuclear physics. Unstable nuclei exibit "exotic" properties which have inspired extensive experimental and theoretical studies. Historically, the existence of such nuclei, is predicted in the early 1972 by Migdal (see Ref. [1]). In this work he showed that in the case of a core and two (interacting) neutrons (e.g. ^{9}Li+$2n$) the attractive force between the two neutrons (itself too weak to form a bound $2n$ system) in the presence of a nucleus (itself unable to bind a single neutron) may lead to a bound state of the three particles (i.e., it is a dineutron coupled to a nuclear core). The nucleon (proton or neutron) halo implies extremely extended radial function of loosely-bound nucleons due to their tunneling effect. The weak binding entails a neutronization of the nuclear surface (neutron halo) and a large nuclear radius. Migdal pointed out also to the possibility that cluster states more complex than dineutrons may exist. Later in the 1985 in the pioneering work of Tanihata *et al* [2] the measured radius of the nucleus ^{11}Li using radioactive ion beams of this nucleus was presented. These experimental results have been interpreted in terms of a neutron halo by Hansen and Jonson in 1987 [3]. The data allowed to study the structure of the nuclei far from the stability, the mechanism of the reactions, and give tests for various theoretical methods for analysis. The best known one-neutron halo nuclei are ^{11}Be and ^{15}C, which can be considered as a ^{10}Be or ^{14}C core, respectively, to which one neutron is loosely bound. Two-neutron halo nuclei are ^{6}He and ^{11}Li, with an α and ^{9}Li core,

IOP Conf. Series: Journal of Physics: Conf. Series **1023** (2018) 012029 doi:10.1088/1742-6596/1023/1/012029

respectively. Although much less probable, proton haloes are also possible and ^{8}B is one of the nuclei that probably exhibit a one-proton halo. There are some evidences that confirm this. The interaction cross section at 790 MeV/nucleon indicates that the rms radius of ^{8}B is different from more tightly bound boron isotopes [4]. The small proton separation energy, only 136.4 keV and the large proton matter radius compared to its neutron matter radius support its halo structure. The narrow momentum distribution of the fragments is another indication of a halo. To study the interesting structure of this nucleus and to investigate theoretically the role of the halo nucleon we can use indirect techniques such as elastic scattering and breakup processes.

In the present work we discuss some features of the nucleus ^{8}B in elastic and breakup reactions (see also [5]). First, we analyze the elastic scattering cross sections of ^{8}B on ^{12}C at 25.8 MeV [6], ^{8}B+^{58}Ni at 20.7, 23.4, 25.3, 27.2, 29.3 MeV [7], ^{8}B+^{208}Pb at 170.3 MeV [8] within the microscopic hybrid model of the OP ([5], [9]-[12]) and we compare the results with the available experimental data using a minimal number of fitting parameters. We test the density distributions of ^{8}B which reflect its proton-halo structure. Second, we estimate important characteristics of the reactions with ^{8}B, such as the breakup cross sections and the momentum distributions of ^{7}Be fragments from the breakup reactions ^{8}B+^{9}Be and ^{8}B+^{197}Au at $E = 41$ MeV/nucleon Ref. [13] and ^{8}B+^{12}C at $E = 36$ MeV/nucleon Ref. [14] within the breakup reaction model. As in our previous works [9]-[12] where neutron-rich He, Li and Be isotopes were considered, we used the hybrid model of OP, in which the real part (ReOP) is calculated by a folding of a nuclear density and the effective nucleon-nucleon (NN) potentials and includes direct and exchange isoscalar and isovector parts. The imaginary part (ImOP) is obtained on the base of the high-energy approximation (HEA) method. In our calculations we use the density distribution obtained within the variational Monte Carlo model [15, 16] and within the framework of the microscopic three-cluster model of Varga *et al.* [17]. Such a complex study based on the microscopic method to obtain the OPs with a minimal number of free parameters and by testing density distributions of ^{8}B which reflect its proton-halo structure would lead to a better understanding of the ^{8}B structure. Also, it would be a test of our microscopic approach to reveal the proton-halo structure of the loosely bound ^{8}B projectile, particularly its density distribution, in the considered elastic scattering and breakup processes.

2. Theoretical formalism and results of the calculations

The microscopic volume OP is constructed in the form:

$$U(r) = N_R V^{DF}(r) + i N_I W^H(r), \tag{1}$$

where the parameters N_R and N_I are related to the depths of the ReOP and ImOP respectively and are fitted by a comparison with the experimental cross sections. The real OP V^{DF} is obtained by a double folding procedure. Generally, the double-folding potential consists of a direct (V^D) and exchange (V^{EX}) part. The V^D and V^{EX} parts of the ReOP have isoscalar (IS) and isovector (IV) contributions. The IS ones of both terms are:

$$V_{IS}^D(r) = \int d^3r_p d^3r_t \rho_p(\mathbf{r}_p)\rho_t(\mathbf{r}_t) v_{NN}^D(s), \tag{2}$$

$$V_{IS}^{EX}(r) = \int d^3r_p d^3r_t \rho_p(\mathbf{r}_p, \mathbf{r}_p + \mathbf{s})\rho_t(\mathbf{r}_t, \mathbf{r}_t - \mathbf{s}) \times v_{NN}^{EX}(s) \exp\left[\frac{i\mathbf{K}(r)\cdot s}{M}\right], \tag{3}$$

where $\mathbf{s} = \mathbf{r} + \mathbf{r}_t - \mathbf{r}_p$ is the vector between two nucleons, one of which belongs to the projectile and another one to the target nucleus. In Eq. (2) $\rho_p(\mathbf{r}_p)$ and $\rho_t(\mathbf{r}_t)$ are the densities of the projectile and the target, respectively, while in Eq. (3) $\rho_p(\mathbf{r}_p, \mathbf{r}_p + \mathbf{s})$ and $\rho_t(\mathbf{r}_t, \mathbf{r}_t - \mathbf{s})$ are the density matrices for the projectile and the target. K(r) is the local nucleus-nucleus momentum and depends on the folding potential V^{DF}. The effective NN interactions v_{NN}^{D} and v_{NN}^{EX} have their IS and IV components in the form of M3Y interaction obtained within g-matrix calculations using the Paris NN potential [18]. The expressions for the energy and density dependence of the effective NN interaction are given, e.g., in Refs. [5],[12]. The ImOP is calculated in the high energy approximation (HEA):

$$W^{H}(r) = -\frac{1}{2\pi^2}\frac{E}{k}\bar{\sigma}_N \int_0^\infty j_0(qr)\rho_p(q)\rho_t(q)f_N(q)q^2dq. \tag{4}$$

In Eq. (4) $\rho(q)$ are the corresponding form factors of the nuclear densities, $f_N(q)$ is the amplitude of the NN scattering and $\bar{\sigma}_N$ is the averaged over the isospin of the nucleus total NN scattering cross section that depends on the energy and accounts for the in-medium effect.

To apply the microscopic OPs to scattering of ^{8}B on nuclei we used realistic density distributions of ^{8}B calculated within the VMC model [15, 16] and from the 3CM in Ref. [17]. In our case, within the VMC method the proton and neutron densities have been computed with the AV18+UX Hamiltonian, in which the Argonne v18 two-nucleon and Urbana X three-nucleon potentials are used [16]. As far as the 3CM densities of Varga *et al.* [17] are concerned, the ^{8}B nucleus has been studied in a microscopic $\alpha+h+p$ three-cluster model (h=^{3}He) using the stochastic variational method, where a Minnesota effective two-nucleon interaction was used. In the calculations of the OPs for ^{8}B+^{12}C the density of ^{12}C was taken in a symmetrized Fermi form. For ^{8}B+^{58}Ni and ^{8}B+^{208}Pb elastic scattering the densities of ^{58}Ni and ^{208}Pb were taken in a form of two-parameter Fermi distributions (for details see Ref. [5]).

In our work we consider a set of coefficients N_R and N_I (see Eq. (1) for the OP) as parameters to be found out from the fit to the experimental data for the cross sections using the χ^2-procedure. We should mention (as it had been emphasized in our previous works ([5],[9]-[12]) that we do not aim to find a complete agreement with the data. The fitted N's related to the depths of the ReOP and ImOP can be considered as a measure of deviations of our microscopic OPs from the case when the values of Ns are equal to unity.

The results of the elastic scattering differential cross sections of ^{8}B using VMC [16] and 3CM [17] densities of ^{8}B are presented in Fig. 1. It can be seen that in both cases of calculations with VMC or 3CM densities the results are in a good agreement with the available data [6]. The differential cross section obtained with VMC density demonstrates more developed diffractional picture. The corresponding values of N_R and N_I parameters obtained in our work for 3CM and VMC densities, as well as the total reaction cross sections σ_R (in mb) are listed in the a Table 1. The elastic scattering cross section of ^{8}B on ^{208}Pb target at 170.3 MeV incident energy is also presented. The theoretical results are compared with the experimental data from Ref. [8]. It can be seen a fair agreement of our microscopic calculations with the experiment. Both VMC and 3CM densities used in the calculations are able to reproduce the data that are restricted in a range of small angles. The values of N_R and N_I and the total reaction cross sections σ_R are also given in Table 1. In Fig. 2 we present our results for ^{8}B+^{58}Ni elastic scattering cross sections at energies 20.7, 23.4, 25.3, 27.2, and 29.3 MeV using the VMC and 3CM densities, respectively. These results are obtained with N_R and N_I which reproduce in a best way the experimental cross sections at considered five energies. Their values from the fitting procedure providing minimal χ^2/N are listed in Table 2. One can see that the results obtained using both densities of ^{8}B are in a good agreement with the data for all energies considered. The values of the renormalization parameters N_R are smaller than the ones deduced for the ^{8}B+^{12}C and ^{8}B+^{208}Pb elastic scattering processes. It is well nown that the couplings to non-elastic channels lead to polarization potentials that can considerably modify the bare potential calculated within the double folding formalism.

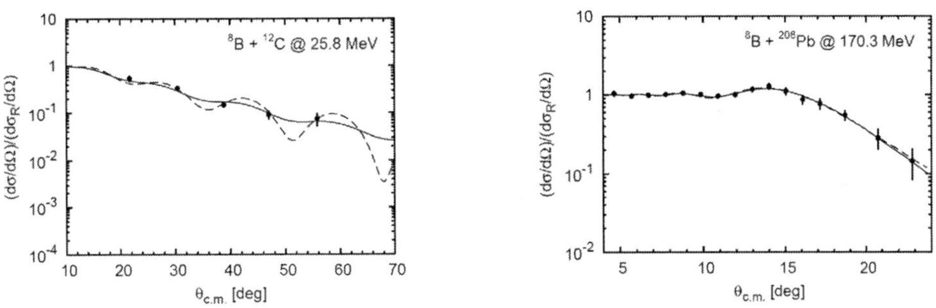

Figure 1. ^{8}B+^{12}C at E = 25.8 MeV (left panel) and ^{8}B+^{208}Pb at E = 170.3 MeV (right panel) elastic scattering cross sections. Solid line: calculations with 3CM density of ^{8}B; dashed line: calculations with VMC density of ^{8}B. The results are compared with the experimental data from Refs. [6] and [8], respectively.

Table 1. The renormalization parameters N_R, N_I, and the total reaction cross sections σ_R (in mb) for the results of ^{8}B+^{12}C and ^{8}B+^{208}Pb elastic scattering at incident energy E (in MeV) presented in Fig. 1 using the 3CM and VMC model densities of ^{8}B, respectively.

Process	Model	E	N_R	N_I	σ_R
^{8}B+^{12}C	3CM	25.8	1.075	0.433	1507.63
	VMC		2.200	0.165	1251.90
^{8}B+^{208}Pb	3CM	170.3	0.661	0.389	3226.73
	VMC		1.358	0.908	3158.15

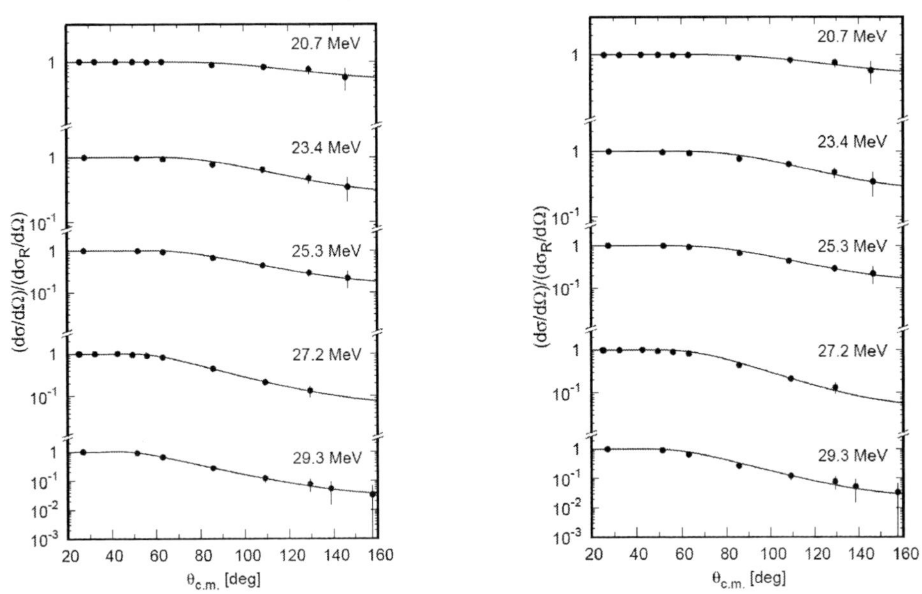

Figure 2. ^{8}B+^{58}Ni elastic scattering cross sections calculated using VMC density of ^{8}B (left panel) and 3CM density of ^{8}B (right panel). The experimental data are taken from Ref. [7].

Table 2. The renormalization parameters N_R, N_I, and the total reaction cross sections σ_R (in mb) for the results of ^{8}B+^{58}Ni elastic scattering at five incident energies E (in MeV) using the VMC and 3CM model densities of ^{8}B, respectively.

Model	E	N_R	N_I	σ_R
VMC	20.7	0.863	2.792	208.69
3CM		0.427	0.393	201.99
VMC	23.4	0.359	1.500	376.08
3CM		0.319	0.266	370.04
VMC	25.3	0.317	1.030	494.86
3CM		0.235	0.212	480.38
VMC	27.2	0.329	1.750	794.03
3CM		0.293	0.252	705.38
VMC	29.3	0.525	1.830	987.29
3CM		0.221	0.248	847.11

Table 3. The parameters V_0 (in MeV), R (in fm), a (in fm) of the Woods-Saxon potentials and the rms radii of the cluster wave function (in fm) obtained by using of the 3CM density of ^{7}Be for three cases (see the text).

V_0	R	a	rms radii
38.22	2.70	0.55	4.51
38.70	2.50	0.20	5.08
38.77	2.48	0.50	6.24

Obviously, for more successful description of cross sections at low energies near Coulomb barrier an inclusion of polarization contributions due to virtual excitations and decay channels of the reactions is necessary to obtain unambiguously the OP renormalization parameters. The good agreement of our results for ^{8}B+^{58}Ni elastic scattering cross sections with the experimental data using both VMC and 3CM densities validates their ability as a reasonable choice to reveal the proton-halo structure of ^{8}B nucleus.

The characteristics of breakup processes of the ^{8}B nucleus on the example of the stripping reaction cross sections and the momentum distributions of the fragments are also considered. We used a simple model in which ^{8}B consists of a core of ^{7}Be and a halo of a single proton. The wave function of the relative motion of the proton and ^{7}Be clusters in ^{8}B is obtained by solving the Schrödinger equation with the Woods-Saxon potential for a particle with a reduced mass of two clusters. The parameters of the WS potentials are obtained by a fitting procedure to reproduce the proton separation energy $S_p = 137$ keV. However, this procedure could provide several sets of potential parameters that satisfy the above condition. The values of WS potential parameters and corresponding rms radii of the cluster formation for $1p$ state in which the valence proton in ^{8}B is mainly bound (see Refs. [19, 20]) are listed in Table 3.

The wave function of the p-state ($l = 1$) of the relative motion of two clusters $\phi_{1m}(\mathbf{s})$ and the corresponding probability density $\rho_0(\mathbf{s})$ of both clusters to be at a mutual distance $\mathbf{s}$ are written as

$$\phi_{1m}(\mathbf{s}) = R_1(s)Y_{1m}(\theta,\varphi), \qquad \rho_0(\mathbf{s}) = \frac{1}{4\pi}|\phi_{1m}(\mathbf{s})|^2, \tag{5}$$

where $R_1(s)$ is the radial wave function and $Y_{1m}(\theta,\varphi)$ are the spherical functions for $l=1$.

For calculations of breakup cross sections and momentum distributions of fragments in the ^{7}Be+p breakup model within the eikonal formalism one needs the expressions of the S-matrix (as a function of the impact parameter b) (Ref.[5]):

$$S(b) = \exp\left[-\frac{i}{\hbar v}\int_{-\infty}^{\infty} U^{(b)}(\sqrt{b^2+z^2})dz\right], \tag{6}$$

where $U^{(b)}=V+iW$ is the OP of the breakup of ^{8}B in its collision with nuclear targets within the ^{7}Be+p cluster model. The longitudinal momentum distribution of ^{7}Be fragments produced in the breakup of ^{8}B in the case of stripping reaction (when the proton leaves the elastic channel) is

$$\left(\frac{d\sigma}{dk_L}\right)_{str} = \frac{1}{3}\int_0^{\infty} b_p db_p\left[1-|S_p(b_p)|^2\right]\int \rho d\rho d\varphi_\rho |S_c(b_c)|^2 \sum_{m=0,\pm1} F_m(\rho), \tag{7}$$

where

$$F_m(\rho) = \left|\int_{-\infty}^{\infty} dz \exp(-ik_L z) Y_{1m}(\theta,\varphi) R_1(r)\right|^2. \tag{8}$$

$|S_c|^2$ and $(1-|S_p|^2)$ give the probabilities after the collision with a target nucleus ($z\to\infty$), the cluster c with impact parameter b to remain in the elastic channel and the proton p to remove the elastic channel. After substituting the spherical functions $Y_{1m}(\theta,\varphi)$ in Eq. (8) one obtains the following expressions

$$F_0 = \frac{3}{\pi}\left[\int_0^{\infty} dz \sin(k_L z)\frac{z}{\sqrt{\rho^2+z^2}} R_1(\sqrt{\rho^2+z^2})\right]^2, \tag{9}$$

$$F_1 = F_{+1} = F_{-1} = \frac{3}{2\pi}\left[\int_0^{\infty} dz \cos(k_L z)\frac{\rho}{\sqrt{\rho^2+z^2}} R_1(\sqrt{\rho^2+z^2})\right]^2. \tag{10}$$

Eq. (7) is obtained when the incident nucleus has spin equal to zero and for the p-state of the relative motion of two clusters in the nucleus with $\mathbf{s}=\mathbf{r}_c-\mathbf{r}_p$, $\rho=\mathbf{b}_c-\mathbf{b}_p$, $\mathbf{s}=\rho+\mathbf{z}$ and

$$b_c = \sqrt{s^2\sin^2\theta + b_p^2 + 2sb_p\sin\theta\cos(\varphi-\varphi_p)} \tag{11}$$

coming from $\mathbf{b}_c=\mathbf{b}_p+\mathbf{b}$, where $b=s\sin\theta$ is the projection of $\mathbf{s}$ on the plane normal to the z-axis along the straight-line trajectory of the incident nucleus.

The stripping cross sections for the reactions ^{8}B+^{9}Be, ^{8}B+^{12}C, and ^{8}B+^{197}Au are calculated from Eq. (7) and are illustrated in Figs. 3, 4, and 5, respectively. The blue dotted, black solid, and red dashed curves in the figures correspond to the three sets of WS parameters given in Table 3.

It is worth to note the relevance between the rms radii of the wave function of the ^{7}Be-p relative motion and the obtained full widths at half maximum (FWHMs) for the considered three cases. Due to the uncertainty principle the widths become smaller with the increase of the distance between two clusters. We note the good agreement with the experimental data from light and heavy breakup targets. It can be seen from Figs. 3 and 4 that the best agreement with the experimental data for the parallel momentum distributions of ^{7}Be fragments in a breakup reaction of ^{8}B on a ^{9}Be target at 41 MeV/nucleon and on a ^{12}C target at 36 MeV/nucleon is achieved when the relative ^{7}Be-proton distance is 5.08 fm or 4.51 fm, respectively, while in the case of ^{8}B breakup on a ^{197}Au target at 41 MeV/nucleon shown in Fig. 5 a larger distance (6.24 fm) is needed to get better coincidence with the data. The calculated FWHMs are 72.07 MeV/c, 108.71 MeV/c and 54.86 MeV/c, respectively and they are closed to the experimentally measured widths.

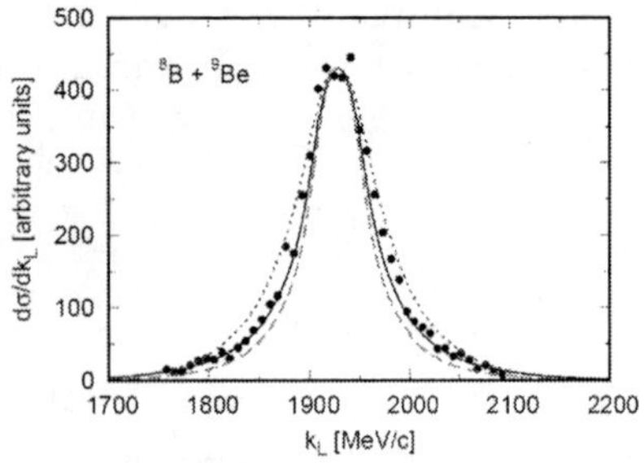

Figure 3. The cross section of stripping reaction in ^{8}B+^{9}Be scattering at E=41 MeV/nucleon. The experimental data are taken from Ref. [13].

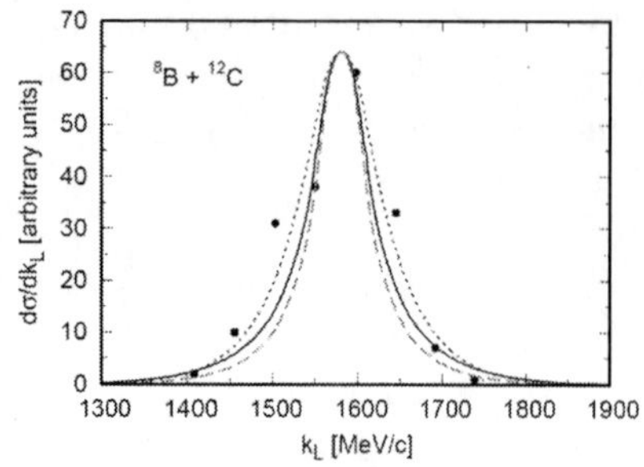

Figure 4. The cross section of stripping reaction in ^{8}B+^{12}C scattering at E=36 MeV/nucleon. The experimental data are taken from Ref. [14].

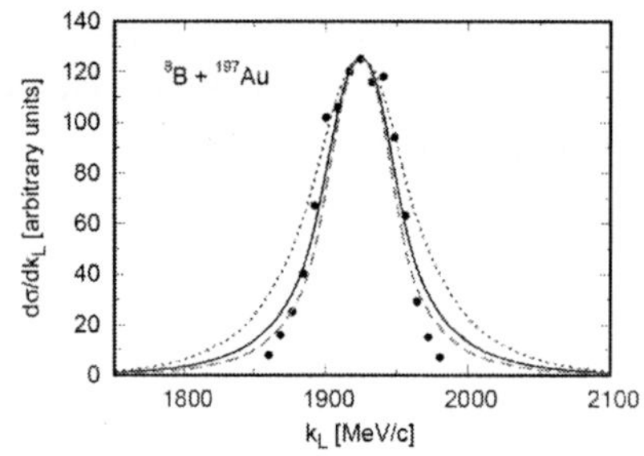

Figure 5. The cross section of stripping reaction in ^{8}B+^{197}Au scattering at E=41 MeV/nucleon. The experimental data are taken from Ref. [13].

3. Conclusions

In the present work microscopic calculations of the optical potentials and the cross sections of elastic scattering ^{8}B+^{12}C, ^{8}B+^{58}Ni and ^{8}B+^{208}Pb in comparison with the available experimental data are performed. We have tested our microscopic model studying the role of the breakup mechanism to analyze properly the whole picture of ^{8}B scattering. We consider the ^{8}B breakup by means of the simple ^{7}Be+p cluster model for the structure of ^{8}B.

We note that from the comparison of the results in the present work with those obtained in our previous works [11, 12] for the cases of ^{11}Li and ^{11}Be breakup, it can be concluded that the halo cluster ($2n$ or n) in these nuclei can be found outside the core with larger probability than the valence proton in ^{8}B. This observation together with the variation of the width with the target in breakup reactions of ^{8}B at almost equal energies show the specific features of the momentum distributions of corelike fragments in breakup of neutron and proton halo nuclei.

In general, we can conclude that our microscopic approach can be applied to reaction studies with exotic nuclei such as ^{8}B. The consistency of our results with the measured elastic cross sections and narrow longitudinal momentum distributions may provide supplemental information on the internal spatial structure of the ^{8}B nucleus supporting its proton-halo nature.

Acknowledgments

The authors are grateful to S.C. Pieper for providing with the density distributions of ^{8}B nucleus calculated within the VMC method and to S.L. Jin for the experimental longitudinal momentum distributions of ^{7}Be fragments from the breakup of ^{8}B on a carbon target. The work is partly supported by the Project from the Agreement for co-operation between the INRNE-BAS (Sofia) and JINR (Dubna). The authors A.N.A., D.N.K., K.S., and M.K.G. are grateful for the support of the Bulgarian Science Fund under the Contracts Nos. DFNI–T02/19 and DNTS/Russia 01/3 and one of them (D.N.K.) under the Contract No. DFNI–E02/6. The authors E.V.Z., K.V.L. and V.K.L. thank to the Russian Foundation for Basic Research (Grant N17-52-18057 bolg-a) for the partial support.

References

[1] Migdal A B 1972 *Yad. Fiz.* **16** 427; 1973 *Sov. J. Nucl. Phys.* **16** 238
[2] Tanihata I *et al.* 1985 *Phys. Lett.* B **160** 380
[3] Hansen P G and Jonson B 1987 *Europhys. Lett.* **4** 409
[4] Tanihata I *et al.* 1988 *Phys. Lett.* B **206** 592
[5] Lukyanov V K *et al.* 2017 *Eur. Phys. J.* A **53** 31
[6] Barioni A *et al.* 2011 *Phys. Rev.* C **84** 014603
[7] Aguilera E F *et al.* 2009 *Phys. Rev.* C **79** 021601(R)
[8] Yang Y Y *et al.* 2013 *Phys. Rev.* C **87** 044613
[9] Lukyanov V K *et al.* 2009 *Phys. Rev.* C **80** 024609
[10] Lukyanov V K *et al.* 2010 *Phys. Rev.* C **82** 024604
[11] Lukyanov V K *et al.* 2013 *Phys. Rev.* C **88** 034612
[12] Lukyanov V K *et al.* 2015 *Phys. Rev.* C **91** 034606
[13] Kelley J H *et al.* 1996 *Phys. Rev. Lett.* **77** 5020
[14] Jin S L *et al.* 2015 *Phys. Rev.* C **91** 054617
[15] Carlson J *et al.* 2015 *Rev. Mod. Phys.* **87** 1067
[16] Pieper S C, private communication.
[17] Varga K, Suzuki Y and Tanihata I 1995 *Phys. Rev.* C **52** 3013
[18] Khoa D T and Satchler G R 2000 *Nucl. Phys.* A **668** 3
[19] Hencken K, Bertsch G and Esbensen H 1996 *Phys. Rev.* C **54** 3043
[20] Esbensen H 1996 *Phys. Rev.* C **53** 2007

IOP Conf. Series: Journal of Physics: Conf. Series **1023** (2018) 012030 doi:10.1088/1742-6596/1023/1/012030

The Nuclotron-based Ion Collider Facility Project. The Physics Programme for the Multi-Purpose Detector

N.S. Geraksiev[1,2] **on behalf of the MPD Collaboration**

[1] VBLHEP, Joint Institute for Nuclear Research, Dubna, Moscow region, Russian Federation
[2] FPT, Plovdiv Univsity Paisii Hilendarski Plovdiv, Bulgaria

E-mail: nikolay.geraksiev@gmail.com

Abstract. The Nuclotron-based Ion Collider fAcility (NICA) is a new accelerator complex being constructed at the Joint Institute for Nuclear Research (JINR). The general objective of the project is to provide beams for the experimental study of hot and dense strongly interacting QCD matter. The heavy ion programme includes two planned detectors: BM@N (Baryonic Matter at Nuclotron) a fixed target experiment with extracted Nuclotron beams; and MPD (MultiPurpose Detector) a collider mode experiment at NICA. The accelerated particles can range from protons and light nuclei to gold ions. Beam energies will span $\sqrt{s} = 12-27$ GeV with luminosity $L \geq 1 \times 10^{30}$ cm^{-2}s^{-1} and $\sqrt{s_{NN}} = 4-11$ GeV and average luminosity $L = 1 \times 10^{27}$ cm$^{-2}s^{-1}$(for ^{197}Au^{79+}), respectively. A third experiment for spin physics is planned with the SPD (Spin Physics Detector) at the NICA collider in polarized beams mode. A brief overview of the MPD is presented along with several observables in the MPD physics programme.

1. Introduction

The Nuclotron-based Ion Collider fAcility (NICA)(Fig.1) is a new accelerator complex being constructed at the Joint Institute for Nuclear Research (JINR), Dubna [1]. The main goal of this facility is to provide heavy-ion collisions in order to investigate the hot and dense barionic matter produced in heavy ion collisions at relatively low energies compared to previous experiments at RHIC and SPS (Fig.2), [2]. Thus formulated the experimental programme at NICA includes simultaneous measurements of observables that are sensitive to high-density effects and phase transitions.

The proposed research programme contains a wide range of collision configurations in both fixed target and collider modes.

2. Nuclotron-based Ion Colider facility NICA

The NICA project in its entirety is going to incorporate several already existing facilities, as well as such still being developed. The injection complex comprises of four sources (for light and heavy ions) and two linacs. The light ion sources are: laser, duoplasmatron sources, source of polarized protons and deutrons, a modernized LU-20 accelerates for beam injection into the synchrotron Nuclotron. The electron string source KRION-6 provides heavy ions of Au^{31+} and delivers them into HILac, which accelerates them to 3.24 MeV/u and injects them to the Booster. The main purpose of the booster is accumulation of $2 \cdot 10^9$ Au^{31+} ions and acceleration of heavy ion bunches to energy required for effective stripping. The Booster is equipped with an electron cooling system for the ion beam cooling in the energy range from injection energy up to 100 MeV/u. The maximum energy of Au^{79+} ions accelerated in the Booster is 600 MeV/u. After acceleration in the Booster ions will be fully stripped and injected into the Nuclotron providing acceleration of $^{197}Au^{79+}$ ions up to the energy of 4.5 GeV/u. The required bunch intensity is about $1 - 1.5 \cdot 10^9$ ions . Two collider rings are constructed one above the other. To provide the beam storage and bunch formation in the collider, three independent RF systems are used. Twenty-two bunches per ring will be stored. For luminosity preservation, an electron and/or a stochastic cooling system will be used [3]. Beam energies will span $\sqrt{s} = 12 - 27$ GeV with luminosity $L \geq 1 \cdot 10^{30}$ cm^{-2}s^{-1} and $\sqrt{s_{AuAu}} = 4 - 11$ GeV and average luminosity $L = 1 \cdot 10^{27}$ cm$^{-2}s^{-1}$(for $^{197}Au^{79+}$), respectively.

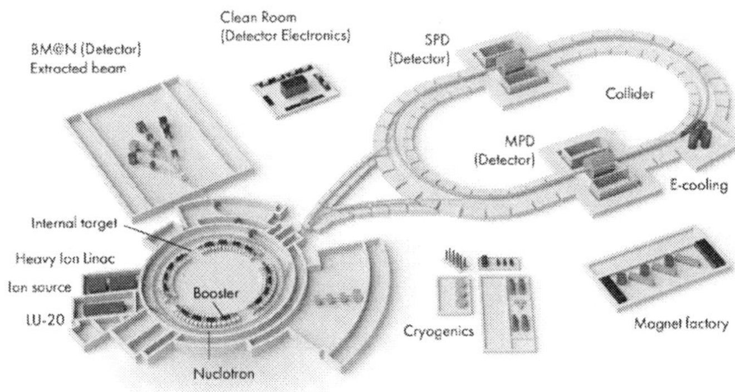

Figure 1. Schematic view of the NICA complex.

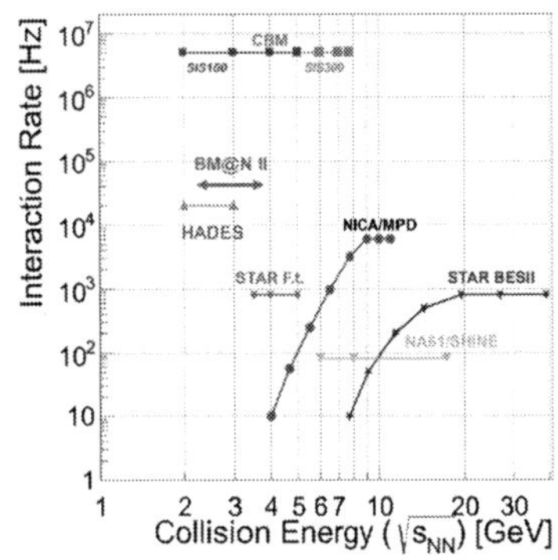

Figure 2. Landscape of present and future heavy-ion experiments.

Two NICA experiments are dedicated to the exploration of the QCD phase diagram: Baryonic Matter at Nuclotron (BM@N)[4] - a fixed target experiment in the beam extracted from the Nuclotron, which will cover the lower energy region, and the Multi-Purpose Detector (MPD)[5] at one of the collider interaction points (IP).

These two experiments will be focused on measurements of particle yields and spectra, ratios, femtoscopy and collective flow; observation of in-medium modification of hadron properties like low-mass dilepton enhancement; study of deconfinement (chiral) phase transition at high baryon density through observation of onset of enhanced strangeness production; search for QCD Critical Point via studies of event-by-event fluctuations and correlations; investigation of the Chiral Magnetic (Vortical) effect through a measurement of Λ-polarization; investigation of hyperon-nucleon interactions in dense nuclear matter via studies of hypernucleus production and their properties.

At the second interaction point of the collider facility, a Spin Physics Detector is proposed to investigate problems related to the origin of spin and a number of scattering problems. The NICA accelerator will be capable of running in polarized beam modes for protons and deutrons. The high intensity and high polarization ($> 50\%$) of the colliding beams present a unique possibility for spin physics research, which is of crucial importance for the solution of the nucleon spin problem (spin puzzle) - one of the main tasks of the modern hadron physics.

3. Multi-Purpose Detector MPD

The Multi Purpose Detector (MPD) (Fig.3) is designed as a 4π spectrometer capable of detecting charged hadrons, electrons and photons in heavy-ion collisions at high luminosity in the energy range of the NICA collider. At the design NICA luminosity, the event rate of minimum bias interactions is of about 7 kHz, and the total charge particle multiplicity exceeds 1000 in the most central Au+Au collisions at $\sqrt{s_{NN}} = 11$ GeV. As the average transverse momentum of the particles produced in a collision at NICA energies is below 500 MeV/c, the detector design requires a very low material budget. Two stages of realization are planned for the MPD. For the first stage a barrel setup is being constructed.

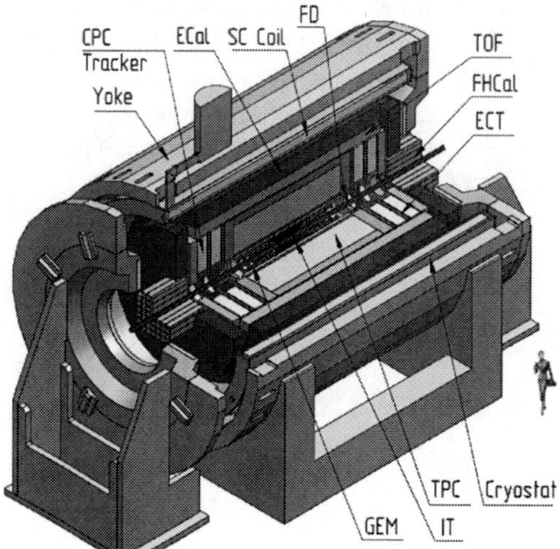

Figure 3. Sschematic view of MPD sub-detector allocation.

The main subdetector systems will be operating inside of a large solenoid with a superconducting NbTi winding and a flux return iron yoke. The solenoid should provide a homogeneous magnetic field of up to 0.5 T. The field inhomogeneity in the tracker area of the detector is about 0.1%. Homogeneity of the radial component of the magnetic induction in the volume of Charged Particle Tracker (TPC) has to be < 0.775 mm.

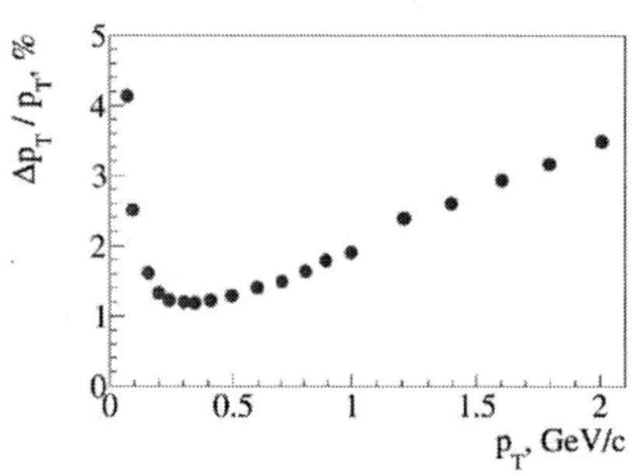

Figure 4. Relative transverse momentum resolution for primary tracks with $|\eta| < 1.3$ reconstructed in TPC.

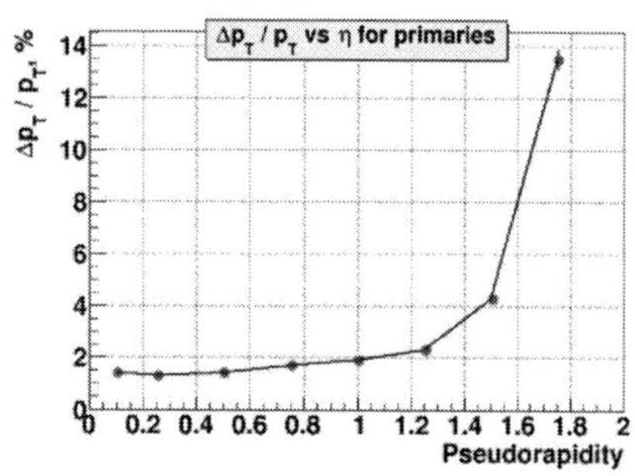

Figure 5. Transverse momentum resolution as a function of pseudo-rapidity for primary tracks with $p_T = 1$ GeV/c.

The main tracking detector in MPD barrel is the Time Projection Chamber (TPC). The TPC is a cylindrical gas detector 2.7 m in diameter and 3.4 m in length. The uniform electric field in the active volume is created by a thin central high voltage electrode, together with a voltage dividing network at the surface of the outer and inner cylinders, and at the readout end-caps. The TPC readout system is based on Multi-Wire Proportional Chambers (MWPC) with cathode readout pads. The main requirements toward the TPC is to provide a sufficient transverse momentum resolution (with a resolution < 1 cm) and energyloss measurements

(dE/dx resolution better than 8 %) for hadronic and leptonic tracks at pseudo-rapidities $|\eta < 1.5$ and $p_T > 100$ MeV/c. Precise primary and secondary track reconstruction and also precise primary (Interaction point) and secondary (decay) vertex reconstruction are of top TPC priority [6].

The Time of Flight System (TOF) outer diameter is 3.4 m, and the total surface is of about 53 m^2. The basic element of the TOF system will be a Multigap Resistive Plate Chamber (MRPC). This system will provide time of flight measurements with geometrical efficiency above 95% and a resolution in the range of 60-80 ps [7], which are complemented by the energy loss (dE/dx) information from the TPC, in order to provide an accurate particle identification (PID). Combined, the measurements from TPC and TOF, should provide an efficient π/K PID separation up to 1.5 GeV/c, π/p separation up to 3 GeV/c, and a very good electron/hadron separation [8].

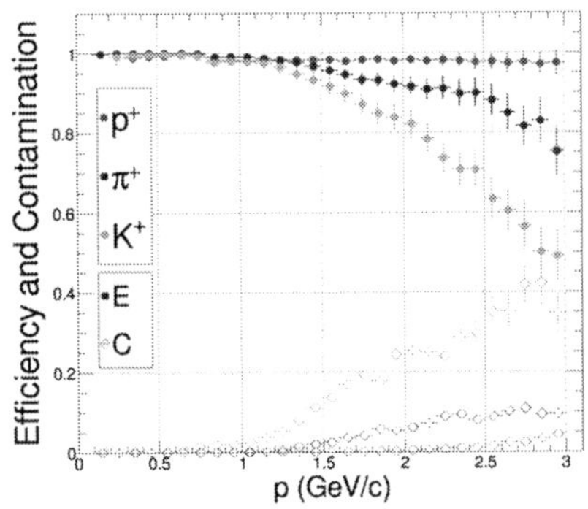

Figure 6. Efficiency(E) and contamination(C) of particle identification for charged hadrons.

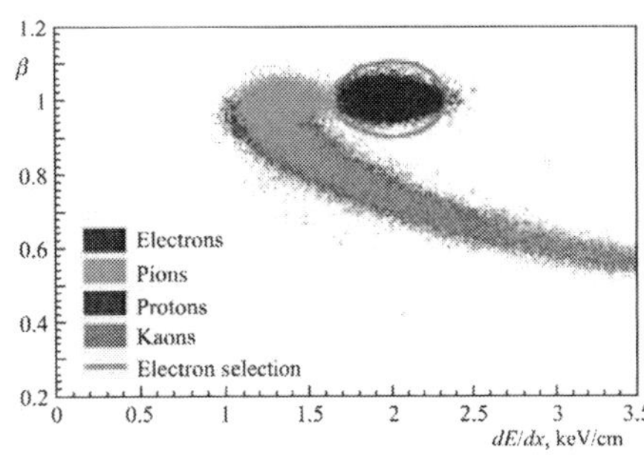

Figure 7. Electron-hadron separation with TPC and TOF measurements.

The Electromagnetic Calorimeter is constructed by shashlyk-type modules of lead+scintilator, each consisting of nine towers of 4x4 cm^2 cross-section. ECal has 15 X_0 radiation lengths with readout wave length shifting fibers. The light collected by these fibers will be read out by avalanche photodiodes (MAPD). The primary role of the electromagnetic calorimeter is to measure the spatial position and energy of electrons and photons. For an increased energy resolution of electromagnetic showers a projective geometry targeting the IP is used. The ECal will provide energy resolution of 5 % at 300 MeV, and time resolution better than 0.5 ns [9].

Two sets of Forward Hadron Calorimeter (FHCal) are allocated at 3.2 m to the right and to the left from the IP, each of which consists of 45 modules providing the necessary transverse granularity. The acceptance of FHCal covers the pseudo-rapidity region from 2.2 to 4.8 The main goal of the FHCal is to provide energy measurements of spectator nucleons for centrality determination and event-plane reconstruction with a resolution of 20-30 degrees [10].

The Forward Detector (FD) consists of two modular sets of Cherenkov detectors placed at a distance of 130 cm to the left and to the right from the interaction point. FD provides provides a trigger system for data taking and start time for TOF with resolution better than 50 ps [11].

Within the NICA timetable, MPD will be commissioned by the end of 2020 and will start physical data taking by 2021. In order to increase low momentum track and primary vertex resolution, the addition of an Inner Silicone Tracker (IT) and Gas Electron Multipliers (GEM) close to interaction point are planned. Also, end-caps on both sides of the barrel are considered: ECT, ETOF, ECAL.

4. Physics Programme for MPD

4.1. Anisotropic Flow Studies

The observation of a azimuthally anisotropic flow is one of the key observables for establishing that a QGP state of matter has formed in the collision. The initial spatial eccentricity in heavy-ion collision geometry leads to a non-uniformity in parton interaction which results in anisotropic particle emissions. The collective flow of hot and dense baryonic matter provides information on the transport properties of the medium and the equation of state. The anisotropic flow measurements of both identified charged particles and reconstructed hyperon decays will be a key measurement in the MPD research programme.

The magnitude of the anisotropic flow is defined using Fourier coefficients of the azimuthal distribution of particles with respect to the collision reaction plane. The reconstructed and generated(true) values of differential flow coefficients are in good agreement. Due to limits in TPC event-plane resolution, FHCal is used to determine the reaction plane.

These results present a good agreement between generated and reconstructed proton differential directed and elliptic flow (Fig.8) [12].

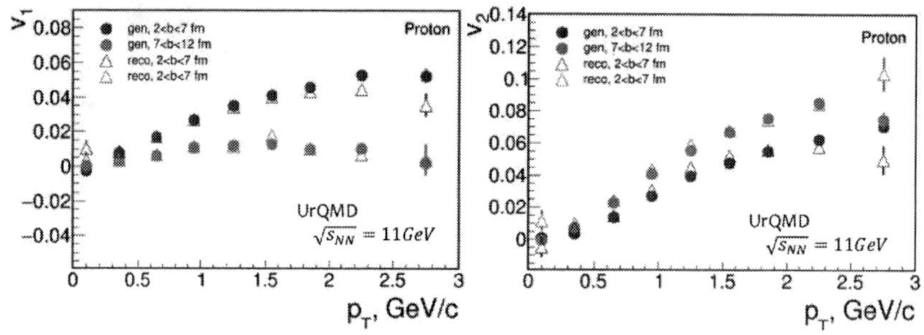

Figure 8. Generated and reconstructed differential anisotropic flow: (a Directed v1 and (b Elliptic v2.

4.2. Hyperon Studies

Production of strange particles is of particular interest because enhanced production of rare strange hadrons(Ξ^-, Ω^-) in A+A collisions (relative to the yields from elementary pp reactions) was predicted as a signal for the QGP formation. The enhancement of the strangeness was experimentally observed at SPS and RHIC, and it is more pronounced for hyperons with larger strangeness content. At present, a complete theoretical description of the (multi)strangeness production mechanism at collision energies of several GeV, has not yet been achieved.

The performance of the MPD detector for reconstruction of hyperons (Fig.9) in central collisions has been evaluated at $\sqrt{s_{AuAu}} = 9$ GeV, using the UrQMD generator and the full MPD reconstruction chain. The combination of particle yield and good signal-to-background ratio will be of key importance in lambda-polarization and hyperon elliptic flow studies [13].

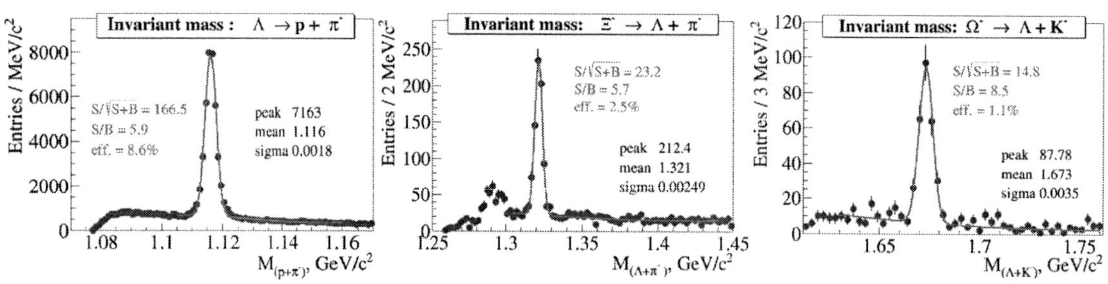

Figure 9. Reconstructed invariant mass of Λ, Ξ^-, Ω^- hyperons.

4.3. Hypernuclei Studies

The NICA project has a good opportunity to investigate hypernuclei in heavy ion collisions as there is a predicted considerable enhancement of nuclear clusters with strangeness in the planned energy ranges. The better understanding of the dynamics of hypernuclei production will provide necessary information on hyperon-nucleon and hyperon-hyperon interactions which is important for the theoretical basis for neutron star models.

The feasibility of hypernuclei reconstruction with the MPD experiment was assessed. Presented are results of simulated $5 \cdot 10^5$ and $6 \cdot 10^7$ central events by the DCM-QGSM generator at $\sqrt{s} = 5$A GeV for the studies of ${}^3_\Lambda H$ and ${}^4_\Lambda He$, respectively. The full reconstruction chain of MPD was used. In the presented invariant mass distribution results with a high signal-to-background ratio and significance are observed (Fig.10). As evident, MPD offers good opportunities for studies of heavy strange probes at the NICA collider [14].

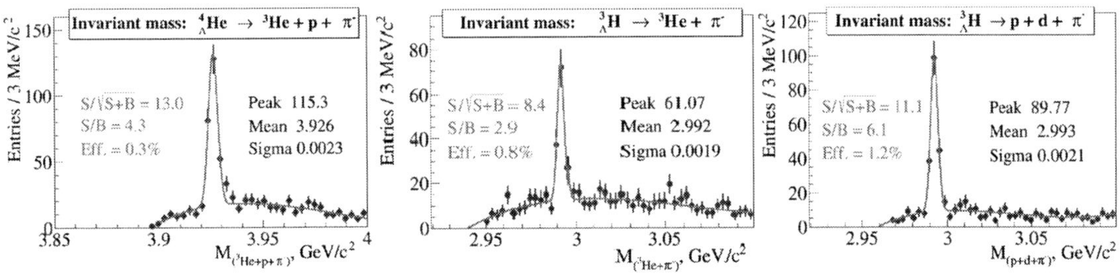

Figure 10. Reconstructed invariant mass of ${}^4_\Lambda He$ and ${}^3_\Lambda H$ hypernuclei.

4.4. Dilepton Studies

The reconstruction of low-mass vector mesons ρ, ω, ϕ by measuring their dileptonic decay channels is one of the top priorities for the MPD research programme (Fig.11). Dileptons are good probes to indicate medium modifications of spectral functions due to chiral symmetry restoration in A+A collisions; the effect is proportional to baryon density.

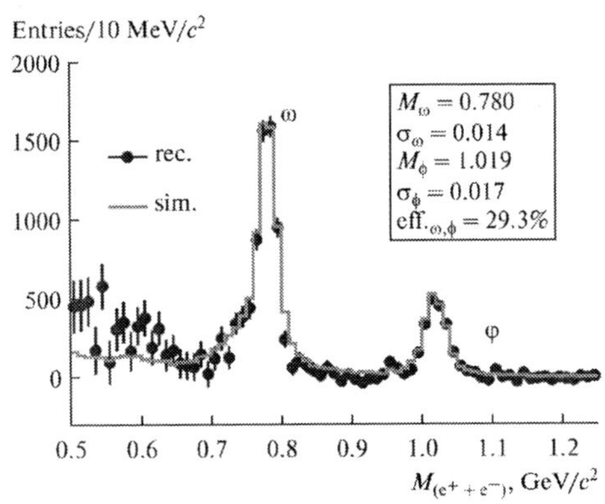

Figure 11. Reconstructed invariant mass of ϕ and ω vector mesons.

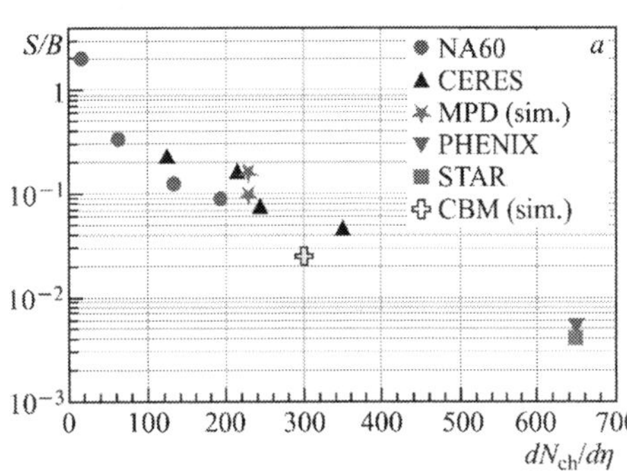

Figure 12. Signal-to-background ratio of dileptons for HIC experiments.

Due to the large combinatorial background, a strict particle selection is required and topological cuts are applied. For the low-mass dielectron study a strict dielectron selection with TPC, TOF, and ECal measurements were used. The signals can be separated from the background within the mass region 0.2-1.5 GeV/c^2. In comparison to other heavy-ion experiments a good signal-to-background ratio is achieved (Fig.12) [15].

4.5. Other Key Observables

The NICA/MPD physics programme entails a vast volume of research possibilities, such as the charge dependence of azimuthal correlations between produced hadrons, baryon stopping power and femtoscopy correlations.

Charged particle azimuthal correlations are an important probe of the QGP matter created in relativistic heavy-ion collisions, as they may be sensitive to a possible effect of local parity violation in the QCD. The azimuthal correlations and their projections are well reconstructed for central and mid-central collisions with MPD stage 1 setup, while stage 2 will afford the ability to study peripheral collisions.

Baryon stopping power is a requisite in describing the conversion of the initial particle collision energy into matter excitation and creation of hot and dense nuclear matter. It was argued that the baryon stopping in nuclear collisions can be a sensitive probe for the onset of deconfinement. At NICA energies, models indicate a peak-deep-peak behavior at mid-rapidity. MPD simulations confirm that this will be observable.

Femtoscopy correlations will provide information on the space-time evolution of a QGP. Such studies provide a constrain on model predictions for the early stage of collision and fireball evolution. The NICA large luminosity along with MPD's precision measurements and large acceptance, will provide a much more detailed space-time information on the emitting system than in previous studies. A more detail overview of the studies above is available [16].

5. Conclusion

A brief description of the NICA project is given, as well as a short overview of the multi-purpose detector and planned physics programme. Models predict that at the NICA energy range a high net-barion density is reached in the interacting QCD matter. For a further understanding of QCD it is important to study the equation of state, to search for phase transitions, chiral symmetry restoration in multiple collision systems and experiments. The experiments on the NICA collider will present competitive and complementary results to those of BNL, CERN and FAIR, and will be able to shed light on multiple theoretical problems for which accurate experimental measurements are not present.

References

[1] Kekelidze V D 2017 *Journal of Instrumentation* **12**
[2] Blaschke D e a 2016 *Eur. Phys. J.* **267** 1277
[3] Kostromin S A e a 2012 *Physics of Particles and Nuclei Letters* **9** 322–336
[4] Kapishin M 2016 *European Physical Journal A* **52** 0–5
[5] Golovatyuk V e a 2016 *European Physical Journal A* **52** 1–7
[6] Cheremukhina G e a 2017 *Journal of Instrumentation* **12**
[7] Babkin V e a 2011 *Bull. Russ. Acad. Sci. Phys.* **75** 1277
[8] Mudrokh A A and Zinchenko A 2017 *J. Phys.: Conf. Ser.* **798** 012071
[9] Tyapkin I e a 2015 *PoS(PhotoDet2015)* 053
[10] Golubeva M e a 2017 *Journal of Physics: Conference Series* **798** 012074
[11] Yurevich V I e a 2015 *Nucl Instrum Methods Phys Res* **A824** 162–164
[12] Svintsov I e a 2016 *Journal of Physics: Conference Series* **755** 011001
[13] Ilieva M e a 2015 *Physics of Particles and Nuclei Letters* **12** 618–627
[14] Ilieva M e a 2016 *Journal of Physics: Conference Series* **668** 012104
[15] Zinchenko A e a 2015 *Physics of Atomic Nuclei* **78** 1497–1503
[16] Batyuk P N e a 2016 *Physics of Particles and Nuclei* **47** 540–566

Energy Production and Transmutation of Nuclear Waste by Accelerator Driven Systems

P K Zhivkov

Institute of Nuclear Research and Nuclear Energy, BAS, 72 Tsarigradsko shouse Blvd, Sofia, Bulgaria

Abstract. There is a significant amount of highly radiotoxic long-life nuclear waste (NW) produced by NPP (Nuclear Power Plants). Transmutation is a process which transforms NW into less radiotoxic nuclides with a shorter period of half-life by spallation neutrons or radiative capture of neutrons produced by ADS (Accelerator Driven System). In the processes of transmutation new radioactive nuclides are produced. ADS is big energy consumer equipment. It is a method for production of a high-flux and high-energy neutron field. All these processes occur in ADS simultaneously. ADS is able to transmute actinides and produce energy simultaneously. The article considers the energy production problems in ADS. Several ideas are developed regarding the solution of the global energy supply.

1. Introduction

In the mid-90s many experiments and theoretical investigations of the parameters and possibilities of ADS (Accelerator Driven System) were started [1, 2, 3, 4, 5, 6]. The investigations may be divided into two parts, experimental and computer simulations. Numerous experiments for determination of the high-energy cross sections and estimation of the high-energy neutron field inside and outside the experimental set-ups have been done. Computer simulations with MCNPX, GEANT4, FLUKA transport codes were used. Most of the codes use group cross sections for calculations, whereas MCNPX uses another method of precise nuclear data tables which are included in the program. Most of the nuclear data tables are defined up to 20 MeV. Some of the tables are extended up to 100 or 150 even up to 200 MeV [7]. The group cross sections do not describe accurately the neutron induced reactions for neutrons with energy E_n<1 MeV because of resonances. The MCNPX transport code is the most powerful code describing nuclear fission. All codes use nuclear models if there are no data tables. MCNPX is the most advanced transport code for calculation of ADS integral energy parameters (described below) with significant amount of fission nuclides such as $^{239,240}Pu$ and $^{233,235}U$. The new cyclotron centre based on TR24 cyclotron at the Institute for Nuclear Research and Nuclear Energy of the Bulgarian Academy of Sciences opens new possibilities for experimental research in the field of accelerator based neutron sources for applied research and nuclear medicine [8].

2. ADS design

There are many suggestions concerning the ADS design. However, first of all the main aims of ADS should be defined. The article aims to investigate the energy production by ADS. The energy is produced by neutron induced fission in isotopes $^{233,235,238}U$, ^{232}Th and $^{239,240}Pu$. All ADS calculations are made by the MCNPX transport code (Bertini model) and (Evaluated Nuclear Data File) ENDF60 data table [9]. The design of ADS is simple because we are interested only in integral parameters such as total neutron production, leakage of neutrons and neutron induced reactions. In general, ADS consists of 4 main parts. The first is the accelerator. A proton beam with energies from 0.66 to 2 GeV is chosen. The second part is a target zone made of ^{nat}U, ^{nat}Pb and a combination of ^{nat}Pb and W. The third part is a reflector and a cooling system, both made of ^{nat}Pb. The forth is nuclear fuel MOX (Mixed Oxide Fuel, 20-30% concentration of isotopes ^{239}Pu+^{235}U). For axial (direction of proton beam) symmetry the ADS has cylindrical shape with approximate dimensions: diameter D=2.5 m, length L=2.5 m, figure 1. The approximate masses of nuclides without shielding (^{nat}U and ^{232}Th) are M(^{nat}U)= 270 kg, M(^{239}Pu) = 109 kg, M(^{240}Pu) = 2 kg, M(^{nat}Pb)= 4.5 t. The masses of the ADS shielding are approximately 100 t and 63 t of ^{nat}U and ^{232}Th, respectively. The effective multiplication factor K_{eff} is 0.98-0.99 for all cases.

Figure 1. Cross sections of ADS. Target zone (^{nat}Pb, ^{nat}U or W), MOX – fuel, reflector ^{nat}Pb and ^{nat}U or Th protection shield (area) or (nuclear fuel production zone). K_{eff} = 0.98 – 0.99. The size is approximately 2.5x2.5 m.

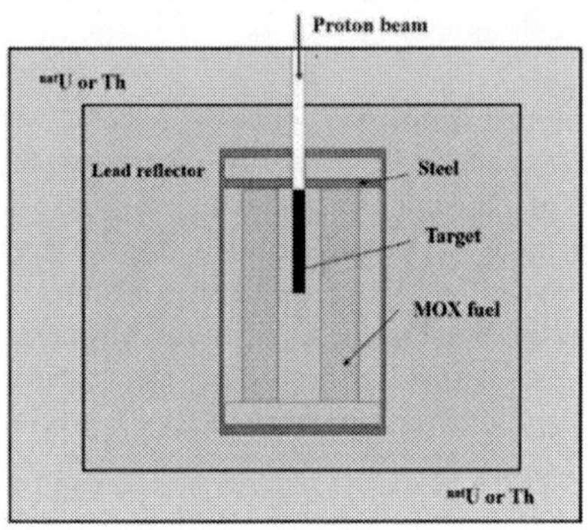

Figure 1a. Axial cross section of ADS, cases 2 and 3.

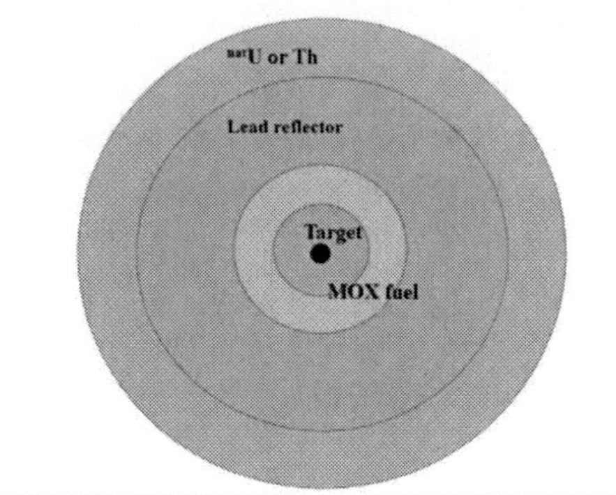

Figure 1b. Cross section perpendicular to the axis of cylinder, cases 2 and 3.

3. ADS simulations

Three cases are chosen for the simulations. The first case is an ADS system without shielding of ^{nat}U or ^{232}Th, but it has a biological shielding and a steel holder of ADS. The biological shielding and the steel holder do not change the neutron field inside ADS, it is important only for the neutron leakage. Three different targets and proton beams with energies E_p=0.66, 1 and 2 GeV (the results are shown in tables 1 and 2) are included in the calculations. The second case is an ADS with ^{nat}U shielding and proton beam E_p=1 GeV and the third case is an ADS with ^{232}Th shielding and proton beam E_p=1 GeV (the results are shown in table 3). For all cases the number of (n,f) reactions for incident neutrons with energies E_n<20MeV is calculated. High energy neutrons and charged particles induced fissions are not included in the calculations. The mass of the lead reflector in cases 2 and 3 is insignificantly higher than that in case 1. The steel holder and the biological shielding (in case 1) are removed and replaced by lead (in cases 2 and 3). The capture cross section of the iron nucleus is high, and because of that, the iron is not included in cases 2 and 3. The aim of ^{nat}U and ^{232}Th shielding is nuclear fuel production by reactions:

$$^{238}U(n,\gamma) \rightarrow {}^{239}U \xrightarrow{\beta^-(23,4\text{min})} {}^{239}Np \xrightarrow{\beta^-(2,4d)} {}^{239}Pu$$

$$^{232}Th(n,\gamma) \rightarrow {}^{233}Th \xrightarrow{\beta^-(22,3\text{min})} {}^{233}Np \xrightarrow{\beta^-(26,94d)} {}^{233}U \qquad (1)$$

The neutron leakage is significant in case 1 and the neutrons will participate in unwanted capture reactions. The best way to avoid this is to place the ^{nat}U and ^{232}Th shielding. For ease of calculations the fission energy per one split is set E_i=200 MeV. This approximation will not change the integral results. In the beginning of the simulations, the total ionization energy deposition of the proton beam in the matter is investigated. The targets (^{nat}U, ^{nat}Pb and ^{nat}P+W) with dimension diameter D=5 cm and length

L=100 cm are irradiated by protons with different energies. The ionization energy deposition E_{ion} for every millimetre in the targets is calculated. The ionization may be calculated with the Bethe equation.

$$-\frac{dE}{dx} = \frac{4\pi n z^2}{m_e \beta^2}\left(\frac{e^2}{4\pi\varepsilon_0}\right)^2 \left[Ln\left(\frac{2m_e c^2\beta^2}{I(1-\beta^2}\right) - \beta^2\right], \text{ where } n = \frac{N_A Z\rho}{A\mu},$$

m_e – electron mass in rest, c – speed light in vacuum, β=v/c, v – speed of the particle, ε_0 – vacuum permittivity, N_A – Avogadro number, Z – charge of the medium, ρ - medium density, A – atom mass of medium, μ – molar mass of the medium. The Bragg-peak of protons is presented (figure 2).

Figure 2. Ionization energy deposition E_{ion} in ^{nat}U, ^{nat}Pb and W targets for proton beams with energies E_p=0.8, 1, 1.2 and 1.5 GeV. The calculations are made for one proton.

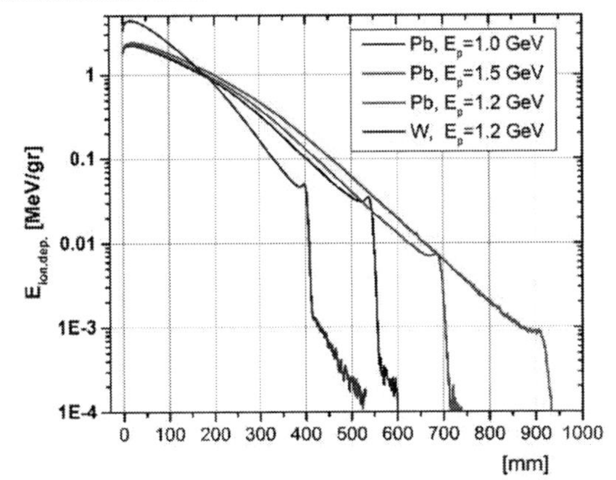

Figure 2a. The E_{ion} deposition in ^{nat}Pb and W targets irradiated by protons with energies E_p=1, 1.2 and 1.5 GeV.

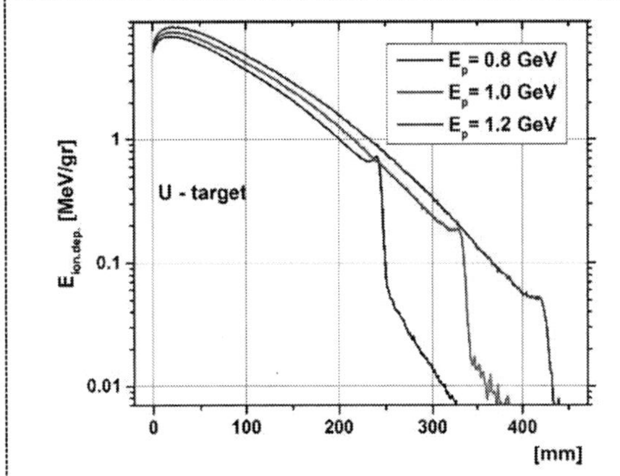

Figure 2b. The E_{ion} deposition in ^{nat}U target irradiated by protons with energies E_p=0.8, 1 and 1.2 GeV.

The next step is K_{eff} calculations. In the real system of energy production the equilibrium concentration of fission nuclides is supposed to be achieved. It means that the number of fissions reactions is less than the number of radiative capture of neutrons by fission nuclides, i.e. the rate of production of nuclear fuel is higher than the rate of consumption of nuclear fuel.

Table 1. Neutron balance in ADS (without ^{nat}U or ^{232}Th shield) for a proton beam with energies E_p=0.66, 1 and 2 GeV and three different targets ^{nat}U, ^{nat}Pb and ^{nat}Pb/W with dimensions D=5 cm and L=50 cm. N_{total} is the total neutron production, N_{escape} is the leakage of neutrons, (n,f) and (n,γ) are the number of reactions and "/" separation of the reactions. The calculations are made for one proton. K_{eff} = 0.98 – 0.99

			^{nat}U target					
E_p GeV	N_{total}	N_{escape}	(n,f)^{235}U	(n,f), (n,γ)^{238}U		(n,f), (n,γ)^{239}Pu		(n,γ)^{240}Pu
0.66	1350	363	6	37	106	330	62	6
1	2630	706	12	70	190	636	134	12
2	5540	1490	24	140	407	1364	255	23
			^{nat}Pb target					

0.66	1392	376	6	36	103	346	65	6
1	2472	668	11	63	183	614	115	10
2	5631	1521	25	145	417	1400	417	24
				$^{nat}Pb/W$ target				
0.66	1050	286	5	27	78	262	50	5
1	2000	540	9	51	148	495	93	8
2	4550	1230	20	116	337	1130	211	20

In that equilibrium stage, the concentrations and distribution of nuclides (fission and reflector) should be such that the K_{eff} is less than 1, for example approximately 0.98-0.99. For all simulations K_{eff} ranges between 0.985 and 0.998. To compare different experimental and computer simulations Beam Power Gain (BPG) is introduced. It is the ratio of the total energy production [MeV] and the beam energy [MeV]. The total energy production is presented by the equation: $E_{tot}=E_{fiss} + E_{ion} + E(n,\gamma)$, where E_{fiss} is energy received by fission, E_{ion} is ionization energy deposition of all ions and $E(n,\gamma)$ is the energy which is produced by the fission of the nuclides ^{233}U and ^{239}Pu, received by reactions (1). In table 2, the beam power gain is calculated by $BPG=(E_{fiss} + E_{ion})/E_p$.

Table 2. Fission E_{fiss} [x10^{4}MeV] and ionization E_{ion} [x10^{4}MeV] energy deposition in ADS for three targets ^{nat}U, ^{nat}Pb and ^{nat}P+W and proton beam energies E_p=0.66, 1 and 2 GeV are presented. The ratio E_{tot}/E_p shows the energy amplifier of the ADS without ^{nat}U or ^{232}Th shield. The calculations are done for one proton. K_{eff} = 0.98 – 0.99

	E_p=0.66GeV			E_p=1GeV			E_p=2GeV		
target	E_{fiss}	E_{ion}	E_{tot}/E_p	E_{fiss}	E_{ion}	E_{tot}/E_p	E_{fiss}	E_{ion}	E_{tot}/E_p
^{nat}U	9.0	0.65	147	15.5	1.1	166	35.5	2.5	190
^{nat}Pb	9.0	0.66	145	16	1.1	170	36.2	2.5	194
^{nat}P+W	6.8	0.5	110	12.8	0.92	127	29.2	2.5	158

Thus, shielding made of ^{nat}U or ^{232}Th is placed to increase nuclear fuel production. The results with two different types of shielding are presented in table 3.

Table 3. Neutron induced reactions (n,f) and (n,γ) in ADS (with ^{nat}U or ^{232}Th shield) for proton beam with energy E_p= 1GeV. The neutron leakage is 2.5% and 5% for ^{nat}U or ^{232}Th shield, respectively. The calculations are made for one proton. K_{eff} = 0.98 – 0.99

			^{nat}U - shield			
	$(n,f)^{235}U$	(n,f),	$(n,\gamma)^{238}U$	(n,f),	$(n,\gamma)^{239}Pu$	$(n,\gamma)^{240}Pu$
MOX	100	144	430	1410	272	6
Target	2	7	30	-	-	-
Shield	140	55	1960	-	-	-
			^{232}Th - shield			
	$(n,f)^{235}U$	(n,f),	$(n,\gamma)^{238}U$	(n,f),	$(n,\gamma)^{239}Pu$	$(n,\gamma)^{240}Pu$
MOX	75	105	310	1020	200	5
Target	2	6	25	-	-	-
	(n,f),	(n,γ) ^{232}Th				
Shield	0	1220		-	-	-

The comparison of nuclear fuel production by reaction $(n,\gamma)^{238}U$ and fission reaction $(n,f)^{239}Pu$ in table 1 shows that this ADS construction cannot achieve an equilibrium stage despite the fact that K_{eff}

IOP Conf. Series: Journal of Physics: Conf. Series **1023** (2018) 012031 doi:10.1088/1742-6596/1023/1/012031

=0.99. The fissions are approximately three times more than neutron capture reactions (nuclear fuel production). The neutron leakage is significant.

Table 4 shows the integral number of (n,f) reactions and the number of production of nuclear fuel ^{239}Pu, ^{233}U by radiative capture and BPG. The equilibrium stage is reached. Moreover, the production rate of nuclear fuel is higher than the fission rate. Most of the new nuclear fuel is generated in the shielding, not in the MOX or target areas. This is good, because the new nuclear fuel can be removed from ADS during irradiation.

Table 4. The total neutron induced fission, production of ^{239}Pu and ^{233}U are shown. The energy amplifier BPGs are presented in the last two columns. The proton beam energy is E_p=1 GeV. E(n,γ) is the fission energy of produced nuclear fuel ^{239}Pu and ^{233}U. K_{eff}=0.99. The calculations are made for one proton.

Shield	$(n,f)_{total}$	$(n,\gamma)^{238}$U	$(n,\gamma)^{232}$Th	BPG=E_{fiss}/E_p	BPG=$(E_{fiss}+E(n,\gamma))/E_p$
^{nat}U	1850	2400		370	850
^{232}Th	1200	310	1220	240	540

Most of the fissions come from $(n,f)^{239}$Pu reaction and the fission cross sections for low energy neutrons are very high. Part of those neutrons are obtained after the interaction with the neutron moderator ^{nat}Pb. Therefore, the additional simulations are done with different data libraries for ^{nat}Pb, the same cross sections for other fission nuclides and the same high energy nuclear model. The total number of fissions in ADS calculated by MCNP6 with ENDF70 data tables [9] (for ^{nat}Pb) are less than that of the fissions calculated with ENDF60. The total number of fissions calculated by MCNP6 and RMCCS or DRMCCS data tables (for ^{nat}Pb) are higher than that of the fissions calculated with ENDF60. Similar behaviour is observed in fuel production in such ADS structure. For ADS calculations the cross sections data tables are very important as well as nuclear models.

4. Discussion

The ADS is supposed to be controlled by the parameters of the proton beam, but the significant presence of MOX (^{239}Pu) in the reactor part requires more calculations and investigations. A sudden increase of the ^{239}Pu fissions to $K_{eff}>1$ is possible, which is out of the control of the accelerating proton beam. The neutron field in ADS is highly inhomogeneous. Close to the target the neutron field has many spallation neutrons. The neutron field in the reactor part (MOX – fuel) consists of fission neutrons, spallation neutrons and reflected neutrons from the lead. The neutron field in the ^{nat}U or/and ^{232}Th shield (nuclear fuel production region) consists of fission and neutrons reflected from the lead. Energy production strongly depends on the Pb cross section neutron moderation. Moreover, the ^{206,207}Pb isotopes can capture low energy neutrons during irradiation and the Pb isotope concentration in the moderator will change.

5. Conclusion

In the reactor part no equilibrium (production rate of nuclear fuel higher than fission rate) can be reached. The equilibrium can be achieved if additional layers of ^{dep}U or ^{232}Th are placed around ADS (case 2 and 3). These layers produce nuclear fuel and reduce the neutron leakage to the minimum.

References

[1] Andriamonje S Angelopoulos A Apostolakis A et al Experimental determination of the energy generated in nuclear cascades by a high energy beam Physics Letters B 348 (1995) 697
[2] http://sckcen.be/en/Technology_future/MYRRHA
[3] https://ntof-exp.web.cern.ch/ntof-exp/
[4] Krivopustov M Pavliouk A Kovalenko A Mariin I et all 2009 *First results studying the transmutation of ^{129}I, ^{237}Np, ^{238}Pu, and ^{239}Pu in the irradiation of an extended ^{nat}U/Pb-assembly with 2.52 GeV deuterons* (Journal of Radioanalytical and Nuclear Chemistry) 279(2):567

IOP Conf. Series: Journal of Physics: Conf. Series **1023** (2018) 012031 doi:10.1088/1742-6596/1023/1/012031

[5] Furman W Adam J Baldin A Berlev A Gundorin N Hushvaktov Zh Kadykov M Kostov L Stoyanov Ch Zhivkov P Bielevicz Westmeier W Manolopoulou M Hashemi-Nezhad SR E&Tr collaboration 2012 *Recent results of the study of ADS with 500 kg natural uranium target assembly QUINTA irradiated by deuterons with energies from 1 to 8 GeV at JINR NUCLOTRON* XXI (International Baldin Seminar on High Energy Physics Problems at Dubna Russia) vol XXI_086

[6] Polański A Zhivkov P Stoynov *Ch Simulation of energy deposition and neutron spectrum of subcritical assembly irradiated with proton beam with MCNPX transport code* 2009 (NATO Security through Science Series C: Environmental Security) pp 333

[7] https://tendl.web.psi.ch/tendl_2015/neutron_html/neutron.html

[8] Tonev D Goutev N and Georgiev L *Cyclotron laboratory of the Institute for Nuclear Research and Nuclear Energy* 2016 *J. Phys: Conf. Ser.* 724 012049.

[9] https://www.oecd-nea.org/janis/

IOP Conf. Series: Journal of Physics: Conf. Series **1023** (2018) 012032 doi:10.1088/1742-6596/1023/1/012032

Double Charge Exchange Reactions and Double Beta Decay.

N. Auerbach

School of Physics and Astronomy, Tel Aviv University

Tel Aviv 69978, Israel

Abstract: The subject of this presentation is at the forefront of nuclear physics, namely double beta decay. In particular one is most interested in the neutrinoless process of double beta decay, when the decay proceeds without the emission of two neutrinos. The observation of such decay would mean that the lepton conservation symmetry is violated and that the neutrinos are of Majorana type, meaning that they are their own anti-particles. The life time of this process has two unknowns, the mass of the neutrino and the nuclear matrix element. Determining the nuclear matrix element and knowing the cross-section well will set limits on the neutrino mass. There is a concentrated effort among the nuclear physics community to calculate this matrix element. Usually these matrix elements are a very small part of the total strength of the transition operators involved in the process. There is no simple way to "calibrate" the nuclear double beta decay matrix element. The double beta decay is a double charge exchange process, therefore it is proposed that double charge exchange reactions using ion projectiles on nuclei that are candidates for double beta decay, will provide additional necessary information about the nuclear matrix elements.

1. Introduction

One of the most important fields of study in modern nuclear physics is double beta decay in nuclei. There are a number of nuclei where such decays are energetically allowed. These are usually ground state to ground state transitions. There is presently an intense activity both in experiment and theory in the nuclear and particle physics to study double beta decays. There are two types of decays, in the first decay, the process is accompanied by two neutrino emission while in the second, more interesting one, the so called neutrinoless decay, there is no emission of neutrinos , (see for example [1].) The observation of neutrinoless decay will put serious constraints on various gauge models beyond the standard one. It would mean that the lepton number conservation symmetry is broken, that the neutrino is a Majorana particle that is, its own anti-particle. The expression of the decay rate of the neutrinoless process involves basically two unknowns, the mass of the neutrino and the nuclear transition matrix element. Having determined the nuclear matrix element one will be able to find the mass of the neutrino. In various experiments double beta decays emitting two neutrinos have been observed [2] however the neutrinoless double beta decay, as of today, was not detected. Only lower limits for the life time of such reaction were set. As mentioned there are many ongoing experiments, and many more are planned to study the neutrinoless double beta decay [1, 3].

The operators that are involved in the beta decay and double beta decay are the Fermi and Gamow-Teller. In the double beta-decay with neutrinos emitted, there is no r dependence and the Fermi type operator does not contribute since the ground state of the final nucleus is not the double isobaric analog of the initial state. (There might be a negligible contribution due to a very small isospin admixture). The important transition is of double Gamow-Teller type. In the case of the neutrinoless process there is essentially a 1/r dependence in the operators and the Fermi transition does contribute, but still it is small compared to the Gamow-Teller transition. The nuclear matrix elements for the double beta decays are very small, 1% or less of the total double Gamow-Teller strength (DGT) [4, 5]. In the past we wrote [4]: "The (ββ-decay) matrix element, is very small and accounts for only a 10−4 to 10−3 of the total DGT sum rule. A precise calculation of such hindered

IOP Conf. Series: Journal of Physics: Conf. Series **1023** (2018) 012032 doi:10.1088/1742-6596/1023/1/012032

transition is, of course, very difficult and is inherently subject of large percent uncertainties. At present there is no direct way to "calibrate" such complicated nuclear structure calculations involving miniature fractions of the DGT transitions. By studying the stronger DGT transitions and, in particular, the giant DGT states experimentally and theoretically, one may be able to "calibrate" to some extent, the calculations of ββ-decay nuclear matrix elements". In double charge exchange (DCX) reactions at least two nuclei are involved in the process. This is also the case in the double beta decays. It is therefore natural to look for DCX reactions to provide some information about the various aspects of nuclear structure involving the dynamics of two nucleons. In the past *pion* DCX reactions were extensively studied because of the existence of intensive pion beams produced in the so called meson factories. Experimental and theoretical studies of the pion DCX reactions were performed, not necessarily in the double beta decay context. One learned a substantial amount of information about the nature of the reaction, about the role of correlations in nuclei and about the various mechanisms that the reaction proceeds [6, 7]. However as the pion programs in the meson factories were declining these studies of DCX declined. Thoughts were given to use some other projectiles in order to perform DCX reactions. Another way to achieve probes that could produce DCX when scattered is to use light or even heavy ions as projectiles. An isospin $T \geq 1$ ion may undergo a double charge exchange reaction from a T_z to $T_z \pm 2$ charge state. The use of $T = 1$ nuclei as probes is the most convenient case [4].

Presently due the interest in the double beta decay new attempts are being made to perform DCX reactions with heavy ions. We mention just a few: In Catania there is a large program to study DCX using ion beams for nuclei that are candidates of double beta decay studies. The large collaboration is called NUMEN. The main motivation is to help to determine the nuclear matrix elements for neutrinoless double beta decay. Reaction with ions O18 to Ne18, etc., are planned [8].There is a DCX program using ions at RIKEN, Japan which is related to double beta decay studies but also emphasizes the exploration of new nuclear structure, as for example observing double Gamow-Telle states, suggested some years ago [4], the study of tetraneutron states [9]. Work at Osaka University [10] consists again of using light ions to study excitations of double charge exchange states, comparing the results to the pion DCX reactions [11] and exploring the possibility that these studies will help to determine better the nuclear matrix elements appearing in the double beta decays.

2. **Isotensor giant resonances**

A "model" giant resonance (state) built on a state $|n\rangle$ is:

$$|Q_\alpha; n\rangle = Q_\alpha|n\rangle / |\langle n|Q_\alpha^+ Q_\alpha|n\rangle|^{-1/2} \quad (1)$$

Q_α is a one-body operator, $Q_\alpha = \sum_i q_\alpha(i)$

For $|0\rangle$, the g.s., we have the "usual" giant resonances.

The operator Q_α might depend on isospin. For example the electric dipole is an isovector and the corresponding operator is:

$$Q_\alpha \equiv D = \sum_i r_i Y_1(\theta_i)\tau_\mu(i), \; with\, \mu = 0, \pm 1 \quad (2)$$

In addition to the "common" $\mu = 0$ dipole one has also the *charge-exchange analogs* $\mu = \pm 1$

The action of the operator with $\mu = -1$ (or $\mu = 1$) twice will lead to states with $\Delta T_z = \pm 2$

These states can be reached in *double* charge-exchange (DCX) processes.

One of the best examples, are the pion DCX reactions (π^+,π^-), (π^-,π^+), studied extensively in the past. Because we deal with isotensor transitions, the selectivity of the DCX is large and enhances the possibilities to observe double giant resonances (states). Examples of isotensor states:

Double isobaric analog state (DIAS): The IAS is defined as:

$$|A_1\rangle = T_-|0\rangle/(N-Z)^{1/2} \tag{3}$$

The DIAS is:

$$|A_2\rangle = T_-^2|0\rangle/[(N-Z)(N-Z-1)]^{1/2} \tag{4}$$

Dipole built on an analog:

$$|D_{-1};A_1\rangle = \sum_i r_i Y_1(\theta_i)\,\tau_{-1}(i)|A_1\rangle/\tilde{N} \tag{5}$$

Yet another example, the double dipole (for the $\Delta T_z = -2$):

$$|D_{-1};D_{-1}\rangle = \sum_i r_i\, Y_1(\theta_i)\tau_{-1}(i)|D_{-1}\rangle/\tilde{N} \tag{6}$$

The operators could also depend on spin, for example the Gamow-Teller operator is:

$$G_\pm = \sum_i \sigma(i)\,\tau_\pm(i) \tag{7}$$

The giant Gamow-Teller (GT) state:

$$|GT\rangle = G_-|0\rangle/N \tag{8}$$

The double Gamow-Teller (DGT) state:

$$|DGT;J\rangle = G_-^2(J)|0\rangle/N(J) \tag{9}$$

J is the total spin of the DGT. For two identical phonons the wave function is symmetric in space-spin-isospin, therefore for $\Delta T = 0{,}2$ $J = 0^+,\ 2^+ \ldots..$

Some properties of these states:

Excitation energies: $E_{Q_1Q_2} \approx E_{Q_1} + E_{Q_2}$

Widths *(spreading widths)* $\Gamma_{n;\lambda} = n\Gamma_{1;\lambda}$. In some theories n is replaced by $\sqrt{n}$

Isospin and intensities: for an isotensor excitation $\Delta T_z = -2$ and for $(N-Z) \geq 2$ there are five isospin values $T' = T+2,\ T+1,\ T,\ T-1,\ and\ T-2.$

For an isotensor of rank k $F_\mu^{(k)}$ the intensities for the various isospin components T' are given by the corresponding Clebsh-Gordan (CG) coefficients and reduced matrix elements:

$$S_{T'} = |\langle T,T,k,\mu|T',T+\mu\rangle|^2\langle T'\|F^{(k)}\|T\rangle^2 \tag{10}$$

The CG coefficient for $T \gg k$ (in our case $T \gg 2$) are dominated by the CG of the aligned isospin $T' = T+\mu$. For $= -k,\ \ (CG)^2 = \frac{2(T-k)+1}{2T+1}$. (11)

In the pion DCX reactions the various spin independent DGRs listed above were excited in several nuclei [7, 11]. These observations were confirmed recently in experiments in the double charge exchange reaction (B_5^{11}, Li_3^{11}) [10].

3. DCX and Correlations in Nuclei.

The DCX process has held out the hope that it would be a means of probing two-body correlations in nuclei. In the initial studies it was found that that uncorrelated nuclear wave functions produced qualitative agreement with experiment at higher energies (pion energies around 300 MeV.) (The cross-sections were proportional to (N-Z) (N-Z-1)). This was the situation when the double giant resonances were studied. However at low pion energies (around 50 MeV) the disagreement with experiment was very large, (sometimes a factor of 50) when uncorrelated wave functions were used to study the double isobaric analog state. It was necessary to include wave functions with correlated nucleons.

The DCX process, as determined in pion charge-exchange reactions to the DIAS involves two basic routes.

1. For uncorrelated n paricles in the transition to a double analog state the route from the initial state to the final state goes via the excitation of the single analog in the intermediate stage, and from there in a charge-exchange to the double analog. This is termed as the analog route or more generally as the *"sequential"* process. The cross section is proportional to the number of pairs one can form from the n nucleons, $n(n-1)/2$.

2. The second process involves correlated nucleons and the process proceeds through intermediate states that are not the analog. This we term the "non-analog route".

An expression that represents this approach was derived [6] for a DCX transition between the J=0^+ and its double analog, for a single configuration j^n, where n is an even number of neutrons occupying the orbit.

The DCX cross section can be written as:

$$\sigma_{DCX} = \frac{n(n-1)}{2}\left|A + \frac{(2j+3-2n)}{(2j-1)(n-1)}B\right|^2 \tag{12}$$

This formula is remarkable in its simplicity. The DCX cross section to the DIAS has two amplitudes A and B (which are angle and energy dependent but j independent). The first represents transitions which occur in the absence of correlations in the nuclear wave function. The cross section due to this term is proportional to the number of neutron pairs to be made into pairs of protons when going from the ground state to the DIAS. This counting rule is independent of the mutual location of the two nucleons in the pair and therefore this dominates the cross section when the DCX operator is of long range. The second term represents the DCX transitions which take place when the nuclear wave function is more than just that of independent particles, that is it involves correlations. This term will contribute when the DCX transition operator is of short range. The first term corresponds to the *"sequential"* while the second one represents the case when the intermediate states are non-analog states [6].

The above formula is extended to the generalized seniority scheme when several orbits are involved [6, 12, 13]. Then:

$$\sigma_{DCX} = \frac{n(n-1)}{2}\left|A + \frac{(\Omega+1-n)}{(\Omega-1)(n-1)}B\right|^2 ;\text{with } \Omega = \textstyle\sum_i (j_i + 1/2) \tag{13}$$

The characteristic behavior of the DCX amplitude to the DIAS as a function of the number of valence neutrons for a single j orbit persists also when there is rather weak configuration mixing. The change is absorbed by the altered values of the A and B coefficients in such a way that the ratio $\frac{B}{A}$ is increased, meaning that the effective DCX operator is of shorter range. But even in the case of stronger mixing the effect could be similar. This may suggest why certain models in which a limited space is employed can provide satisfactory results when effective operators are used. (An example is the IBM description of the double beta decay).

Double beta decay and DCX

The process of double beta decay is very similar to the nuclear structure part of DCX reactions. We will not present the specifics of double beta decay. One can find this in the extensive literature. We will concentrate on common features of these processes. In particular the similarities are close when we deal with DCX reactions involving the Fermi and Gamow-Teller strength. For example the two-neutrino amplitude can be written as:

$$A_{GT}^{2\nu} = \sum_n \frac{\langle f|G_-|n\rangle\langle n|G_-|i\rangle}{E_i-E_n-\varepsilon} \tag{14}$$

Where G_- is the Gamow-Teller operator defined above, eq. (7). (The 0-neutrino amplitude includes also a Fermi amplitude and the operator has an r dependence.)

The total coherent DGT strength is:

$$S_{DGT} = \sum_f |\sum_n \langle f|G_-|n\rangle\langle n|G_-|i\rangle|^2 \tag{15}$$

$$S_{2\beta} = |\sum_n \langle f_0|G_-|n\rangle\langle n|G_-|i\rangle|^2 \tag{16}$$

is the double beta decay strength (in the closure approximation) and is a ***coherent*** sum also. Thus:

$$S_{2\beta} \sim |\langle f|G_-(1)G_-(2)|i\rangle|^2 \tag{17}$$

One can approach this problem of the distribution of DGT strength in a straightforward way [14, 15, 16]. One calculates the shell-model states in an extended model space in the parent nucleus, (N, Z), in the intermediate nucleus (N - 1, Z + 1), and in the final nucleus (N- 2, Z + 2). Having determined the nuclear wave functions one then evaluate directly the *incoherent* sum and *coherent* sum. In figure 1 we show the coherent running sum of the two neutrino beta decay of ${}^{48}_{20}Ca$ as a function of the intermediate states. We note thát this is not a monotonic function and some of the contribution are positive and some negative. This happens for all nuclei we calculated and all the different interactions used.

Some (n, p) studies were performed in which one chooses as the target states the final states of the relevant double beta decay process. In these (n, p) experiments one measures the GT strength to the various intermediate single charge exchange states. One can then take and multiply the two GT strengths obtained in (p,n) and (n,p) to the same intermediate states obtained in each of the two reactions and try to sum over the intermediate states observed. For example, let us take the A = 48 nuclei. The (p, n) and (n, p) reactions are performed correspondingly on the 48Ca and 48Ti targets. The intermediate states measured are in both reactions are located in the 48Sc nucleus. Actual experiments were performed at the initial nucleon energies of 200 MeV. At these energies the spin states and in the forward direction GT states in particular are excited predominantly. Using the DWIA analysis, one was able to extract the GT strength for the various intermediate states observed in these two reactions. Note that only the squares of the matrix elements of the type $\langle J_i|\sigma\tau|n\rangle$ $and\langle n|\sigma\tau|J_f\rangle$ can be deduced from these two one-body experiments. The relative signs are not determined in such processes. By measuring transitions to a great number of intermediate states, one in principle can determine the incoherent sum for the g.s. to g.s. transition. There are several difficulties in this procedure even when one tries to determine the incoherent sum. First, because of final experimental resolution on the one hand and large density of intermediate states on the other, one is not always sure that it is the same matching intermediate state that is excited when the (p,n) and (n, p) experiments are compared. Second, and related to it is the fact that when a strong GT transition in (p, n) is observed to a given intermediate state, the (n, p) transition to the same intermediate state might be (and often is) weak, and vice versa. This of course makes the experimental exploration of the double beta decay amplitude difficult. The ground states in the two target nuclei are not related by any simple transformation and therefore the action of the $\sigma\tau_{\pm}$ operators leads to mismatched distributions of strength in the intermediate nucleus. The combined studies of (p, n) and (n, p) may provide only limited information concerning the nuclear structure aspects of the 2β decay matrix elements. In view of the things discussed above it is clear that because of the coherence of the 2β decaymatrix elements, the use of (n, p) and (p, n) strength cannot determine the value of such matrix elements. Also, quite obviously in such experiments, one cannot excite all DGT strengths. In order to be able experimentally to study the DGT strength (including the 2β decay g.s. to g.s. transition), one must employ processes in which the leading terms are genuine two-body transitions. Double charge-exchange (DCX) reactions are the natural candidates for such a study.

We should point out that one of the difficult problems in the theoretical studies of double beta decay is the "universal" quenching of the single Gamow-Teller strength. Experiments show that 30-40% of GT strength is missing in the main GT peaks. This affects the double beta decay transition matrix elements. The source of the above quenching is not certain. There are several ideas. One is that the missing strength is due to the $\Delta - h$ configurations interacting with the GT states and the missing strength is 300 MeV above the GT peak [17]. Another theory is that the GT strength is fragmented and the missing strength is several tens of MeV above the main GT peak [18, 19, 20]. These two different theories will affect the double beta decay matrix element differently. The uncertainty, (because of this quenching), could be as large as a factor of 2 for double beta decay matrix elements.

Acknowledgement

We wish to thank the Bulgarian Academy of Science for the hospitality during authors visit in Bulgaria.

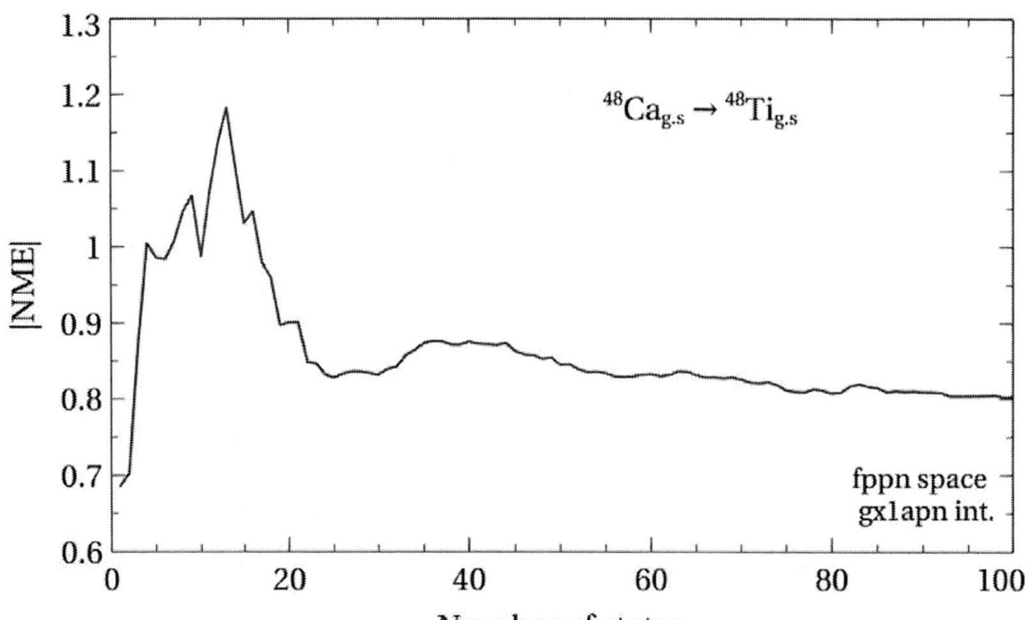

Figure1. The running sum of the DGT contribution to the neutrininoless double beta decay matrix element as a function of the number of intermediate states.

IOP Conf. Series: Journal of Physics: Conf. Series **1023** (2018) 012032 doi:10.1088/1742-6596/1023/1/012032

References

1. Avignone III FT, Elliot SR, and Engel J, 2008 Rev. Mod. Phys. **80**, 481

2. Elliot SR, Hahn A, and. Moe MK, 1987 Phys. Rev. Lett. **59**, 2020

3. Barabash AS, 2011 Physics of Atomic Nuclei, **74**, 603

4. Auerbach N, Zamick L and. Zheng DC, 1989 Annals of Physics (N.Y.) **192**, 77

5. Shimizu N, Menendez J, and Yako K, 2017 arXiv: 1709.01088v1

6. Auerbach N, Gibbs WR and Piasetzky E, 1987 Phys. Rev. Lett. **59**, 1076 ; Auerbach N, Gibbs WR, Ginocchio JN , and Kaufmann WB, 1988 Phys. Rev. C **38** 1277

7. Mordechai S, Auerbach N et al., 1988 Phys. Rev. Lett. **60**, 408; Mordechai S, Auerbach N et al., 1988 Phys. Rev. Lett. **61**, 531

8. Capuzzello F et al., 2015 Eur., Phys. J. A**51,** 145

9. Uesaka T, et al., 2015 RIKEN, RIBF, NP-PAC, NP21512, RIBF141 and private communication.

10. Takahisa A, Ejiri H, et al., 2017 preprint.

11 Mordechai S et al., 1994 Int. Jour. of Mod. Phys. E, Vol.**3** No.**1** 39.

12. Wu HC, Ginocchio JN, Dieperink AEL, and Scholten O,1996 Phys. Rev. C**54** 1208

13. Barea J and Iachello F, 2009 Phys. Rev. C **79** 044301

14. Zheng, DC, Zamick L and Auerbach N, 1990 Annals of Physics (N.Y.) **197**, 343

15. Loc BM, Auerbach N, 2017 private communication

16. Brown BA, Horoi M, and Sen'kov RA, 2014 Phys. Rev. Lett. **113**, 262501

17 Bohr A and Mottelson BR, 1981 Phys. Lett. B **100,** 10

18. Bertsch GF and Hamamoto I, 1982 Phys. Rev. C **26**, 1323

19. Klein A, Love WG and Auerbach N, 1985 Phys. Rev. C **31**, 710

20. Zelevinsky VG, Auerbach N and Bui Minh Loc, 2017 Phys. Rev. C, **96**, 044319

IOP Conf. Series: Journal of Physics: Conf. Series **1023** (2018) 012033 doi:10.1088/1742-6596/1023/1/012033

Simulations of negative hydrogen ion sources

A Demerdjiev, N Goutev and D Tonev

Institute for Nuclear Research and Nuclear Energy, Bulgarian Academy of Sciences, BG-1784, Sofia, Bulgaria

Abstract. The development and the optimisation of negative hydrogen/deuterium ion sources goes hand in hand with modelling. In this paper a brief introduction on the physics and types of different sources, and on the Kinetic and Fluid theories for plasma description is made. Examples of some recent models are considered whereas the main emphasis is on the model behind the concept and design of a matrix source of negative hydrogen ions. At the Institute for Nuclear Research and Nuclear Energy of the Bulgarian Academy of Sciences a new cyclotron center is under construction which opens new opportunities for research. One of them is the development of plasma sources for additional proton beam acceleration. We have applied the modelling technique implemented in the aforementioned model of the matrix source to a microwave plasma source exemplifying a plasma filled array of cavities made of a dielectric material with high permittivity. Preliminary results for the distribution of the plasma parameters and the φ component of the electric field in the plasma are obtained.

1. Introduction

The negative hydrogen/deuterium (H^-/D^-) ion sources are an integral part of devices as proton accelerators [1] and magnetic confinement fusion reactors of the tokamak type (ITER - currently under construction at Cadarache France). In the field of fusion applications the Negative Ion (NI) sources are a vital component of the systems for additional heating - Neutral Beam Injection (NBI) - of tokamak fusion plasmas, as they complement the ohmic one. Development of these systems for the ITER tokamak is an extensive task allocated in fusion research institutes worldwide [2]. The H^- ion sources found their first application in the Van der Graaf accelerators, which dates back to the 50's of the 20-th century, and later on in the cyclotron accelerators. The currently used and developed H^- ion sources in proton accelerators are of the same type as the ones in the systems for additional heating of thermonuclear plasma. Regarding the discharge type the sources are: DC (direct current, based on glow discharge or arc discharge), Radio Frequency (RF) and microwave discharges.

There are two mechanisms for production of H^- ions - volume (in the plasma volume) and surface production (on the discharge walls). The H^- ions are created in the plasma volume through dissociative attachment of electrons to vibrationally excited molecules ($e + H_2(\nu) \rightarrow H + H^-$). This is a two step process: excitations of the vibrational levels (ν) of the ground electron state through collisions with fast electrons (energy above 10 eV), followed by dissociative attachment of slow electrons (energy $\simeq 1$ eV) to vibrationally excited molecules. The effectiveness of H^- surface production ($H, H^+ + e_{wall} \rightarrow H$) depends on the surface material. Since a low work function material is required, usually Cs or Ba are used.

The two main types of H^- ion sources are DC and RF. In both of them the volume production of H^- could be optimised by employing a magnetic filter (MF) (Figure 1 and 2). It consists of

IOP Conf. Series: Journal of Physics: Conf. Series **1023** (2018) 012033 doi:10.1088/1742-6596/1023/1/012033

magnetic cusps with fields that effectively separate the volume of the source in one region with fast electrons and another region with slow electrons, i.e. it reduces the electron temperature. NI sources with MF are called tandem sources. The plasma in the DC source is ignited by a hot filament (Figure 1), whereas the typical RF sources are inductive discharges and the plasma is generated by the RF field introduced in the driver region by RF coil (Figure 2). The other two regions in the RF NI sources are, respectively, expansion (the plasma generated in the driver region diffuses into it) and extraction regions. The NI are extracted via an extraction system which consists typically of 3 multi-aperture grids: the plasma grid (PG), the extraction grid (EG), the grounded grid (GG). The NI are extracted from the plasma by applying a positive voltage (5-10 kV) between the PG and the EG. Subsequently, electrons are accompanying the NI extraction, hence they need to be deflected. For this purpose secondary a magnetic field, created by permanent magnets embedded in the EG (Figure 2), deflects the electrons out of the beam on the grid components.

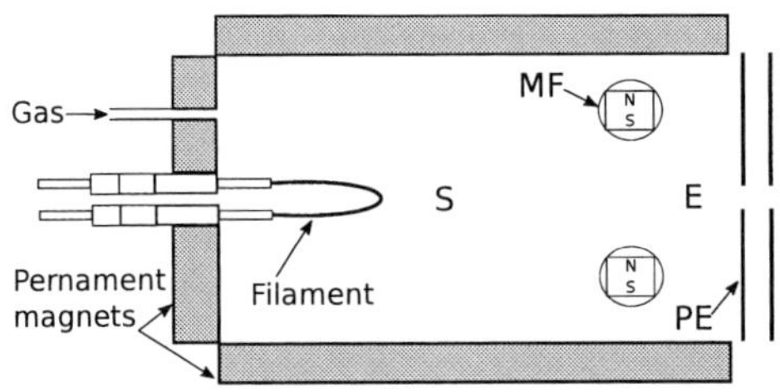

Figure 1. Schematic representation of a DC NI tandem source. S - plasma volume, E - extraction region, MF - magnetic filter, PE - plasma electrode of the extraction device.

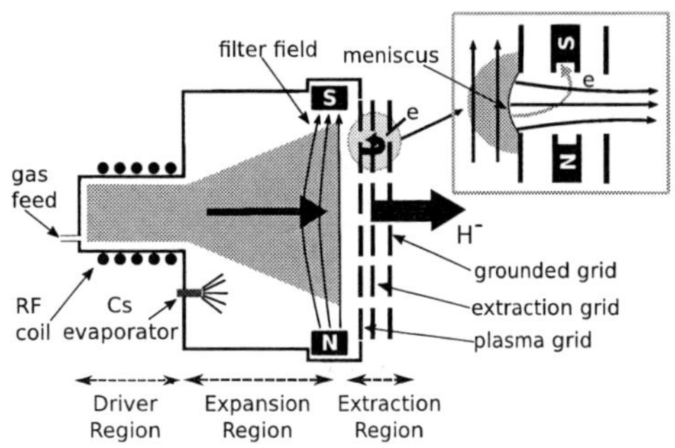

Figure 2. Schematic representation of RF NI tandem source.

Depending on their application, NI sources vary in design and beam parameters. For example, the beam requirements for the full size ITER neutral beam injection experiment Source for Production of Ions of Deuterium Extracted from Radio-frequency plasma (SPIDER) are: 40 A beam current; 3600 s pulse length; 1 MeV beam energy [2]. NI source for the spallation neutron source project at the Oak Ridge National Laboratory has the following characteristics: RF source with surface production; 40 mA beam current; 1 ms pulse length; 2.5 MeV beam energy [3].

In the research and development of H^- ion sources the key issues are summarised as follows: improvement of the H^- ions production efficiency under low pressure conditions; development of large volume H^- ion sources with spatial uniformity; optimisation of the H^- ions extraction from the source; high energy acceleration and good beam optics. The rapid development of computer resources has boosted the computational methods in plasma physics. In the next part the two theories for plasma description (respectively, the kinetic and fluid theories) shall be presented and, respectively, simulation models in which the Particle In Cell (PIC; for collisional models the Monte Carlo Collision (MCC) scheme is applied [4]; we shall use the PIC-MCC notation for such models) or Finite Element Method (FEM) are implemented. Both of them are well established tools in the field of plasma physics simulations, more information could be found in the literature [4–6]. These models will show part of the progress that has been made in the field of simulations of H^- ion sources. The main topic will be the H^- matrix ion source which concept [7] is built upon self-consistent model coupling gas discharge part within the fluid plasma theory with electrodynamics. The modelling equations are numerically solved by applying FEM. In the current study an example of the aforementioned modelling technique will be presented as we applied it in the development of a two-dimensional, axially symmetric, self-consistent model of a new plasma source. The source exemplifies a plasma filled linear array of cavities made of material with high dielectric permittivity. The discharge is in argon. The gas pressure is 10 Torr. We have obtained preliminary results for the distribution of the electron density and temperature; phi (φ) component of the electric field. The plasma is generated by a transverse electric (TE) wave with a frequency of 2.4 GHz, and 160 W deposited power. The application of this source is in the field of a plasma source for additional acceleration of a proton beam. The development of such a plasma source could be a part of the research program of the cyclotron center which is under construction at the Institute for Nuclear Research and Nuclear Energy of the Bulgarian Academy of Sciences [8].

2. Plasma description theories, H^- ions source modelling

There are two theories for plasma description, respectively - kinetic and fluid. The kinetic theory is based on obtaining the distribution function $f(\boldsymbol{r},\boldsymbol{v},t)$, which gives us the distribution of the particles in the six dimensional phase space (position $\boldsymbol{r}$ and velocity $\boldsymbol{v}$ space of the particles) at a given moment of time t. The distribution function has to satisfy the Boltzmann equation:

$$\frac{\partial f}{\partial t} + \boldsymbol{v}.\nabla f + \frac{\boldsymbol{F}}{m}\frac{\partial f}{\partial \boldsymbol{v}} = \left(\frac{\partial f}{\partial t}\right)_c \tag{1}$$

Where $\boldsymbol{F}$ is the force acting on the particles and $\partial/\partial\boldsymbol{v}$ - is the velocity space gradient. The right hand side term of the equation is the collisional term. In a sufficiently hot plasma the collisions can be neglected and furthermore the force is usually entirely electromagnetic (Lorentz force: $\boldsymbol{F} = q(\boldsymbol{E} + \boldsymbol{v}\times\boldsymbol{B})$; q - electrical charge of the particles; $\boldsymbol{E}$-electric field; $\boldsymbol{B}$-magnetic field), then the Boltzmann equation takes it's special form - the Vlassov equation. Having obtained the distribution function by solving the Boltzmann or respectively the Vlassov equation we can obtain the macroscopic parameters. Typically we need the distributions of the following (macroscopic) parameters: the density of charged particles of certain species $n_\alpha(\boldsymbol{r},t)$; their velocity $v_\alpha(\boldsymbol{r},t)$ and their temperature $T_\alpha(\boldsymbol{r},t)$ (α-electrons or ions). The kinetic theory is applied in the cases when non-Maxwellian plasmas and non-local effects are studied like the electron plasma oscillations and the Landau damping (particle-wave interactions).

Fortunately in many cases the plasma species have Maxwellian velocity ditribution. Thus a more crude model (than the kinetic theory) for description of plasma phenomena is applicable - the fluid theory for plasma description. In the fluid description typically the first three moment equations are included, respectively: the continuity equation (2); the fluid equation of motion (momentum transport equation) (3); the electron energy balance equation (4):

$$\frac{\partial n_\alpha}{\partial t} + \nabla.(n_\alpha \boldsymbol{u}_\alpha) = \frac{\delta n_\alpha}{\delta t} \tag{2}$$

$$m_\alpha n_\alpha\left[\frac{\partial \boldsymbol{u}_\alpha}{\partial t} + (\boldsymbol{u}_\alpha.\nabla)\boldsymbol{u}_\alpha\right] =$$

$$= Z_\alpha e n_\alpha \boldsymbol{E} + Z_\alpha e n_\alpha\left[\boldsymbol{u}_\alpha \times \boldsymbol{H}\right] - \nabla p_\alpha - n_\alpha \mu_{\alpha\beta}\nu_{\alpha\beta}\left(\boldsymbol{u}_\alpha - \boldsymbol{u}_\beta\right) - m_\alpha \boldsymbol{u}_\alpha \frac{\delta n_\alpha}{\delta t} \tag{3}$$

$$-\frac{3}{2}\frac{\partial n_e T_e}{\partial t} + \nabla.\boldsymbol{J}_e = Q_{ext} - P_{coll} - e\boldsymbol{\Gamma}_e.\boldsymbol{E}_{DC} \tag{4}$$

Where, in these equations $n_\alpha, m_\alpha, T_\alpha$ and p_α are, respectively, the density, mass, temperature and partial pressure for given type of species. $\boldsymbol{u}_\alpha$ is the fluid velocity, $\mu_{\alpha\beta} = m_\alpha m_\beta/(m_\alpha + m_\beta)$ is the reduced mass for elastic collisions between particle species α and β, $Z_\alpha = \pm 1$ respectively

IOP Conf. Series: Journal of Physics: Conf. Series **1023** (2018) 012033 doi:10.1088/1742-6596/1023/1/012033

for ions and electrons, $\mathbf{\Gamma}_{\rm e}$ is the electron flux, $\boldsymbol{J}_{\rm e} = -(5/2)n_{\rm e}D_{\rm e}\nabla T_{\rm e} + (5/2)n_{\rm e}T_{\rm e}\boldsymbol{u}_{\rm e}$ - electron energy flux, consisting of convective and heat fluxes ($D_{\rm e}$ - electron diffusion coefficient); $Q_{\rm ext} = (1/2)({\rm Re}\sigma)|E|^2$ is the deposited power by the external RF field, the $\delta n_\alpha/\delta t$ term is the collisional generation and loss of charged particles through the processes of ionization and recombination. On the right hand side of (4) despite the collisional losses $P_{\rm coll}$ the losses for generation of constant electric field are also noted ($-e\mathbf{\Gamma}_{\rm e}.\boldsymbol{E}_{\rm DC}$, where $\boldsymbol{E}_{\rm DC}$ is the constant electric field inside the discharge).

Since the two theories for plasma description are being discussed we have to note that the plasma is a dynamic system and respectively its charged particles are in constant motion. The charged particle fluxes induce currents inside the plasma volume, which interact with each other. Those interactions on their turn cause modifications to the macroscopic plasma properties. Thus in order to have a self-consistent decription of the plasma we need to include in our set of equations in a proper manner the Maxwell equations. Depending on the geometry and the manner of power deposition in the discharge the Maxwell equations could be reduced, respectively, to: Poisson's equation (simulation of DC discharges; extraction region of the NI sources [9, 10]); equations for the azimuthal components of the electric or magnetic field (simulation of cylindrical RF inductive discharges), other alternative is the vector potential equation of the RF magnetic field; electromagnetic wave equations (in the case of surface microwave heated discharges).

The NI sources are sophisticated devices, thus a number of issues should be taken into account in the research and development process. Here we shall consider issues related to the extraction region and the RF power deposition in the NI sources. RF power deposition models are the tools which give us insight on the physics of the different states and operating regimes of the discharge, and thus help us to optimize their design. The physics of the NI extraction region requires detailed modelling by itself, since many processes and parameters should be taken into account in order to insure the beam quality.

An example of electromagnetic PIC-MCC models for studying the RF power deposition are the models employed for the plasma modelling of CERN Linac4 H^- and The J-Parc Linac sources [11, 12]. The Linac4 H^- ion source model was further developed [13, 14] giving insight on the effect of design and operational parameters on the plasma dynamics in the source, and finally the plasma heating in the Linac4 H^- source in the nominal operational conditions was simulated for the first time, taking into account the neutral populations as well as the external static magnetic fields [15]. The transition between the two stable operating regimes (capacitive and inductive [16]) of the inductive discharges as well as their ignition and stable state are other main issues when RF heating of inductive discharges is studied. Some recent works employing PIC models [12, 14, 17, 18] consider these issues.

The complexity of the physics involved in the extraction process is a result of: sophisticated shape of the 3D electromagnetic field; kinetic reactions and plasma surface processes. The key parameter is the meniscus (Figure 2) - the 3D frontier between the plasma and the beam - as its shape and position determine to a great extent the beam optics. Concerning the beam quality one of the main issues are the co-extracted electrons which induce thermal loads on the electron dumping system, which contributes to increased beam emittance. Formation of beam halo (part of the beam with poor optics) and top-bottom assymetry [19] (caused by the magnetic field filter) are other issues which should be taken into account. Recent works take into account and/or study these effects, as some of the models for simulation of the extraction region are capable to simulate the extraction systems of ITER like NBI in three dimensions and even a first attempt for simulation of the full-size ITER negative ion source is made [9, 10, 20–22].

The aforementioned models are based on kinetic plasma description, but as it was noted a great deal of the plasma phenomena could be described accurately enough by the fluid theory. Here the concept for matrix source of volume produced H^- ions [7], based on fluid models, shall

IOP Conf. Series: Journal of Physics: Conf. Series **1023** (2018) 012033 doi:10.1088/1742-6596/1023/1/012033

be presented briefly.

The small discharge radius is in the basis of the matrix source, theoretically motivated by results obtained within two-dimensional fluid models of discharges inductively driven by a cylindrical coil [23, 24]. The small radius (2-3 cm) ensures high density of the negative ions in the region with the highest value of the plasma potential [23] as a result of the negative ion flux in the DC electric field in the discharge. Increasing the radius decreases the electric field and respectively the ion flux, and thus the accumulation effect of NI on the discharge axis is diminished [24]. The measured [25] higher density of the NI in the driver region (a small-radius quartz tube) of a two chamber, tandem small-size plasma source, inductively driven by a cylindrical coil confirms the theoretical results. The source design - a matrix of gas-discharge tubes - has orientated the work to planar-coil driving of the matrix [24]. Thus, the suggested construction of the source is for a matrix of small-radius inductively driven discharges with: a single aperture extraction from each discharge; RF inductive driving by a planar coil (a large area planar coil, common for the whole matrix). Finding RF driving which would ensure both efficient RF power transfer to the plasma and equal plasma parameters (including their spatial distribution) in the separate discharge tubes is an issue which comes to the fore in the development of matrix plasma sources. For the purpose a three-dimensional model coupling the gas discharge part within the fluid plasma theory with electrodynamics was developed. Applying the model several tests [26] of different coil designs and source constructions were made. Finally a zigzag coil with an Ω-shaped conductor on the bottom of each discharge tube proved to be an effective coil for RF driving of a source completed as a matrix of small-radius discharges [27].

3. Microwave excitation of linear array of plasma filled cavities

In this section will be presented the obtained by us preliminary modelling results, for macroscopic plasma parameters and the φ component of the electric field, of microwave heated plasma source. The chosen gas as discharge medium is argon due to its simple chemistry in discharges at low pressure. The intended application of the plasma source is in the field of additional acceleration of cyclotron proton beam. The simulated source represents a plasma filled linear array of cavities made of a diellectric material with high permittivity ($\varepsilon = 25$) (Figures 3 and 4). In the current study it is shown that this structure is capable of sustaining a surface microwave discharge with high plasma density, i.e. surface microwaves heating the plasma in the cavities. For the purpose we have implemented the same modelling techniques, as in the study of RF heating of the matrix source [26, 27].

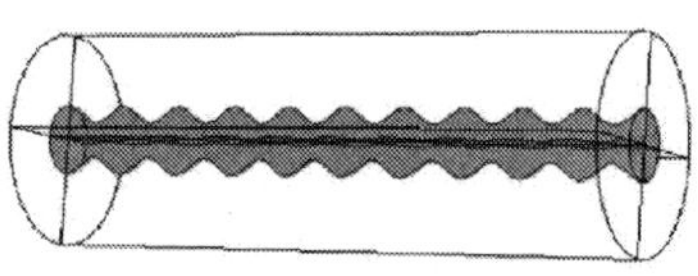

Figure 3. Shematic representation of the source.

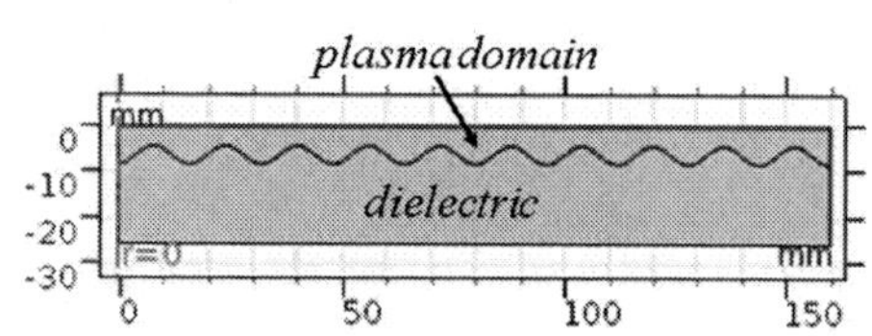

Figure 4. Modelling domain.

The model is implemented in the commercial software Comsol Multiphysics ®, which uses the FEM for solving the modelling equations. We have developed a self-consistent model by applying the built-in interface for microwave plasma. The electron density and mean electron energy are computed by solving a pair of fluid equations (electron continuity and energy balance equations). The model considers the following heavy species: Ar atoms, Ar^+ ions, Ar^* excited atoms (electronically excited states are lumped into a single species). The electrodynamics consists of: solving the Poisson equation in the plasma domain to obtain the distribution of the

DC potential; and solving the wave equation for the high frequency electric field in the entire modelling domain.

Since the geometry of the source is intrinsically axially-symmetric (Figure 3), we reduced the modelling domain to a two-dimensional domain with axial symmetry (Figure 4). On the plasma domain a wall boundary condition (for the fluid equations) is applied to account for the loss of electrons due to their net and random flux to the walls and, respectively, ground ($V = 0$) for the Poisson's equation.

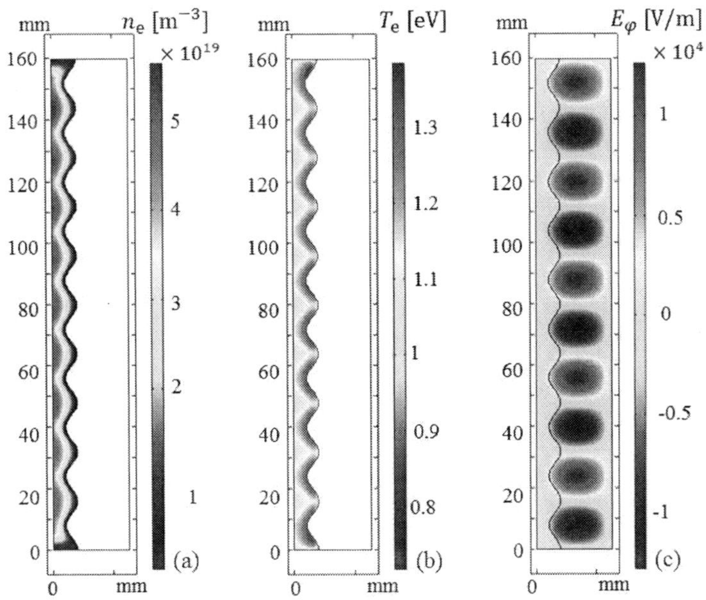

Figure 5. Two-dimensional distributions of: (a) n_e; (b) T_e; (c) E_φ.

The electric field wave equation has the following boundary conditions: input port - microwave deposited power of 160 W ($f = 2.4$ GHz TE-wave) fed through the top side of the dielectric; output port on the bottom side; perfect electric conductor - lateral side of the dielectric and on the top and botom sides of the plasma domain. The chemical mechanism in the plasma volume is simplified as it consists only of 3 species and 7 reactions: elastic collisions; direct ionzation via collision with electrons (15.8 eV); all electronically excited levels lumped into a single excited level (11.5 eV); excited argon atoms are consumed via superelastic collision with electrons ($-11.5\, eV$); stepwise (4.24 eV) and Penning ionization; quenching with neutral argon atoms. Two surface reactions (particle-wall interactions) are taken into account: excited atoms convert to ground state argon atoms and of the argon ions convert to neutral argon atoms. The cross-sections and rate coefficients of the argon volume and surface reactions are taken from the available data within the Comsol Multiphysics plasma interface.

The gas discharge part of the implemented model solves electron continuity and energy balance fluid equations, thus the results obtained for the macroscopic plasma properties are for the distribution of the electron density and temperature Figure 5 (a) and (b). Solving the wave equation of the high frequency electric field we obtain the distribution of the φ component of the electric field Figure 5 (c). The calculations are done in two steps: first Eigenfrequency study is performed in order to obtain a proper eigenfrequency of the studied stucture ($f = 2.4$ GHz) which is input for the second step of the study - Frequency-Transient simulation. The electron density obtained is high $\sim 10^{19}$ m^{-3}, thus the electron plasma frequency is also high - $f_e = (1/2\pi)(e^2 n_e/m_e\varepsilon_0)^{1/2} = 63$ GHz (where e is the electron charge; n_e - the electron density; m_e - the electron mass; ε_0 - the vacuum permittivity). The wave frequency is much lower than the electron plasma frequency ($f << f_e$), thus the wave does not propagate in the plasma, it is absorbed and its energy is dissipated in the region near the dielectric wall. As a result the electron temperature has a maximum in this region (Figure 5 (b)). The electrons gain energy in the region with maximum electron temperature and carry it in the plasma volume where it is dissipated through collisions, therefore the electron density has its maximum in the plasma volume (Figure 5 (a)). The well known formula from the literature [28] $E_{max} = (m_e c\omega_p)/e$ (c - speed of light (in vacuum); $\omega_p = 2\pi f_e$ - plasma frequency) gives us the maximum electric field in plasma, i.e. the maximum electric field which could be waked inside the plasma. Since it is a function of the electron density ($\omega_p \propto n_e^{1/2}$) the value of electric field in the plasma is high $E_{max} \sim 680$ MV/m. The main conclusion of these results is that this structure of the plasma source and the chosen manner of plasma heating ensure high enough electron density which is a requirement for plasma acceleration sources. Next step in the development of the model is to change the discharge gas to hydrogen.

4. Acknowledgments

The research has been supported by Bulgarian Science Fund under Contract No. DN 08/6, 13.12.2016 and Contract No. DM 18/2, 12.12.2017.

References

[1] Dinev D 2006 *Particle accelerators* Marin Drinov Academic Publishing House ISBN - 13:978-954-322-067-0
[2] Bacal M and Wada M 2015 *Appl. Phys. Rev.* **2** 021305
[3] Keller R 2002 *Rev. Sci. Instrum.* **73** 914
[4] Verboncoeur J P 2005 *Plasma Phys. Control. Fusion* **47** A-231-A260
[5] Jardin S 2010 *Computational methods in plasma physics* Chapman & Hall/CRC ISBN - 13: 978-1-4398-1095-8
[6] Birdsall C K, Langdon A B 2004 *Plasma physics via computer simulations* Taylor & Francis
[7] Lishev St, Paunska Ts, Shivarova A and Tarnev Kh 2012 *Rev. Sci. Instrum.* **83** 02A702
[8] Tonev D, Goutev N, Georgiev L S 2016 *J. Phys. Conf. Proc.* **724** 012049
[9] Nishioka S, Miyamoto K, Okuda S, Goto I, Hatayama A, and Fukano A 2014 *Rev. Sci. Instrum.* **85** 02A737
[10] Mochalskyy S , Lettry J and Minea T 2016 *New J. Phys.* **18** 085011
[11] Mattei S, Ohta M, Hatayama A, Lettry J, Kawamura Y, Yasumoto M, and Schmitzer C 2013 *AIP Conf. Proc.* **1515** 386
[12] Shibata T and et al 2017 *AIP Conf. Proc.* **1869** 030017
[13] Ohta M, Mattei S, Yasumoto M, Hatayama A, and Lettry J 2014 *Rev. Sci. Instrum.* 85 02B113
[14] Mattei S, Ohta M, Yasumoto M, Hatayama A, Lettry J, and Grudiev A 2014 *Rev. Sci. Instrum.* 85 02B115
[15] Mattei S, Nishida K, Onai M, Lettry J, Tran M Q, and Hatayama A 2017 *AIP Conf. Proc.* **1869** 030018
[16] Kortshagen U, Gibson N D and Lawler J E 1996 *J. Phys. D: Appl. Phys.* **29** 1224
[17] Nishida K, Mattei S, Mochizuki S, Lettry J, and Hatayama A 2016 *J. Appl. Phys.* **119** 233302
[18] Shibata T and et al 2016 *Rev. Sci. Instrum.* **87** 02B128
[19] Schiesko L, McNeely P, Franzen P, Fantz U and the NNBI Team 2012 *Plasma Phys. Control. Fusion* **54** 105002
[20] Revel A, Mochalskyy S, Montellano A , Wnderlich D, Fantz U, and Minea T 2017 *J. App. Phys.* **122** 103302
[21] Ichikawa M and et al 2017 *AIP Conf. Proc.* **1869** 030024
[22] Taccogna F and Minelli P 2017 *New J. Phys.* **19** 015012
[23] Paunska Ts, Shivarova and Tarnev Kh 2011 *AIP Conf. Proc.* **1390** 165
[24] Paunska Ts, Shivarova S and Tarnev Kh 2010 *J. Appl. Phys.* **107** 083301
[25] Lishev St, Shivarova A and Tsankov Ts 2010 *J. Phys: Conf. Series* **223** 012001
[26] Demerdzhiev A, Lishev St, Tarnev Kh and Shivarova A 2015 *AIP Conf. Proc.* **1655** 040011
[27] Tarnev Kh, Demerdjiev A, Shivarova A and Lishev St 2016 *Rev. Sci. Instrum.* **87** 02B123
[28] Seryi A 2016 *Unifying Physics of Accelerators, Lasers and Plasma* CRC Press, Taylor & Francis Group ISBN - 13: 978-1-4822-4059-7

IOP Conf. Series: Journal of Physics: Conf. Series **1023** (2018) 012034 doi:10.1088/1742-6596/1023/1/012034

Nuclear and analytical methods for investigation of high quality wines

D Tonev[1], E Geleva[1], T Grigorov[1], N Goutev[1], H Protohristov[1], Ch Stoyanov[1], V Bashev[1], I Tringovska[2,3] and D Kostova[2,3]

[1]Institute for Nuclear Research and Nuclear Energy, Bulgarian Academy of Sciences "Tzarigradsko shaussee" 72 Blvd, 1784 Sofia, Bulgaria
[2]Maritsa Vegetable Crops Research Institute, ul."Brezovsko shosse " 32, 4003 Plovdiv, Bulgaria
[3]Center of Plant Systems Biology and Biotechnology, blvd. "Ruski" 139, 4000 Plovdiv, Bulgaria

dimitar.tonev@inrne.bas.bg

Abstract. Nuclear and analytical methods can help to determine the year of production – vintage and the geographical provenance of high quality wines. A complex analytical investigation of Melnik fine wines from "Artarkata" vineyards, Vinogradi village near Melnik in Southwest Bulgaria using different methods and equipment were performed. Nuclear methods, based on measured gamma-ray activity of ^{137}Cs and specific activity of ^{3}H can be used to determine the year of wine production. The specific activity of ^{137}Cs was measured in wines from different vintages using Low-Background High-Resolution Gamma-Spectrometry. Tritium measurements in wine samples were carried out by using a low level liquid scintillation counting in a Packard Tri-Carb 2770 TR/SL liquid scintillation analyzer. The identification of the origin of wines using their chemical fingerprints is of great interest for wine consumers and producers. Determination of 16 chemical elements in samples from soil, wine stems, wine leaves and fine wine from the type Shiroka Melnishka, which are grown in typical Melnik vineyard was made, using Inductively Coupled Plasma-Optical Emission Spectrometry (ICP-OES).

1. Introduction.

Fine wines in the world are nowadays subject to increased demand driving up prices and creating environment of limited supply and high value that make these products attractive for counterfeiting. Recent studies estimated counterfeit wine market between 6 and 30 million US $ [1]. Due to their high quality, Bulgarian fine wines are eventually also objects of counterfeiting.

The Melnik region is world renowned for its red wine, produced since 1346 from the endemic grape Broad-Leaved Melnik (Shiroka Melnishka loza). The special flavour of this grape, the aging in unique dry underground caverns and the secrets of the local vintners gives the Melnik wine its extraordinary quality. Wines from this particular grape variety were the most popular Bulgarian wines in Western Europe in the 19th and 20th century. One of the most famous connoisseurs of the red wine from Melnik is said to be Sir Winston Churchill.

A nuclear method, based on the measurement of the gamma-ray activity of the radioisotope ^{137}Cs have been utilized to control the authenticity of an old wine [2]. It occurs in the wine from fruit contamination, which can result from direct deposition on fruit surfaces, absorption by the fruit skin and transport to the pulp, deposition to soil, root uptake and transfer to fruit [3]. ^{137}Cs decays into ^{137}Ba, which emits a single gamma-ray with an energy of 661 keV. It was released into the environment in the 20th century as a result of nuclear weapons testing (1946-1961) and the Chernobyl power plant accident in 1986. In both cases, the radioisotope is accumulated in soil and measurable amount of it can be detected by means of high-resolution gamma-ray spectrometry. ^{137}Cs with a half-life of 30 years is well suited for radiometric determination of age of wine because its activity is strongly correlated with the year of production of wine form certain geographic region [2].

The present lack of any research facility for irradiation (a research reactor or/and a research accelerator) in Bulgaria opens a possibility for such interesting experiments like the present one. The new TR24 cyclotron project in Bulgaria [4] will give many new possibilities for experiments in Bulgaria in the field of nuclear physics.

In Bulgaria, as in all countries in South-Eastern Europa, the ^{137}Cs deposition originates mainly from fallout from Chernobyl accident. The total ^{137}Cs fallout in Bulgaria was estimated to 1.3 x 10^{15} Bq, or 1/30 of the total released ^{137}Cs [5]. Later studies of ^{137}Cs in soil from South-West Bulgaria have shown mean value of 30 Bq/kg at the time of investigation [6].

It is possible to date a wine through the activity measurement of the radioactive isotope ^{3}H using the low level liquid scintillation spectrometry technique. Dating wines with ^{3}H dates back to 1954 when it was investigated by Kaufman and Libby [7]. Measurement of wine with ^{3}H by Schöhofer was also performed in the early 1990s [8]. However, tritium measurements are hampered by nuclear fallouts and the spreadout of nuclear power plants.

^{3}H, is a low-energy beta emitter ($E_{\beta max}$ = 18.6 keV) with a period of half live equal to 12.43 years. Its presence in the nature is a consequence of the interactions of cosmic rays in the Earth atmosphere. It can also be produced in human activities. It is released into the atmosphere from weapons manufacturing, operation of nuclear power plants, and reprocessing of nuclear fuels. The ^{3}H environmental level increased in the period from 1945 to 1963 as a result of nuclear weapon tests [9-11]. This radionuclide shows slight differences in the chemical and physical characteristics in comparison with hydrogen, and for this reason can replace it in water samples. Tritiated water (HTO) moves through the environment like ordinary water, cycling readily through the hydrologic components in the environment and as a consequence it is in abundance in the atmospheric fallout [9].

Recent studies were directed on methods indicating the geographical provenance or designation of origin of high quality wines using their chemical „fingerprints“ (unique combination of elements present in wine). Individual vineyard areas have unique trace element "signatures" that permit the identification of the provenance of a wine. This suggests that trace elements could play some role in determining the distinctiveness of wine that differentiates wines of one region from those of another. The analysis of trace metals in wines is of great importance for quality control of wine, authenticity control of wine, wine metals toxicity, bioavailability etc [12]. The identification of the geographical origin of wines is of great interest for wine consumers and producers, since it may provide determinant criteria for wine price and guarantees of quality [13].

Provenance establishment of wine is based on the principle that elements in wine are derived mainly from the soil and environment, and that the concentration of the majority of elements is not significantly changed during the wine production. Additionally, the element content depends on grape variety, weather and agricultural practices. Therefore, these methods involve analysis of the element and microelement concentrations, not only of certain wine, but also of vineyard soil [14].

The present study aimed to investigate the possibility for control of vintage and designation of origin of high quality Melnik wines by means of radiological methods and chemical „fingerprints“.

2. Material and methods

The object of the research is fine Melnik wine – „Baba Ivanka Wine" produced from local Broad-Leaved Melnik grape variety in a single vineyard near the village Vinogradi (5 km south of Melnik),

IOP Conf. Series: Journal of Physics: Conf. Series **1023** (2018) 012034 doi:10.1088/1742-6596/1023/1/012034

community Sandanski, province Blagoevgrad, in „Artarkata“ site with coordinates 41°29’ 49.45” N, 23° 23’ 20.99” E, altitude 390 m.

The radiological study includes measurements of gamma-ray natural and artificial radioactive isotopes and tritium in studied wine.

2.1. *Gamma activity*

For the Gamma measurements wine samples of fine Melnik wine were placed in standard 0.5 L Marinelli beakers. The measurement took 24 to 48 hours depending on the specific activity of the samples. The activity measurement was carried out using a low-background lead shielded gamma-spectrometer, lined with cadmium, copper and plexiglass, equipped with High Purity Ge-detector of 20 % relative efficiency and 1.9 KeV resolution for the 1332 keV peak of ^{60}Co was used. For data acquisition a multichannel analyzer was used at 8192 channels with dead time/pileup correction.

2.2. *^{3}H activity*

Wine samples were purified by distillation as a pre-treatment step to remove the content of organic compounds and reduce quenching, which might adulterate the results. An aliquot of 10 ml from the second fraction of the distillate was transferred in a polyethylene vial, mixed to 10 ml of ULTIMA GOLD LLT scintillation cocktail for measurement. The samples were measured by us using a Liquid Scintillation Counter, Packard TRI-CARB 2770 TR/SL. All measurements were performed using 300 minutes counting time. Efficiency calibration was performed automatically during the procedure of normalization with internal unquenched standard – ^{3}H and ^{14}C.

2.3. *Inductively Coupled Plasma – Optical Emission Spectrometry*

For the geographical characterization of the wine 16 elements (Al, Fe, Ba, Ca, Co, Cr, Cu, Pb, K, Li, Mg, Mn, Na, Ni, Zn and Ga) were analyzed. Chemical analysis was also performed on samples from the vineyard soil. The wine sample was prepared by mixing wines from several recent vintages. Soil samples of 15 points within the vineyard were collected from a depth of 0 to 20 cm. The representative soil sample was then prepared by removing plant residue, air drying, mixing, finely grounding and passing through a 20 mesh filter to obtain very fine particles. Samples were subjected to microwave digestion in a closed PTFE vessel (Anton Paar) according to U.S. EPA Method 3052 and subsequently analyzed by Inductively Coupled Plasma – Optical Emission Spectrometry. Measurements are made using ICP-OES Optima 7000 model of the company Perkin Elmer with dual-view configuration. ICP Multi Element Standard Solution IV CertiPUR® (Merck) was used as reference material.

3. Results and discussion

3.1. *Gamma activity*

The results showed that samples of Melnik fine wine contain the isotopes ^{40}K at the level of 20 Bq/l and ^{137}Cs. ^{40}K is a natural isotope whose percentage in the total potassium content is estimated at 0.0119 [14]. The gamma-ray activities of ^{137}Cs and ^{40}K for different vintages Melnik fine wine are given in table 1.

The data of radio-cesium in samples of Melnik fine wine show a reduction of the ^{137}Cs activity with time, with highest values for the vintage year 1986 (figure 1). The 1986 vintage was strongly affected by radioactive contamination because in May the vines were at an advanced stage of development. Rapid decrease of the ^{137}Cs activity is observed for 1987. During 1986-1987 the cesium fallout is partly washed out or deposited and fixed in soil mostly in the upper 10 cm layer, therefore in later years only a limited amount of cesium is absorbed by the plants.

Table 1. Results from gamma-ray measurements on single grape Melnik wine from different vintages [Bq/l]

Vintage	^{137}Cs [Bq/l]	^{137}Cs [Bq/l][b]	^{40}K [Bq/l]
1985	< 0.10[a]	-	15.50 (0.88)
1986	24.9 (0.37)	44.6 (0.70)	15.40 (1.50)
1987	3.95 (0.08)	7.20 (0.14)	17.00 (1.00)
1988	1.22 (0.05)	2.08 (0.08)	16.80 (0.90)
1989	0.70 (0.04)	1.17 (0.08)	18.30 (0.90)
1990	0.43 (0.05)	0.71 (0.08)	20.75 (1.10)
1992	0.15 (0.04)	0.24 (0.07)	15.30 (0.78)
2001	< 0.15[a]	-	19.30 (1.80)

[a] minimum detected ^{137}Cs gamma-activity
[b] ^{137}Cs gamma-activity corrected for the year of production

A strong correlation was established between ^{137}Cs gamma activity of different vintages and the year of production (figure 2).

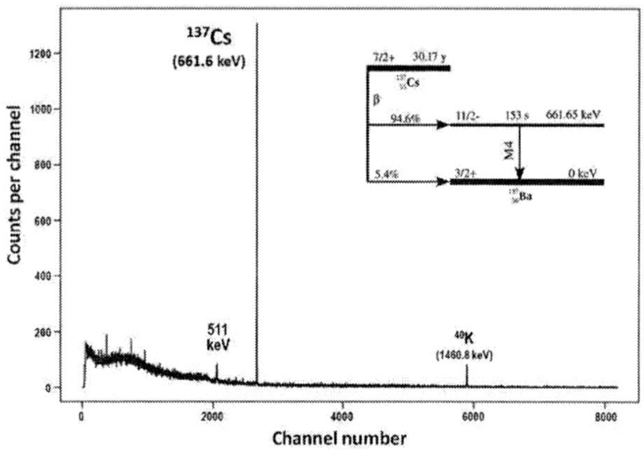

Figure 1. Gamma-ray spectrum of Melnik fine wine, 1986 vintage

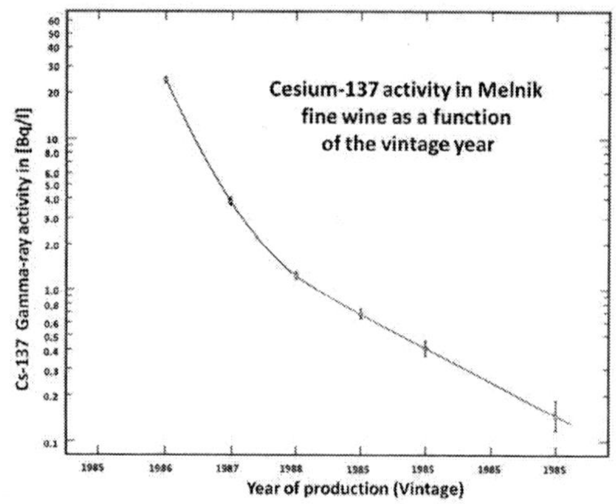

Figure 2. ^{137}C [Bq/l] in Melnik fine wine as a function of the year of production

3.2. ^{3}H *activity*

^{3}H determinations can provide information about the natural origin of many substances including wine, which can be used to determine the age of wine of a certain vintage [8]. The tritium content in the wine is that of the rain which nourished the vineyard. The specific activity of ^{3}H for different vintages of Melnik fine wine are given in table 2.

Table 2. Tritium specific activity (Bq/l) of Melnik wine from different vintages

Vintage	^{3}H [Bq/l]	^{3}H [Bq/l][a]
1985	7.36 (0.91)	32.4 (4.0)
1986	16.0 (1.6)	63.0 (6.3)
1990	13.1 (1.4)	54.6 (5.8)
2000	8.75 (1.03)	16.6 (1.9)
2001	4.37 (0.60)	7.83 (1.1)

[a] ^{3}H activity corrected for the year of production

It was found that ^{3}H activity of Melnik wine was the highest during the period 1986-1990. The minimum value (7.83 Bq/L) was for a wine produced in 2001. There is no strong dependence between ^{3}H activity of different vintages and the year of production, mainly due to atmospheric variations, which prevents us to carry out precise dating. In that sense, activity measurement of the hydrogen isotope ^{3}H is complementary to the nuclear method, based on *in situ* measured gamma-ray activity of the radioisotope ^{137}Cs.

3.3. Chemical examination

The presence of trace elements in wine is the consequence of the atmospheric deposition of airborne particulate matter on grapes and/or of the intake of these elements by the grapevine from ground water and soil, pesticides and fertilizers [12].

The ICP-OES element analysis provides the content of sixteen elements, all available in soil, active and capable of migration in stems, leaves and wine. The concentration of some other elements in the wine samples are below the detection limit. The preliminary results for chemical elemental analysis have been presented in the works [16, 17].

The data could help to identify chemical "fingerprint" for Melnik wines, because Melnik fine wines are produced mostly from the same variety – the endemic Broad-Leaved Melnik grape vine, grown in the same isolated geographical region in Southwest Bulgaria. The measured element concentrations are shown in table 3. For comparison, the concentrations measured in soil, wine stems and wine leaves, are also indicated.

Table 3. Concentrations of detected chemical elements in samples from soil, wine stems, wine leaves and fine wine from a single vineyard in the Vinogradi village near Melnik

	Element	Al	Fe	Ba	Ca	Co	Cr	Cu	Pb
	λ (nm)	396.15	238.20	233.53	317.93	228.61	267.72	327.3	220.35
Sample	Soil	22801.6	18455.3	102.8	8615.1	6.0	26.1	51.9	8.2
	Stem	10.1	9.8	4.8	2668.5	0.0	0.2	3,6	0.3
	Leave	113.8	85.5	13.9	23876.6	0.1	0.6	3,7	0.2
	Wine	9.63	1.41	0.12	21.5	0.0	0.0	0.045	0.0

	Element	K	Li	Mg	Mn	Na	Ni	Zn	Ga
	λ (nm)	766.49	670.78	285.21	257.61	589.59	231.60	213.8	417.21
Sample	Soil	3704.2	31.2	4974.9	479.8	382.4	10.2	72.6	67.7
	Stem	3430.8	0.2	1165.7	17.8	17.8	0.0	6.7	0.0
	Leave	3298.1	0.5	4957.5	90.7	52.0	0.0	10.2	0.1
	Wine	89.31	0.45	69.70	0.71	4.80	0.0	0.12	0.01

All chemical elements detected in the soil and in the wine samples are considered for the study. From these elements, Ca, K and Na levels in wine can be influenced by regional variations in the soil fertilization practices and in the wine making process and are therefore not sufficient reliable markers for fingerprinting. Furthermore, Fe concentration can have mostly technological origin. We consider Li, Mg, Mn and Zn as best markers for fingerprinting, because of the good correlations obtained between the element concentrations in wine and soil. These concentrations, in the relative proportion measured, can be used for determining with extreme precision the region of origin of wine.

IOP Conf. Series: Journal of Physics: Conf. Series **1023** (2018) 012034 doi:10.1088/1742-6596/1023/1/012034

4. Conclusions

A complex analytical investigation of Melnik fine wine from "Artarkata" vineyards, Vinogradi village near Melnik in Southwest Bulgaria using different methods and equipment, were carried out.

A correlation was established between the ^{137}Cs activity and year of production for Melnik fine wine allowing determining the vintage in the period 1986-1992.

It was found that ^{3}H activity of Melnik wine has higher values in the period 1986-1990, decreasing during the following years to average values.

Sixteen chemical elements were investigated, using ICP-OES. The obtained specific concentrations of detected chemical elements in soil and wine could be used as initial database for geographical fingerprinting of Melnik fine wines.

5. Acknowledgements

The authors express their gratitude to Ivanka Necheva and Jordanka Toneva for their outstanding support. This research has been supported by Bulgarian Science Fund under Contract No.DN08/6.

References

[1] Pender R 2010 *Master of Wine Dissertation*, The Institute of Masters of Wine, London
[2] H u b e r t Ph, Perrot F, Gaye J, Médina B and Pravikoff M 2009 *C. R. Physique* **10** 622
[3] Desideri D, Roselli C and Meli M 2010 *Food and Chemical Toxicology* **48** 650
[4] Tonev D, Goutev N and Georgiev L 2016 *J. Phys: Conf. Ser.* **724** 012049
[5] V a p i r e v E, Georgiev G, Jordanov T and Hristova A 1996 *Bulgarian Journal of Physics*, **3/4** 129
[6] K o s t o v L, Kobilarov R, Kostova L, Mladenov M, Protohristov H and Stoyanov C 2007 *Nuclear Methods for Non-Nuclear Applications* Heron Press Sofia 59
[7] Kaufman S and Libby W 1954 *Phys. Rev.* **93** 1337
[8] Schönhofer F 1992 *Radiocarbon* **34** (3) 768
[9] Madruga M, Sequeira M and Gomes A 2008 *Advances in Liquid Scintillation Spectrometry* 353
[10] Palomo, M, Penalver A, Aguilar C and Borrull F 2007 *Applied Radiation and Isotopes* **65** 1048
[11] S´anchez A, Montero M, Escobar V and Vargas M 1999 *Appl. Radiat. Isotopes* **50** 1049
[12] Stafilov, T, Cvetković J, Arpdjan S and Karadjova I 2002 *BAÜ Fen Bil. Enst. Dergisi* 4.2 90
[13] Perez-Trujillo J, Barbaste M and Medina B 2012 *Journal of Wine Research*, **13** (3) 243
[14] Bentlin F, Pulgati F, Dressler V and Pozebon D 2011 *J. Braz. Chem. Soc.* **22** (2) 327
[15] Królak E and Karwowska J 2010 *Polish J. of Environ. Stud.* **19** (3) 599
[16] Tringovska I, Tonev D, Damjanova A, Grigorov T, Gutev N, Protohristov H and Kovachev N. 2016 *Acta Horticulturae* **1115** 63 DOI: 10.17660/ActaHortic.2016.1115.10
[17] Tonev D *et al.* 2016 *Comptes rendus de l'Acade'mie bulgare des Sciences*, **69** (6) 707

National cyclotron centre at the Institute for Nuclear Research and Nuclear Energy

D Tonev, N Goutev, G Asova, A Artinyan, A Demerdjiev, L S Georgiev, M Yavahchova, V Bashev, S G Genchev, E Geleva, M Mincheva, A Nikolov, D T Dimitrov

Institute for Nuclear Research and Nuclear Energy, BAS, Tzarigradsko chaussee 72, Blvd., BG-1784, Sofia, Bulgaria

dimitar.tonev@inrne.bas.bg

Abstract. An accelerator laboratory is presently under construction in Sofia at the Institute for Nuclear Research and Nuclear Energy. The laboratory will use a TR24 type of cyclotron, which provides a possibility to accelerate a proton beam with an energy of 15 to 24 MeV and current of up to 0.4 mA. An accelerator with such parameters allows to produce a large variety of radioisotopes for development of radiopharmaceuticals. The most common radioisotopes that can be produced with such a cyclotron are PET isotopes like: ^{11}C, ^{13}N, ^{15}O, ^{18}F, ^{124}I, ^{64}Cu, $^{68}Ge/^{68}Ga$, and SPECT isotopes like: ^{123}I, ^{111}In, ^{67}Ga, ^{57}Co, ^{99m}Tc. Our aim is to use the cyclotron facility for research in the fields of radiopharmacy, radiochemistry, radiobiology, nuclear physics, materials sciences, applied research, new materials and for education in all these fields including nuclear energy. Presently we perform investigations in the fields of target design for production of radioisotopes, shielding and radioprotection, new ion sources etc.

1. Introduction

The National cyclotron center project was launched in 2012 at the initiative of the Council of Ministers of the Republic of Bulgaria and the management of the Institute for Nuclear Research and Nuclear Energy (INRNE) to solve a very old problem of the country - the lack of domestic production of short lived medical cyclotron radioisotopes. This problem has existed for decades despite repeated efforts of several generations of researchers from INRNE and the medical community to persuade the Government that such a center should be built in Bulgaria as was done a long time ago in the other Eastern European countries.

The project envisions a new cyclotron laboratory as a part of INRNE consisting of: a specialized building that meets the regulatory requirements in the field of radiation safety and GMP (good manufacturing practices) in the pharmaceutical industry; a bunker with a TR-24 cyclotron; a sector for applied research and development in radiopharmacy with enhanced educational functions; a sector for the production of 18F-FDG and, in future, of other radiopharmaceuticals [1]. Presently Bulgaria has no research nuclear facility, neither a research reactor, nor an accelerator. With the new facility we will restart our experimental research program not only in the field of nuclear physics, but also in many interdisciplinary fields related to nuclear physics. An excellent presentation of the accelerator physics in given in the recent work [2].

The TR24 cyclotron is an accelerator which has recently been installed at CYclotron pour la ReCherche et lEnseignement (Cyrce‘) of the Institut Pluridisciplinaire Hubert Curien (IPHC) in

Strasbourg, France or will be operated soon at the Institute of Radiopharmaceutical Cancer Research of Helmholtz-Zentrum Dresden Rossendorf (HZDR), Germany; Rzez, Czech Republic etc. The Laboratori Nazionali di Legnaro at INFN, using a 70 MeV cyclotron, plans to perform similar to our activities [3].

Figure 1. A picture of the open cyclotron TR24 during the test in Vancouver in August 2015.

2. Challenges of the INRNE cyclotron project

One of the main objectives of the project is to allow for regular supply of 18F-FDG to the regional PET/CT center at a low price, providing "full cost recovery", with a small profit margin used to sustain the laboratory. This will enable more hospitals in Bulgaria to purchase new PET/CT equipment as well as to increase the throughput of the existing scanners. The wider access to PET/CT will improve the health and extend the life of the patients suffering from cancer.

Figure 2. The Y-line of the Sofia cyclotron.

The availability of a production sector, meeting all modern regulatory GMP requirements will ensure high quality training of nuclear pharmacists and will close the chain from research through R&D to end products with applications in medicine.

The TR24 cyclotron has been selected because in the next twenty years INRNE will be able to produce a wide range of radioisotopes with applications in medicine with a relatively low initial investment and moderate maintenance costs.

At these proton energies and beam current it is possible to produce commercial quantities of radioisotopes with traditional and expected future applications in medicine as: PET radioisotopes – 18F, 124I, 64Cu, 68Ge/68Ga; SPECT (Single photon emission computed tomography) radioisotopes – 123I, 111In, 67Ga, 99mTc; therapeutic alpha emitters – 225Ac/213Bi, 230U/226Th.

Some part of the equipment for the production of radioisotopes will be designed and constructed at INRNE with the help of the Design and Production Workshop - Physics. The radioisotopes produced by the cyclotron will be used for the development of new radiolabelled molecules for diagnostics [4-6] and therapy purposes [7,8].

The aim of the INRNE cyclotron laboratory is to provide a high intensity and a high quality proton beam to perform forefront research in interdisciplinary fields like medical, chemical, biological and material sciences, nuclear physics and applied research. The proton beam will be used for innovative investigations in the field of new materials, like for example these used for the ITER project. The cyclotron can also produce very low currents (10nA) for radiobiology applications. A longer term goal of the laboratory is to become involved in preclinical studies and clinical trials through collaborations and research projects with academic partners and private companies. The laboratory in Sofia will open a new possibility for experimental nuclear structure research in Bulgaria. We plan to investigate level-schemes of excited nuclei as well as to measure lifetimes of excited nuclear states. A proton beam will be used to bombard a target and with a multidetector system to investigate the nuclear structure. Lifetimes in the femtosecond region can be measured using Doppler-shift attenuation method (DSAM) following the inelastic proton scattering reaction. Such an approach is presented in details in the work of Hennig et al. [9]. In the setup used in the Cologne University, the proton in the exit channel of the $(p, p'\gamma)$ reaction is detected with charged particle detectors placed close to the target. This approach uses the centroid-shift version of the DSAM method and profits from the coincident detection of scattered protons and de-exciting γ-rays.

Figure 3. Factory acceptance tests in Vancouver, Canada in 2015.

A similar setup we plan to install at the cyclotron laboratory of INRNE in order to measure lifetimes for low-lying states in nuclei which could be approached by (p,p'γ) reaction. We would like to continue our research in the fields of investigation of phase transitions in nuclear physics [10], chirality in nuclei [11,12] and nuclear structure of light nuclei [13]. A parallel activity of the cyclotron laboratory of INRNE includes education and training of physicists, chemists, pharmacists and biologists for radioisotopes production, radiochemistry, quality control of radiopharmaceuticals and their uses in medical imaging, as well as training of specialists for nuclear energy. More general practical topics like radiation protection and dosimetry will also be covered by the training and education program of the laboratory. Most of the training and education activities will be carried out by involving the students in the ongoing research projects of the laboratory. Special attention will be paid to the training of students which plan to work in the field of nuclear energy. The education of these students will guarantee the high level of nuclear safety of the Bulgarian nuclear power plant.

3. The accelerator TR24

The TR24 cyclotron is produced and commercialized by Advanced Cyclotron Systems Inc., located in Vancouver, Canada. The TR-24 cyclotron of INRNE, Figure 1, has the following characteristics: accelerates negatively charged hydrogen ions; variable energy of the proton beam; minimum energy of the proton beam of 15 MeV; maximum energy of the proton beam of 24 MeV; proton beam current of 400 μA, upgradeable to 1000 μA; simultaneous extraction of two proton beams; external CUSP ion source. With a very high current of up to 400 μA upgradeable to 1 mA and proton energy ranging from 15 to 24 MeV, the cyclotron can produce commercial quantities of more than 15 radioisotopes by irradiating simultaneously two high current targets placed at the exit ports or along high current beam lines. At the beginning one Y-like beam line, shown in Figure 2, will be mounted at one of the exit ports of our cyclotron for greater flexibility. Presently we have two target stations and three targets for the production of 18F-FDG.

Figure 4. Delivery and storage of the cyclotron in INRNE.

4. ^{67}Ga and ^{68}Ga – Radioisotopes with long history and bright future

Among the radioisotopes with wide use in the clinical practice that will be produced at the cyclotron center are Gallium-67 (^{67}Ga) and Gallium-68 (^{68}Ga).

Radiopharmaceuticals with ^{67}Ga and ^{68}Ga are used primary for imaging techniques. One of these techniques for ^{67}Ga is scintigraphy and the result is 2D images captured by gamma cameras. Another

imaging approach which is using gamma cameras is SPECT which results in 3D images. The imaging of radiopharmaceuticals with ^{68}Ga is based on PET technique.
^{67}Ga (T1/2 = 3.26 d) decays100% through electron capture and emits γ-rays with energies of 93.3 keV (37%), 184.6 keV (20.4%), and 300.2 keV (16.6%). In addition to its use for diagnostic applications, interest in the effectiveness of Auger emissions from ^{67}Ga for some specific therapeutic applications has also been reported. The ^{67}Ga can be produced by a reactor and the most common ^{67}Ga cyclotron production methods include proton irradiation of isotopically enriched ^{68}Zn targets by the $^{68}Zn(p,2n)^{67}Ga$ reaction [14]. Separation of pure gallium radioisotopes produced by cyclotrons from the target material and target holder can be achieved by ion exchange. Recovery of gallium from aqueous solutions is achieved by chemical precipitation, complexation and ion exchange [15].
^{67}Ga-citrate is indicated for detecting cancer and some acute inflammatory lesions after patient injection. The half-life of 3 days has the advantage of performing delayed patient scans after significant accumulation of the radiopharmaceutical. The ^{67}Ga is predominantly supplied as the citrate complex in a multidose vial. The isotope decays by electron capture to yield detectable energies, hence it requires thick shielding. $T_{1/2}$ of ^{67}Ga also allows for delayed scintigraphy [16].
^{67}Ga-citrate is an FDA-approved SPECT agent for imaging of various cancers and inflammatory lesions. Once injected into a patient, ^{67}Ga is easily transmetallated from the weak citrate chelate and becomes ~99% transferrin bound. Because of this, the radionuclide imaging with ^{67}Ga-citrate is not specific and has the additional disadvantages of long imaging time, low resolution and high patient radiation dose. The uptake of the radiometal in the liver, the kidneys and other organs has limited its use for diagnosis of various diseases [17,18].
^{68}Ga in clinical trials proves itself to be a good contender for replacement of some of the currently used radionuclide pharmaceuticals and presents a great opportunity for optimization of treatments and imaging methods. Its shorter half-life of 68 minutes allows for faster metabolic clearance and it makes it possible to use on the same patient with a shorter waiting period in between usage. It has a high positron abundance of 90%, and robust coordination chemistry with many ligands.
The applications of ^{68}Ga are various and one of them is as a diagnostic agent for managing the treatment of gastroenteropancreatic neuroendocrine tumors (GEP-NETs) and neuroendocrine tumors in general. Neuroendocrine tumors are malignant and benign tumors that develop in the hormone-producing cells of the body's neuroendocrine system and are located in organs such as the stomach, intestines, pancreas, and lungs. The idea is to be able to locate NETs with a PET/CT scanner in adult and pediatric patients, utilizing the tumors big number of somatostatin receptors and the radiopharmaceuticals ^{68}Ga-DOTA-TATE, ^{68}Ga-DOTA-TOC, ^{68}Ga-DOTA-NOC [19,20].
Another application is for imaging of pancreatic cancer with a PSMA (prostate-specific membrane antigen) scan. PSMA has considerable overexpression in most of the prostate cancer cells. Several compounds for labelling PSMA have been developed and are currently being investigated. Out of them PSMA inhibitor Glu-NH-CO-NH-Lys(Ahx)-HBED-CC labeled with ^{68}Ga presents itself as the most popular for studying. Even more, in patients who have prostatic cancer which is biochemically recurrent but with low serum PSA values, the usage of ^{68}Ga-PSMA–PET imaging has been shown to increase detection of metastatic sites, compared with conventional imaging or PET examination with different tracers [21,22].
^{68}Ga can be used as radioisotope for Intravascular Irradiation (IVRT) of Coronary Vessels in the occurrence of restenosis. Liquid-filled balloons for coronary brachytherapy provide significant advantages over solid sources when it comes to dose homogeneity. However, they carry the risk of life-threatening radiointoxication after balloon rupture and/or laboratory contamination in case of a spill. The positron emitter ^{68}Ga is well suited to overcome these safety obstacles while providing full therapeutic efficacy [23].
The development of inhibitors of key oncogenic signaling pathways as targeted therapies of cancer would be expedited by the ability to assess the effect of a drug on its target *in vivo*. Radionuclide-based molecular imaging, as illustrated by preclinical PET studies with the use of ^{68}Ga, is a promising approach to such pharmacodynamic assessment of anticancer drugs [24].

Acknowledgments
The project has been supported by the Department of Energy of USA, the Nuclear Power Plant "Kozloduy", the Bulgarian Ministry of Education and Science, the Bulgarian Council of Ministers. This research has been supported by Bulgarian Science Fund under Contract No. DN 08/6, 13.12.2016. The authors acknowledge the support and encouragement for the success of the project from colleagues and friends.

References

[1] Tonev D, Goutev N and Georgiev L S 2016 *J. Phys. Conf. Proc.* **724** 012049
[2] Dinev D 2006 *Particle accelerators* Marin Drinov Academic Publishing House ISBN - 13:978-954-322-067-0
[3] de Angelis G, Prete G and Andrigetto A, for the SPES collaboration, et al. 2016 *EPJ Web of Conferences A* **107** 01001
[4] Marchand P, Bekaert V, Ouadi A, Laquerriere P and Brasse J 2013 *J. Nucl. Med.* **54** 15N-17N
[5] Marchand P, Ouadi A, Pellicioli M, Schuler J, Laquerriere P, Boisson F and Brasse D 2016 *Nucl. Med. and Biol.*, **43** 520-527
[6] Pellicioli M, Schuler J, Marchand P and Brasse D 2017 *AIP Conf. Proc.* **1845** 020018
[7] Pellicioli M, Ouadi A, Marchand P, Foehrenbacher T, Schuler J, Dick-Schuler N and Brasse D 2017 *AIP Conf. Proc.* **1845** 020017
[8] Brasse D and Nonat A 2015 *Radiometals, Dalton Transactions* **44** 4845-4858
[9] Hennig A and et al. 2015 *Nucl. Instrum. Meth. Phys. Res. A* **794** 171
[10] Tonev D and et al. 2004 *Phys. Rev. C* **69** 034334
[11] Tonev D and et al. 2006 *Phys. Rev. Lett.* **96** 052501
[12] Tonev D and et al. 2014 *Phys. Rev. Lett.* **112** 052501
[13] Tonev D and et al. 2002 *Phys. Rev. C* **65** 034314
[14] Little F E and Lagunas-Solar M C 1983 *Int. J. Appl. Radiat. Isot.* **34** 631-637
[15] Ahmed I M, El-Nadi Y A and El-Hefny N E 2013 *Hydrometallurgy* **131–132** 24-28
[16] Tsopelas C 2015 *The Scientific World Journal* **2015** 676719
[17] Dumarey N, Egrise D, Blocklet D, Stallenberg B, Remmelink M, del Marmol V, Van Simaeys G, Jacobs F and Goldman S 2006 *J Nucl Med.* **47** 625–632
[18] Parisi M T 2011 *Pediatr Radiol.* **41** 803–810
[19] Graham M M, Gu X, Ginader T, Breheny P and Sunderland J J 2017 *J. Nucl. Med.* **58** 1450-1451
[20] Wild D, Bomanji J B, Benkert P, Maecke H, Ell P J, Reubi J C and Caplin M E 2013 *J Nucl Med.* **54** 364-372
[21] Maurer T and et al. 2016 *J. Urol.* **195** 1436-1443
[22] Fendler W P and et al. 2016 *J Nucl Med.* **57** 1720-1725
[23] Stoll H P, Hutchins G D, Winkle W L, Nguyen A T, Hou D, Appledorn C R, Romeike B and Marc K L 2001 *Circulation.* **103** 1793-1798
[24] Smith-Jones P M, Solit D, Afroze F, Rosen N and Larson S M 2006 *J Nucl Med.* **47** 793-796